350
Topics in Current Chemistry

Aims and Scope

The series Topics in Current Chemistry presents critical reviews of the present and future trends in modern chemical research. The scope of coverage includes all areas of chemical science including the interfaces with related disciplines such as biology, medicine and materials science.

The goal of each thematic volume is to give the non-specialist reader, whether at the university or in industry, a comprehensive overview of an area where new insights are emerging that are of interest to larger scientific audience.

Thus each review within the volume critically surveys one aspect of that topic and places it within the context of the volume as a whole. The most significant developments of the last 5 to 10 years should be presented. A description of the laboratory procedures involved is often useful to the reader. The coverage should not be exhaustive in data, but should rather be conceptual, concentrating on the methodological thinking that will allow the non-specialist reader to understand the information presented.

Discussion of possible future research directions in the area is welcome.

Review articles for the individual volumes are invited by the volume editors.

Readership: research chemists at universities or in industry, graduate students.

More information about this series at
http://www.springer.com/series/128

Jay S. Siegel • Yao-Ting Wu
Editors

Polyarenes II

With contributions by

I. Agranat · P.U. Biedermann · J.L. Delgado · S. Filippone ·
F. Giacalone · Mª.Á. Herranz · B. Illescas · T. Iwanaga ·
N. Martín · R.R. Parkhurst · M. Prato · E.M. Pérez ·
T.M. Swager · S. Toyota · E. Vazquez · V.A. Zamolo

Editors
Jay S. Siegel
School of Pharmaceutical Science and Technology
Tianjin University
Tianjin
China

Yao-Ting Wu
Department of Chemistry
National Cheng Kung University
Tainan
Taiwan

ISSN 0340-1022 ISSN 1436-5049 (electronic)
ISBN 978-3-319-07301-9 ISBN 978-3-319-07302-6 (eBook)
DOI 10.1007/978-3-319-07302-6
Springer Cham Heidelberg New York Dordrecht London

Library of Congress Control Number: 2014955786

Printed on acid-free paper

Springer is part of Springer Science+Business Media (www.springer.com)

Preface

The coming year (2015) will commemorate an important event in the history of chemistry, and – in particular – aromatic chemistry.

Exactly 150 years ago, on January 27, 1865, Adolphe Wurtz, formally presented August Kekulé´s first benzene paper at a meeting of the Chemical Society of Paris. It was this event which was celebrated only 25 years later at the famous *Kekulé-Feier* or *Benzolfest* on March 11, 1890 in Berlin, in which Kekulé – after the plenary address of the vice-president of the Deutsche Chemische Gesellschaft, Adolf Baeyer, who was also Kekulé´s first student – revealed his dreams to the public in which he had arrived at structural theory and the hexagonal structure of benzene (There is a vast amount of literature on the (two) dreams and the *Benzolfest* – from the psychological and historical viewpoints, from the importance of the benzene formula in organic chemistry and industry to the question of whether it was really Kekul*é* who was the first to propose this breakthrough. None of this will be repeated here; I just want to recommend to the reader the – in my view – most significant and clairvoyant monograph about this fascinating area of early organic chemistry: [1]).

The report on the Benzolfest in *Berichte der Deutschen Chemischen Gesellschaft* [2] is still a marvelous and highly enjoyable piece of chemical literature and – not the least – a tribute to international co-operation and friendship. It contains many deep insights into the nature of the scientific process and distinguishes itself by the humbleness and honesty of the speaker. Still, there are some conclusions which, I believe, are erroneous.

The most notable of these reads (in German): *Etwas absolut Neues ist noch niemals gedacht worden, sicher nicht in der Chemie.* (Absolutely new ideas have never been conceived, certainly not in chemistry). The first half of this quote, a paraphrase of a citation from the Old Testament (Ecclesiates 1:9) has been cited on many occasions, but it is – at best – only partly true.

To be sure, during most of the first century after the discovery of benzene (Faraday 1825) and Kekulé´s hexagon structure (1865) the terms aromatic chemistry and benzene chemistry were practically synonyms, especially if the enormous success of the coal tar and dyestuffs industry is taken into account. It is a safe assumption that Kekulé´s dreams during these decades would have focused on the countless derivatives which can be prepared from this icon of organic chemistry. But, alas, Kekulé did nor dream any more, at least not in public.

The situation began to change when, as a consequence of quantum theory, and in particular the Hückel theory, nonbenzenoid aromatic molecules came onto the scene. These indeed represented many new aromatic (and antiaromatic) entities, whether neutral or charged. Clearly, this development, which ended in the 1960s and 1970s and was strongly driven by the introduction of new physical methods into organic chemistry (NMR spectroscopy!) as well as the development of computational chemistry represents the high point of aromatic chemistry of the twentieth century.

Interestingly, both benzenoid and nonbenzenoid aromatic compounds share a *structural* concept; namely, that all these systems were – by theory had to be! – planar. The origin of this obsession with planarity is manifold. It certainly goes back to benzene itself which in the minds of many chemists was a *rigid* entity. It is the tile-like chemical compound par excellence: just by putting these hexagonal moieties together one can build up an endless series of compounds: naphthalene, anthracene, phenanthrene, pyrene, coronene, the polyaromatic hydrocarbons (PAHs), etc., ad infinitum. All of them completely flat. The notion of rigidity of these compounds was also supported by the stiff molecular models usually employed by chemists in their daily work. And, of course, the Hückel rule seemed to demand planarity with all the axes of the overlapping p-orbitals oriented in parallel fashion for optimal effect.

Had Kekulé had access to an IR spectrometer (and also known the corresponding theory!) he would have noticed that the rigidity of his hexagon and its derivatives is rather limited, the out-of-plane bending vibrations being clearly recognizable in the region of ca. 1,000 cm^{-1} and below in the IR spectrum, i.e., in a low energy section.

That in real life – in the laboratory of the synthetic chemist – the benzene rings do not have to be planar to display their aromaticity (as shown by their chemical and spectroscopic properties) became increasingly obvious when Cram´s seminal studies on cyclophanes began to appear, hydrocarbons which are characterized by "bent and battered" rings (Cram).

Other deliberate steps "out of the plane, into three-dimensional space" were also undertaken in the 1960s (see the corannulene synthesis by Lawton and Barth 1966). The real death blow to planar thinking, which is just another word for rigid thinking, came with the fullerene work of Kroto, Smalley, Krätschmer, and many others. I am sure that even Kekulé would have called these results "new."

From this paradigm change onwards, the benzene ring and all the aromatic and nonbenzenoid compounds derived therefrom – in whatever configuration or conformation – have become the building blocks for an endless variety of π-systems with planar or three-dimensional structures. The endless chemical space which can be constructed with double and triple bonds (in the simplest case) is becoming structurally as complex as the C–C single bond compounds. That many of these artifacts do not survive under the usual laboratory conditions most often has to do with their high reactivity, not their structural complexity.

To return to the title question: a modern day Kekulé would most likely not sit in front of a fancy fireplace or in a bus, but look at the marvelous aromatic molecules discussed in every single chapter of this book – and then start dreaming.[1]

Braunschweig, Germany Henning Hopf

References

1. Rocke AJ (2010) Image and reality. The University of Chicago Press, Chicago
2. (1890) Ber Deutsch Chem Ges 23:1302. The speech is available on the net (in German): http://www.sgipt.org/wisms/geswis/chem/kek1890.htm

[1] Advice to younger readers: Kekulé in his Benzolfest speech did not recommend "dreaming" as a scientific method. In fact, he specifically addressed this problem by saying:

> *"Let us dream, gentlemen (chemists at that time were practically exclusively male), then we will possibly find the truth. But beware of publishing our dreams, before we have tested them with a wide-awake mind."*

Contents

Top Curr Chem (2014) 350: 1–64
DOI: 10.1007/128_2012_414

Published online: 29 March 2013

Buckyballs

Juan L. Delgado, Salvatore Filippone, Francesco Giacalone, Mª Ángeles Herranz, Beatriz Illescas, Emilio M. Pérez, and Nazario Martín

Abstract Buckyballs represent a new and fascinating molecular allotropic form of carbon that has received a lot of attention by the chemical community during the last two decades. The unabating interest on this singular family of highly strained carbon spheres has allowed the establishing of the fundamental chemical reactivity of these carbon cages and, therefore, a huge variety of fullerene derivatives involving [60] and [70]fullerenes, higher fullerenes, and endohedral fullerenes have been prepared. Much less is known, however, of the chemistry of the uncommon non-IPR fullerenes which currently represent a scientific curiosity and which could pave the way to a range of new fullerenes. In this review on buckyballs we have mainly focused on the most recent and novel covalent chemistry of fullerenes involving metal catalysis and asymmetric synthesis, as well as on some of the most significant advances in supramolecular chemistry, namely H-bonded fullerene assemblies and the search for efficient concave receptors for the convex surface of fullerenes. Furthermore, we have also described the recent advances in the macromolecular chemistry of fullerenes, that is, those polymer molecules endowed

J.L. Delgado and E.M. Pérez
IMDEA-Nanoscience, Campus de Cantoblanco, 28049 Madrid, Spain

S. Filippone, Mª.Á. Herranz and B. Illescas
Facultad de Química, Departamento de Química Orgánica, Universidad Complutense, 28040 Madrid, Spain

F. Giacalone
Universitá degli Studi di Palermo, Dipartimento di Scienzie e Tecnologie Biologiche, Chimiche e Farmaceutiche, Palermo, Italy

N. Martín (✉)
Facultad de Química, Departamento de Química Orgánica, Universidad Complutense, 28040 Madrid, Spain

IMDEA-Nanoscience, Campus de Cantoblanco, 28049 Madrid, Spain
e-mail: nazmar@quim.ucm.es

with fullerenes which have been classified according to their chemical structures. This review is completed with the study of endohedral fullerenes, a new family of fullerenes in which the carbon cage of the fullerene contains a metal, molecule, or metal complex in the inner cavity. The presence of these species affords new fullerenes with completely different properties and chemical reactivity, thus opening a new avenue in which a more precise control of the photophysical and redox properties of fullerenes is possible. The use of fullerenes for organic electronics, namely in photovoltaic applications and molecular wires, complements the study and highlights the interest in these carbon allotropes for realistic practical applications. We have pointed out the so-called non-IPR fullerenes – those that do not follow the isolated pentagon rule – as the most intriguing class of fullerenes which, up to now, have only shown the tip of the huge iceberg behind the examples reported in the literature. The number of possible non-IPR carbon cages is almost infinite and the near future will show us whether they will become a reality.

Keywords Asymmetric synthesis · Endohedral fullerenes · Fullerenes · Macromolecular chemistry · Molecular wires · Non-IPR fullerenes · Organic photovoltaics · Supramolecular chemistry

Contents

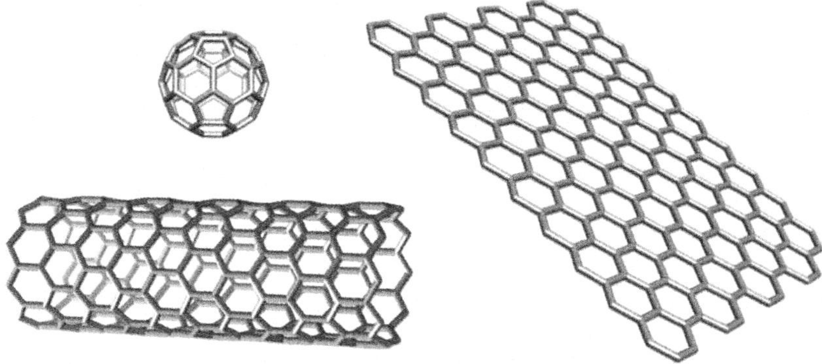

Fig. 1 Chemical structure of C_{60}, a single wall carbon nanotube and graphene

1 Introduction

From the 118 elements which have been identified according to the periodic table, only 94 elements are naturally occurring on earth while the remainder are considered artificial elements. Among the natural elements, carbon is the only one providing the basic requirements for life. Its ability of hybridization of atomic orbitals to produce sp^3, sp^2, and sp hybrid orbitals confers upon it the singular property of having a variety of allotropic forms. Surprisingly, it is less than three decades since only two allotropes of carbon, namely diamond – constituted by sp^3 carbon atoms – and graphite – formed by sp^2 carbon atoms – were known by the scientific community. Both allotropes show a reticular structure with carbon atoms spreading infinitely through the three space directions.

A new scenario emerged in 1985 with the advent of fullerenes, the third and only molecular allotropic form of carbon, formed by highly symmetric closed cages of a well-defined number of carbon atoms [1]. Interestingly, fullerenes have been present on our planet from its very beginning as well as in outer space (the presence of fullerenes C_{60} and C_{70} has recently been detected by IR in huge amounts in a young planetary nebula (Tc 1). The fullerene content is around 1.5% of the carbon present in the nebula, roughly corresponding to the mass of three moons [2]).

Soon after the discovery of fullerenes, other important different forms of carbon were found, namely, in chronological order, multiwall [3] and singlewall [4, 5] carbon nanotubes, and most recently graphenes [6] which have provoked great excitement and expectation in the scientific community (Fig. 1). In the meantime, a wide variety of other less common nanoforms of carbon also emerged such as nanohorns, nanoonions, nanotori, nanobuds, peapods, etc., whose properties and chemical reactivity are less well known to date [7]. Furthermore, fullerenes have been skillfully combined with other elements allocated to their inner empty space, affording a large, singular, and promising family of *so-called* endohedral fullerenes – those containing an atom, molecule, or complex in their inner cavity – whose properties and chemical reactivity are strongly influenced by the elements inside the ball [8].

The increasing number of nanoforms of carbon gives rise to a first taxonomic problem. Should the aforementioned nanoforms of carbon be considered as different allotropes? Answering this question could be accomplished by considering the IUPAC definition of allotrope as “the different structural modifications of an element” (IUPAC Compendium of Chemical Terminology, 2nd edition, 1997). Although at a first glance they could be considered as different allotropes, the scientific community considers fullerenes as the third allotropic form of carbon in which the sp^2 carbon atoms are bonded to form spherical, tubular, or ellipsoid structures, thus gathering all the above forms within the same allotrope of “fullerenes.”

The above considerations do not, however, affect “buckyballs,” a nickname which refers only to the former fullerenes, that is, those constituted exclusively by closed cages of a precise number of carbon atoms, the most representative examples being C_{60} and C_{70} molecules, and the most abundant and easy to obtain fullerenes. In this chapter we will discuss the properties, chemical reactivity, and some of the most realistic applications of fullerenes, including endohedral fullerenes and those rare fullerenes which do not follow the “isolated pentagon rule” (see below).

1.1 Brief History of Fullerenes

The discovery of fullerenes represents one of the most recent examples of serendipity in chemistry. Fullerenes were discovered by Robert F. Curl, Sir Harold W. Kroto, and the late Richard E. Smalley in 1985 [1] during carbon nucleation studies under red giant star conditions. Only 11 years later, in 1996, these scientists were awarded the Nobel Prize in Chemistry for “the discovery of fullerenes” [9–11]. Just a year before, C_{60} had been declared the molecule of the year by the scientific journal *Science*.

This important finding has resulted in a new field with a broad impact in science, thus provoking great excitement in the scientific community, ranging from chemistry, physics, and engineering to practical applications in materials science and biomedical applications [12]. Actually, the impact of the new fullerenes goes beyond the scientific world and, since this molecule was first found in Texas, the State Parliament declared Buckminsterfullerene C_{60} the molecule of Texas State in May 1997.

However, a major breakthrough in fullerene science occurred in 1990 when Wolfgang Krätschmer and Donald Huffman (two astrophysicists) prepared fullerene C_{60} for the first time in multigram amounts [13] thus opening the fullerene world to chemical functionalization and, therefore, to the unlimited imagination of chemists for synthesizing new and sophisticated fullerene architectures. The importance of this achievement was pointed out by some of the Nobel laureate scientists. In Smalley’s own words: “Had there not been a method to make it in measurable amounts, it would not have had an impact.” Curl also recognized this scientific

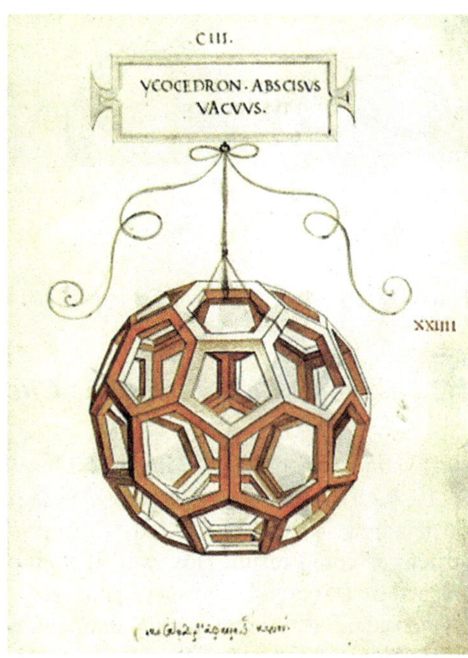

Fig. 2 Ycocedron Abscisus Vacuus by Leonardo da Vinci

contribution, stating: "Huffman's work took it from mass spectrometers to the laboratory. It must have been a close decision by the Nobel committee over who should get it."

However, the history of fullerenes started many years before. Actually, the possible existence of the C_{60} molecule was proposed 15 years before its discovery by Eiji Osawa from Kyoto University. Kroto who did not know this work at the time of discovery, in part due to its publication in Japanese, has given great credit to this Japanese scientist.

The idea of a closed carbon cage was initially proposed by David E.H. Jones in 1966 in an article written in the *New Scientist* under the name "Daedalus" [14]. He suggested that giant empty cages could be formed by distorting a planar net of hexagonal carbons (a vision of graphene?) by adding impurities. However he was never able to explain how it could be done. It is interesting to note, nevertheless, that in situ TEM experiments correlated with quantum chemical modeling have demonstrated that flat graphene sheets undergo a direct transformation to fullerene cages under 80-keV electron beam irradiation [15].

Some more romantic chemists have seen in Leonardo da Vinci's work the first scientific "modeling" of the C_{60} molecule in his famous illustration for the book by Luca Pacioli entitled "De Divina Proportione," published in Venice in 1509. In this illustration the truncated icosahedron called "Ycocedron Abscisus Vacuus" by Leonardo is shown (Fig. 2).

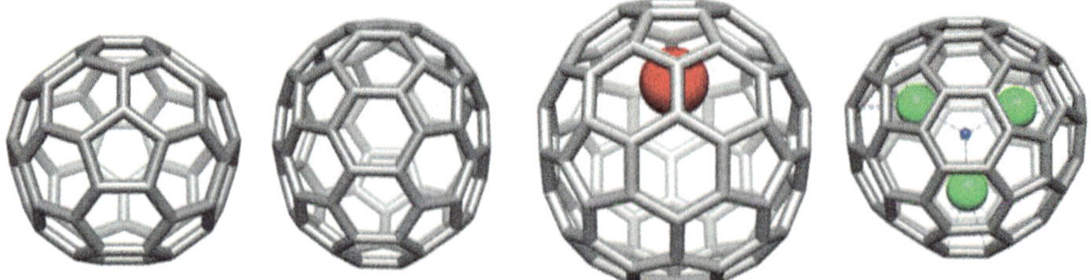

Fig. 3 Empty fullerenes C_{60} and C_{70} (*left*) and endohedral fullerenes La@C_{82} and Sc_3N@C_{80} (*right*)

1.2 General Remarks on the Chemical Reactivity of Fullerenes

Buckyballs constitute a family of closed cage carbon allotropes that contains 2 (10 + *H*) carbon atoms, where *H* is the number of hexagons, while the number of pentagons is always fixed at 12. In principle, an unlimited number of fullerene structures could result. However, the simplest and most abundant is C_{60}, which is formed by 60 carbon atoms –12 pentagons and 20 hexagons–, followed by C_{70}. C_{60} has icosahedral symmetry and a diameter of 7.8 Å. An important structural motif of fullerenes is the so-called "isolated pentagon rule," which means that all pentagons in the molecule must be isolated from other pentagons, since structures with fused pentagons are highly destabilized due to the increase in strain energy and resonance destabilization [16] (Fig. 3).

In contrast to diamond and graphite, which are sparingly soluble in organic solvents, fullerenes are soluble in some organic solvents. They undergo a variety of chemical reactions in solution to afford a huge number of fullerene derivatives which, in general, preserve the outstanding chemical, physical, and electrochemical properties of pristine fullerenes. The study of the chemical reactivity of fullerenes has experienced an unprecedented development during the last two decades and is expected to continue on this steep slope.

The singular 3D geometry of fullerenes containing 30 or more highly reactive double bonds constitutes a new scenario where a variety of different chemical reactions can be tested. The convex surface of fullerenes offers new possibilities for the study of new reactions and mechanisms under severe geometrical constraints on carbon atoms showing a singular $sp^{2.3}$ hybridization [17].

The contributions stemming from well-established and versatile protocols to functionalize fullerenes chemically have yielded a broad spectrum of intriguing, tailor-made fullerene derivatives. The remarkable properties of the latter are continuously under investigation and form the basis in the search for practical applications of fullerenes.

As mentioned above, the C_{60} molecule is formed from 12 pentagons and 20 hexagons linked by single and double carbon–carbon bonds. The calculated bond distances reveal subtle differences between the [5,6]- and [6,6]-bonds with values of 1.45 and 1.38 Å, respectively. Due to the mixed character of 1,3,5-cyclohexatrienes and [5]radialenes, C_{60} behaves as a highly strained electron-poor alkene. The chemical reactivity is mainly driven by strain relief and, therefore, addition reactions have been widely used [18]. Interestingly, although similar reactivity patterns have also been observed for higher fullerenes, chemical reactivity tends to decrease significantly with their size [19–21].

A variety of chemical reactions, namely nucleophilic additions, cycloaddition reactions, free radical additions, halogenations, hydroxylation, and metal transition complexations, have been reported for C_{60}. However, addition reactions, electron transfer reactions, and reactions involving the opening of the fullerene cage (molecular surgery) have been studied in more detail. It is worth mentioning the ease with which fullerenes are reduced by means of electron-rich chemical reagents as well as electrochemically. Their oxidation, however, is considerably more difficult to achieve. These experimental findings are in agreement with former theoretical calculations which predicted that C_{60} has a low energy LUMO which is triply degenerated and, therefore, accepts up to six electrons in solution to form up to the hexaanion [22]. The theoretical predictions were later confirmed by electrochemical measurements recording from the monoanion to the hexaanion using a toluene/acetonitrile 5:1 by volume solvent mixture at −10°C [23].

For a wider and more detailed study of the basic reactivity of fullerenes, the reader is referred to the aforementioned monographs that comprehensively cover the properties and chemical reactivity of fullerenes [19–21].

2 New Covalent Chemistry of Fullerenes

Significant effort is still being devoted to the chemical modification of fullerenes. Even though most of them are based on the chemistry of electron-poor olefins, fullerene curved double bonds have given rise to a quite peculiar fullerene chemistry. Remarkable examples of this reactivity have been provided by the use of fuller-1,6-enynes, fullerene analogues of 1,6-enynes involving a highly reactive fullerene double bond as the "ene" moiety. Thus, fulleropyrrolidines **1** bearing a propargyl group on the C-2 of the pyrrolidine ring undergo an unusual thermal [2+2] cycloaddition reaction affording regioselectively a cyclobutene-fullerene derivative **2** (Scheme 1) [24]. A different change in chemoselectivity is observed when an internal alkyne is used in the fullerenynes. In that case, allenofullerene derivatives (**3**) are obtained as a result of a formal "ene" reaction where the alkyne moiety with the α CH group acts as an "ene" component, despite the unfavorable geometry (Scheme 1) [25].

Another example of the intriguing behavior of fullerene double bonds has recently been reported by Bazan et al. in which fullerenes behave as a neutral carbon based Lewis acid [26]. Thus, when C_{60} reacts with the N-heterocyclic carbene **4**, that acts as a Lewis base, a thermally stable zwitterionic Lewis acid–base adduct **5** is formed. The bulk of the substituents of carbene species, along with the delocalization of its positive charge, prevent the expected cyclopropanation reaction and a C–C single bond, with a length of 1.506 Å is formed instead (Scheme 2).

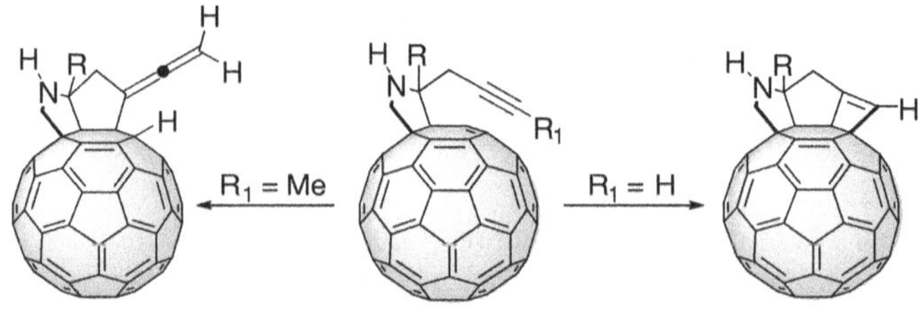

Scheme 1 Fullerenynes (**1**) bearing an alkyne unit give rise to cyclobutene derivatives (**2**) regioselectively, and allenic structures (**3**)

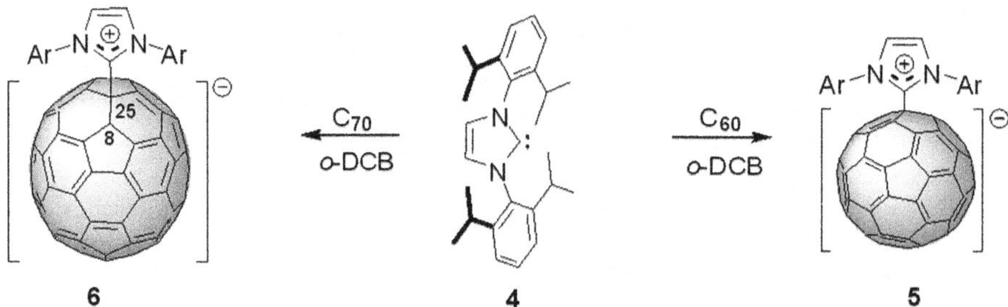

Scheme 2 Fullerene-carbene Lewis acid–base adducts

The addition also occurs onto C_{70} with the regioselective formation of a new bond between the carbene and the carbon atom C-25 of [70]fullerene [27].

Among the numerous methods for chemical functionalization of fullerenes reported during the last decade, some important trends have been outlined for achieving new properties or major control over factors like reactivity, selectivity, and compatibility with a wider range of functional groups.

2.1 *New Reactions on Fullerenes Involving Metals*

The nucleophilic addition of organometallic reagents, such as organolithium [28] or Grignard salts [29], is one of the classical methods for fullerene functionalization. [30–32]. More recently, the use of transition metals has expanded the arsenal of chemical tools, achieving new structures with better control of reactivity and selectivity. Thus, fullerenes' double bonds, despite their electron-poor character, are able to act as the alkene component in a Pauson–Khand (PK) reaction. Therefore, when a 1,6-fullerenyne is treated with $Co_2(CO)_8$, highly efficient and regioselective intramolecular PK products showing three [33] (or five [34]) fused pentagonal rings on the same hexagon of the fullerene surface were formed (Scheme 3).

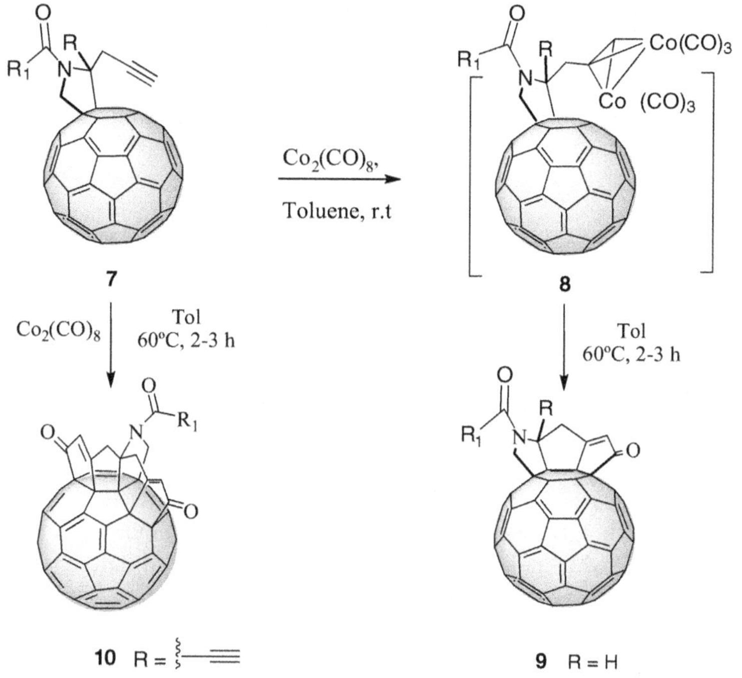

Scheme 3 Pauson–Khand reaction of fullerenynes

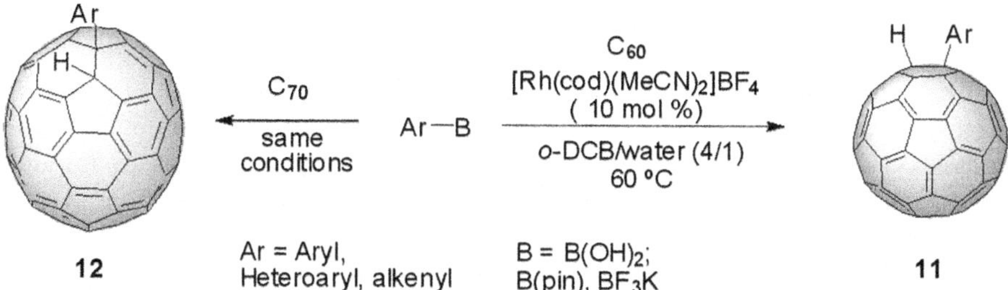

Scheme 4 Rhodium and palladium catalyzed additions of organoborane reagents onto fullerenes

A further step has been the use of transition metal catalysis in fullerene chemistry as a smart alternative to avoid high loading of organometallic reagent and to achieve remarkable levels of reactivity and selectivity. An interesting example of this approach has been the arylation and alkenylation of fullerenes catalyzed by a rhodium complex reported by Itami [35]. Similarly to the reaction of organoboron compounds with electron-deficient alkenes and alkynes, rhodium (I) complexes catalyze the hydroarylation of C_{60} (or C_{70}) with arylboronic acid in aqueous solution. The reaction proceeds with high regioselectivity and in a mono-addition selective manner (Scheme 4)

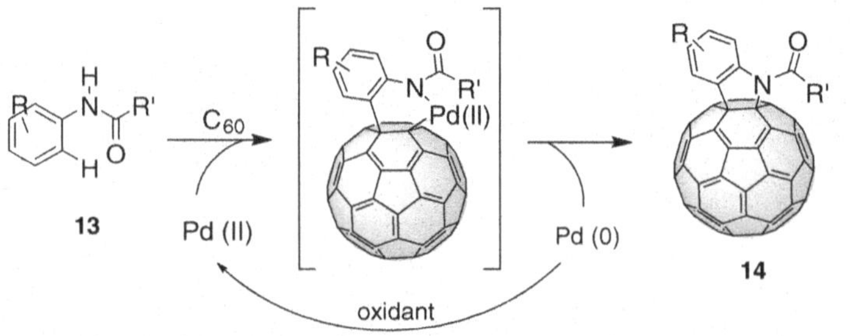

Scheme 5 Fulleroindolines prepared by Pd(II) catalysis

The use of $[Rh(cod)_2]BF_4$ gave rise to an optimal combination of good yield (61%) and excellent selectivity (>95%) showing an important effect of the counteranion of the rhodium complexes in sharp contrast with the reported example of conventional olefins. The authors claimed a catalytic cycle reaction where cationic Rh complex and water produce Rh–OH species. After transmetalation of the thus-formed Rh–OH with $RB(OH)_2$, the Rh–R species undergoes addition on the C_{60} double bond. Finally, protonolysis of the formed fullerenyl Rh species affords the product R–C_{60}–H (**11**, **12**) with regeneration of the cationic Rh species.

Shortly after, the same authors also developed a palladium(II) catalyst Pd(2-PyCH=NPh)$(OCOC_6F_5)_2$ for the hydroarylation of fullerene with boronic acids that, along with good catalytic activity (reaction generally occurs at room temperature), presents a bench stability in the solid state and efficiency under air conditions. Single crystal X-ray diffraction analysis confirmed unequivocally the addition of the aryl moiety and hydrogen in a 1,2-fashion at the α bond of C_{70} with the phenyl group attached at the position close to the pole of the C_{70} unit [36].

Analogously, Co-catalyzed hydroalkylation of C_{60} with reactive alkyl bromides in the presence of Mn reductant and H_2O at ambient temperature gave the monoalkylated C_{60} in good to high yields. The reaction probably occurs through a reduced Co(0 or I) complex that promotes generation of a radical ($R^{\bullet}$) and the addition to C_{60} [37].

An intriguing copper-catalyzed radical reaction that involves a formal C–H bond activation has been reported by Nakamura. The reaction efficiently couples an arylacetylene or enyne to a penta(aryl)[60]fullerene bromide in a formal [4+2] fashion to form a dihydronaphthalene ring fused to a fullerene sphere [38].

Palladium acetate catalyzes cycloaddition onto C_{60} of a variety of anilides through a C–H bond activation, affording fulleroindolines (**14**) in a highly regioselective manner (Scheme 5) [39].

2.2 Asymmetric Catalysis on Fullerenes

Chirality is an important but undeveloped topic in fullerene science [40, 41]. Along with inherently chiral pristine fullerenes, optically active derivatives have been

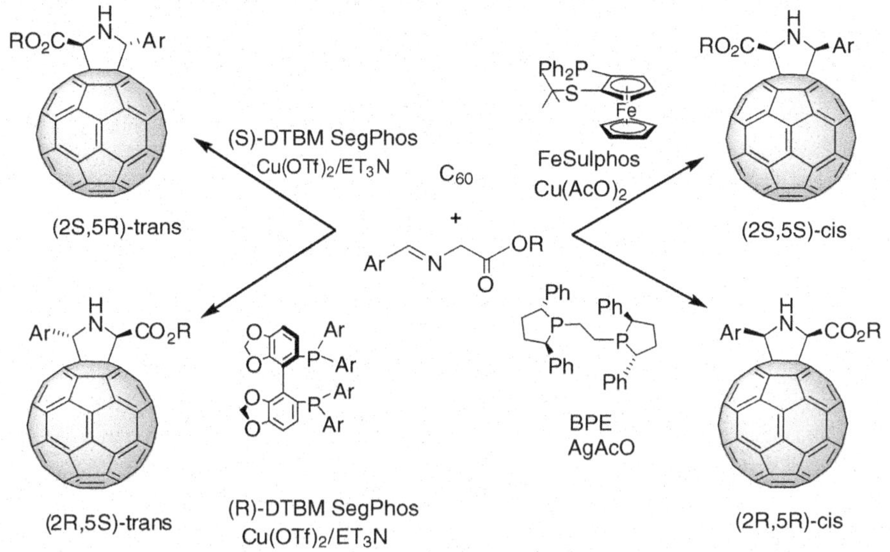

Scheme 6 Stereodivergent 1,3-dipolar cycloaddition of chiral N-metalated azomethine ylides onto C_{60}

obtained by chiral induction from chiral starting reagents or after tedious and expensive HPLC separation from the racemic mixture [40].

The non-coordinating nature of fullerenes has hampered the use of asymmetric metal catalysis and, therefore, the employ of enantiopure fullerene derivatives has been limited to a few examples [42–44].

In this respect, a major breakthrough has been the chiral activation of a 1,3-dipole in the cycloaddition of N-metalated azomethine ylides onto C_{60} (Scheme 6) [45]. By using catalytic amounts of transition metals and the suitable ligand, the cycloaddition of a series of α-iminoesters occurs under very mild conditions and good yields, affording pyrrolidinofullerenes (probably the most important class of fullerene derivatives) with complete control of diastereoselectivity.

More important, chiral complex formed by copper(II) acetate and (*R*)-Fesulphos led the cycloaddition toward the formation of (2*S*,5*S*)-*cis* pyrrolidinofullerenes, whereas the use of silver acetate and chiral (−)-BPE ligand switches the enantioselectivity toward the opposite (2*R*,5*R*)-*cis* pyrrolidinofullerene enantiomers.

Shortly after, complete control on the stereochemical outcome and a fully stereodivergent synthesis of all the possible stereoisomers of disubstituted fulleropyrrolidines was achieved [46]. The use of the complex Cu(II) triflate/(*R*) or (*S*)-DTBM segphos switches the diastereoselectivity towards both enantiomers of the unusual *trans* pyrrolidine with high enantiomeric excess. For this latter process, the authors invoke the presence of a stepwise mechanism that justifies a *supra–antara* formal [4+2] cycloaddition (Scheme 6).

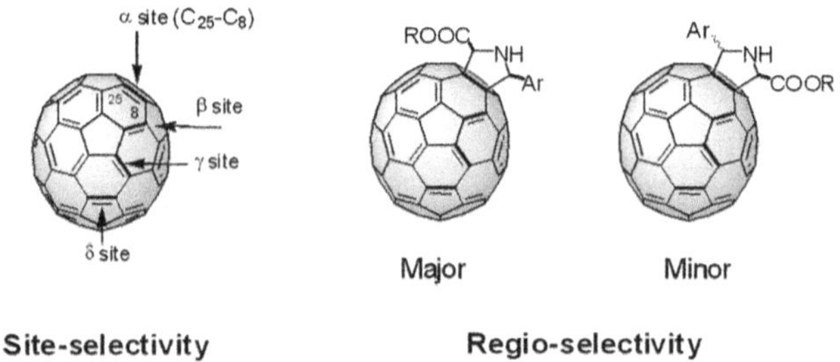

Fig. 4 Site-, regio-, diastereo-, and enantio-selective cycloaddition onto C_{70}

The application of this methodology to higher fullerenes has also been achieved with excellent selectivity. Indeed, higher fullerenes are characterized by a more complex covalent chemistry due to the minor degree of symmetry. In C_{70}, the number of different 6,6 double bonds is increased compared to the single kind of 6,6 double bond in C_{60} and, therefore, even higher control is necessary to face the distinct levels of selectivity encountered.

The very mild conditions used for this transformation allowed direction of cycloaddition of chiral N-metalated azomethine ylides toward the more reactive α double bond (C_{25}–C_8 according IUPAC nomenclature) with an almost complete site-selectivity. Despite the use of unsymmetrical dipoles, all the complexes used afforded the pyrrolidines bearing methoxycarbonyl group in the polar region of C_{70} as major product with good levels of regio-selectivity (Fig. 4) [47].

Finally, for this regioisomer all four possible stereoisomer were obtained by the use of the suitable metal chiral complex with high values of diastereo- and enantio-selectivity [43–45].

Theoretical calculations (B3LYP/LANL2DZ) indicated a stepwise mechanism for this cycloaddition where the first step is critical for the stereochemical outcome. Furthermore, the high regioselectivity has been accounted for by the calculated nucleophilic and electrophilic Fukui indexes.

Chiral functionalization of endofullerenes represents one step further in the application of chiral metal catalysis. The first chiral endohedral metallofullerenes were recently prepared by using such methodology on a racemic mixture of a non-IPR metallofullerene derivative, La@C_{72}($C_6H_3Cl_2$) [48]. This mono-functionalized metallofullerene was chosen due to the calculated energy level of the LUMO orbital being suitable to give rise to 1,3-dipolar cycloaddition. Despite this complexity, eight optically pure bis-adducts of La@C_{72} (four from the clockwise starting material enantiomer and four from the anticlockwise enantiomer) were isolated using non-chiral HPLC. These endohedral fulleropyrrolidines resulted from the addition onto only 2 sites among 108 possible addition sites. For each site, two 2,5-disubstituted pyrrolidine regioisomers are formed with fixed (2*S*,5*S*) and optical purities as high as 98% (Scheme 7). All the isomers feature a strong Cotton effect that is dominated by the inherent chirality of the carbon core. The four isomers

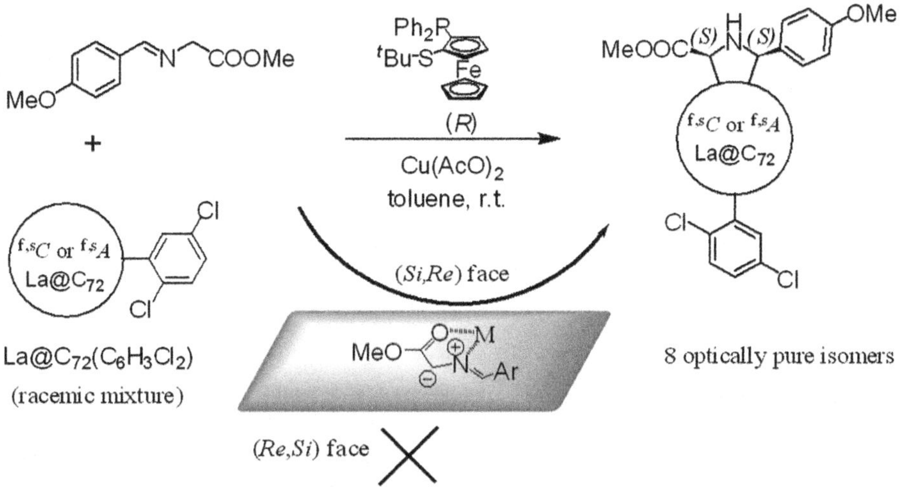

Scheme 7 The 1,3-dipolar cycloaddition reaction of N-metalated azomethine ylide on $La@C_{72}(C_6H_3Cl_2)$ (as a racemic mixture), using Cu(II) Fesulphos complex as a catalyst, affords eight different optically active pyrrolidines with a fixed (2*S*,5*S*) configuration

formed from the clockwise enantiomers exhibit circular dichroism (CD) spectra that are opposite in sign to those derived from the anticlockwise enantiomer.

2.3 Retro-Cycloaddition Reactions of Fullerene Cycloadducts

Among the many well-known exohedral reactions developed on the fullerene sphere, cycloaddition reactions have played a prominent role with applications in fields such as medicinal chemistry [49] and materials science [50]. The fullerene derivatives obtained through this functionalization method display, in general, an acceptable degree of stability; however, in the last few years a number of studies have reported efficient retro-cycloaddition methodologies for the most important fullerene cycloadducts. In this section we will describe the different retro-cycloaddition conditions for each type of fullerene derivative, and the applications of these methodologies to carry out protection–deprotection protocols that could selectively add or remove addends from fullerenes while leaving others unperturbed.

2.3.1 Retro-Diels–Alder Reaction

Fullerenes are excellent dienophiles that can undergo [4+2] cycloaddition reactions with different dienes such as anthracene or cyclopentadiene. This reaction is controlled by the properties of the dienes and can proceed at room temperature, at reflux or under microwave irradiation. The rate of Diels–Alder reaction is affected

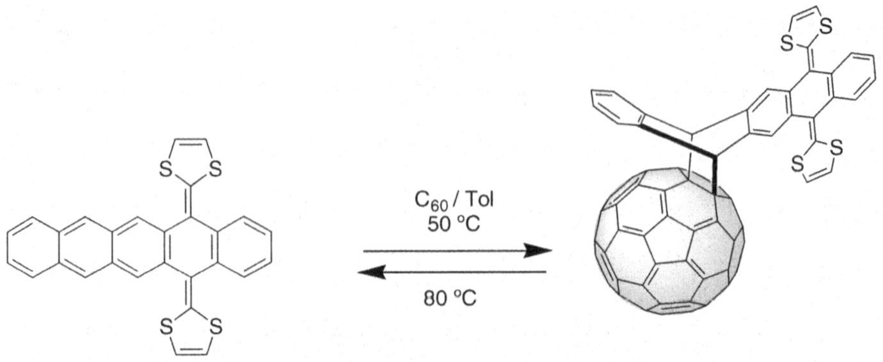

Scheme 8 Thermally reversible C_{60}-based donor–acceptor ensembles

by the overall gain or loss of aromaticity of the dienophile (fullerene core) and/or the diene [51]. Most of these Diels–Alder adducts are thermally unstable and can undergo efficient retro-Diels–Alder upon mild heating [51]. Even though the regiochemistry of Diels–Alder additions is low, a few exceptions have been reported. In this regard, the reaction of C_{60} with an excess of 2,3-dimethyl-1,3-butadiene at elevated temperatures yields a hexakis-adduct with T_h symmetry with an average effective selectivity >80% for each addition step [52].

In our research group we have investigated the Diels–Alder cycloaddition of anthracene derivatives bearing fused π-extended tetrathiafulvalenes (TTFs) to C_{60} to yield thermally reversible donor–acceptor materials (Scheme 8) [53]. NMR and cyclic voltammetry experiments allowed the determination that the retro-cycloaddition process starts around 50°C and continues during the 50–80°C range. Thus, taking advantage of this finding, this donor–acceptor compound is able to act as an ON/OFF switch, using non-fluorescing Diels–Alder adducts of C_{60} which, upon heating, revert to the starting materials.

2.3.2 Retro-Cyclopropanation Reaction

In the Bingel–Hirsch reaction, the deprotonation of an α-halomalonate leads to a nucleophilic anion which attacks the fullerene core to yield methanofullerenes [54]. Using this methodology a large variety of fullerene derivatives have been described and, in general, they are stable in air and under high thermal and oxidative conditions. However, these derivatives can efficiently undergo a retro-cyclopropanation reaction under reduction conditions.

The electrochemical retro-Bingel reaction of the (alkoxycarbonyl)methanofullerenes of C_{60}, C_{70}, C_{76}, and *ent*-C_{76} at the second reduction potential by controlled potential electrolysis (CPE) has been reported by Echegoyen, Diederich and co-workers [55]. The retro-cyclopropanation by CPE is selective for the methano-addend, which must have at least one strong electron-withdrawing group. The presence of other groups such as cyclohexene, pyrrolidine, and

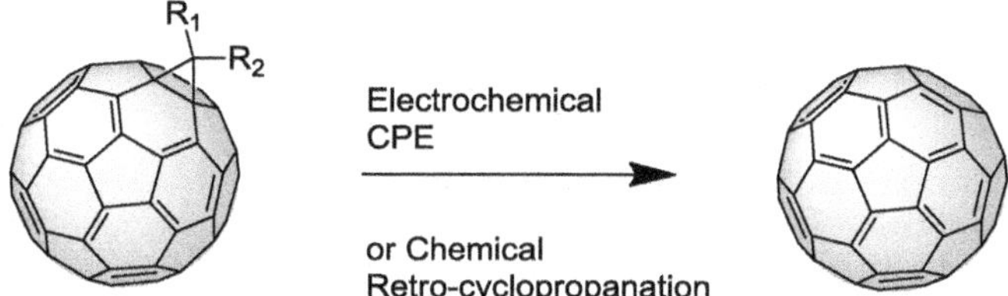

Scheme 9 Different approaches to induce the retro-cyclopropanation reaction

benzocyclobutene rings fused to the [6,6]-bond were not affected by CPE [56]. This singularity offered a new and versatile protecting/deprotecting group strategy. The chemical retro-cyclopropanation of C_{60} and C_{70} mono-adducts was reported in 2000 by Diederich, Echegoyen and co-workers [57]. In this case, the methano-addend was removed from C_{60} and C_{70} after heating at reflux with amalgamated magnesium powder (10% $HgBr_2$) for 3 days, yielding 73% of C_{60} and 63% of C_{70}, respectively. In the case of bis-adducts of C_{60}, the yield of recovered C_{60} varied between 13% and 48% and no isomerization reactions were detected. These chemical conditions can selectively remove only the methano addends in the presence of other functional groups such as pyrrolidines, offering a versatile protecting/deprotecting group strategy (Scheme 9) [57].

2.3.3 Retro-1,3-Dipolar Cycloaddition Reactions

Among the 1,3-dipolar cycloaddition reactions, the addition of azomethine ylides is considered as one of the simplest and most efficient procedures for the functionalization of fullerenes [58]. Azomethine ylides are reactive intermediates that can be generated in several ways, although the decarboxylation of iminium salts, derived from condensation of α-amino acids with aldehydes or ketones, is the easiest and most general procedure commonly followed. Pyrrolidinofullerenes are stable compounds, although in the last few years a number of methodologies have been described to promote the retro-cycloaddition reaction of these derivatives to afford pristine C_{60} fullerene.

Martín et al. [59] have recently described the thermally induced retro-cycloaddition of pyrrolidino[3,4:1,2]fullerenes. The authors studied the retro-cycloaddition process on a series of pyrrolidinofullerenes under a variety of experimental conditions. The best results were obtained heating the corresponding fulleropyrrolidine in the presence of a dipolarophile such as maleic anhydride and copper triflate. Under these conditions, the reaction led to the quantitative formation of the parent unsubstituted C_{60} in all cases. This methodology was also effective in inducing the retrocycloaddition for the mono-adduct mixture of three isomers of [70] fulleropyrrolidine which afforded pristine C_{70} in 95% yield.

Another important finding was the use of C_{60} as a dipolarophile. A mixture of a fullerene bis-adduct was heated to reflux in *o*-DCB in the presence of C_{60} and

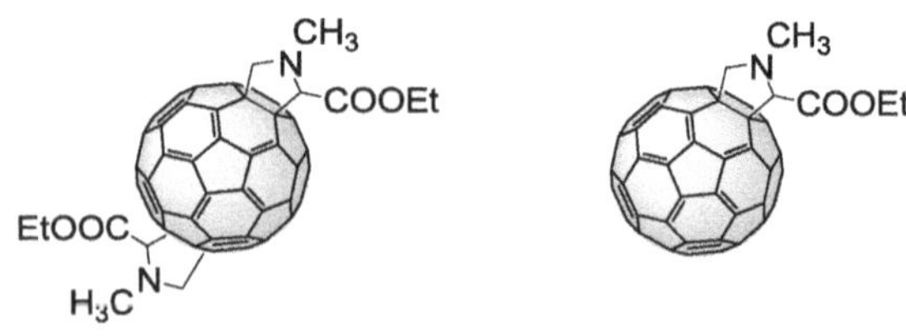

Fig. 5 A mixture of bis-adducts (*left*) in the presence of C_{60} as dipolarophile produces monoadduct (*right*)

HPLC analysis of the reaction confirmed the quantitative formation of the corresponding monoadduct (Fig. 5) [59]. This interesting result opened new avenues for improving the yields of mono-adduct formation from the usually undesired bis-cycloadducts obtained as byproducts in 1,3-dipolar cycloaddition reactions of azomethine ylides to C_{60}. This observation has recently been used as an indirect proof of the covalent attachment of pyrrolidine fragments to single-walled carbon nanotubes (SWCNT). The authors applied the same conditions described by Martín et al. [59] by heating a sample of pyrrolidino-SWCNT in the presence of C_{60} to act as dipolarophile and the corresponding pyrrolidinofullerene compound was detected, thus confirming the efficient trapping of the thermally generated azomethine ylide [60].

Guryanov et al. [61] described an alternative protocol to achieve the retro-cycloaddition of pyrrolidinofullerenes. The authors reported the quantitative retro-cycloaddition of pyrrolidinofullerenes under microwave irradiation in an ionic liquid (1-methyl-3-*n*-octyl-imidazolium tetrafluorborate) without further additives. The combination of microwave irradiation in an ionic liquid offers the unique opportunity for very efficient flash-thermal activation in conjunction with a strong stabilization of ionic intermediates or reactants. In this case, the ionic liquid served as an ideal medium to solvate the incipient 1,3-dipole whose release in solution was likely assisted by electrostatic interactions with the complementary ions of the solvent. In agreement with the mechanism proposed by Filippone et al. [62], cycloreversion occurred through the formation of a reactive 1,3-dipolar intermediate which was expected to be stabilized by the ionic liquid environment.

Lukoyanova et al.[63] had reported an alternative way to induce the retro-cycloaddition of pyrrolidinofullerenes using electrochemical techniques. The authors induce the retro-process by controlled potential electrolysis (CPE) at an applied potential determined from cyclic voltammetry experiments.

In order to determine whether the experimental conditions previously used for the retro-cycloaddition of fulleropyrrolidines and fulleroisoxazolines [64] are suitable for 2-pyrazolinofullerenes, Delgado et al. [65] followed the above-mentioned protocol: excess of dipolarophile, as well as copper triflate to facilitate the retro-cycloaddition process. According to the experimental findings, *C*-aryl-*N*-aryl-2-pyrazolino[60]fullerenes do not undergo an efficient retro-cycloaddition process under a variety of experimental conditions, which reveals that these compounds are thermally stable fullerene derivatives. In contrast, the presence of an alkyl chain in the carbon atom of the pyrazole ring results in an easier cleavage of the 1,3-dipole, leading to pristine C_{60} in good yields (72%). These results show the importance of thermal stability in order to prepare new C_{60}-based materials, as well as the nature of the substituents (alkyl or aryl) which has a strong influence on the thermal stability of the cycloadducts. In particular, this is a key issue in photovoltaic cells, where long exposure to sunlight results in drastic temperature increases of the photo- and electro-active materials [65].

3 Macromolecular Chemistry of Fullerenes

Soon after the first protocols for the chemical functionalization of fullerene were set up, the rush for its incorporation in polymeric backbones started [66]. The main aim was to combine the well established processability, ease of handling, and toughness of polymers with the rather unique pool of properties of fullerenes in order to achieve new materials with combined features or even with unprecedented properties. Obviously, the final scope of the preparation of such hybrids is their application in cutting edge technologies. Nevertheless, due to the incomplete disclosure of the chemistry of fullerene, the firsts attempts to achieve fullerene-polymers led to uncharacterisable or inutile materials often obtained by employing empirical synthetic methods. Fortunately, these problematics were overcome with the development of well established methodologies for the modification of fullerenes, and nowadays the macromolecular–fullerene hybrids can be designed and tailored at will and fully characterized. Thanks to the synthetic versatility of polymers, several examples of polyfullerenes have been reported to the date, which can be classified accordingly to their chemical structure or to their properties as well as to their applications [67–69]. Herein we use a “classical” classification in which each family is structurally homogeneous and few common protocols can be followed to prepare their corresponding members.

3.1 Classification and Synthetic Strategies

An easy way to order polyfullerenes is accordingly to their increasing chemical complexity and difficulty in preparing them (Fig. 6). The synthesis of macromolecular fullerenes may be as simple as the mixing of C_{60} and a polymeryzing reactant, or may require several carefully controlled reaction steps leading to unprecedented superarchitectures.

Even simpler, in some ways, is the preparation of all-carbon fullerene polymers [70, 71], which are all those materials constituted exclusively by fullerene units covalently linked to each other without any additional linking groups or side groups. The members of this family are prepared by exposing pristine fullerene to a strong external stimulus such as visible light [72], pressure [73], and plasma irradiation [74], with no problems in getting the final structure. During the polymerization, [2+2] cycloaddition reactions between two double bonds of two neighboring C_{60} molecules take place generating new cyclobutane rings [75].

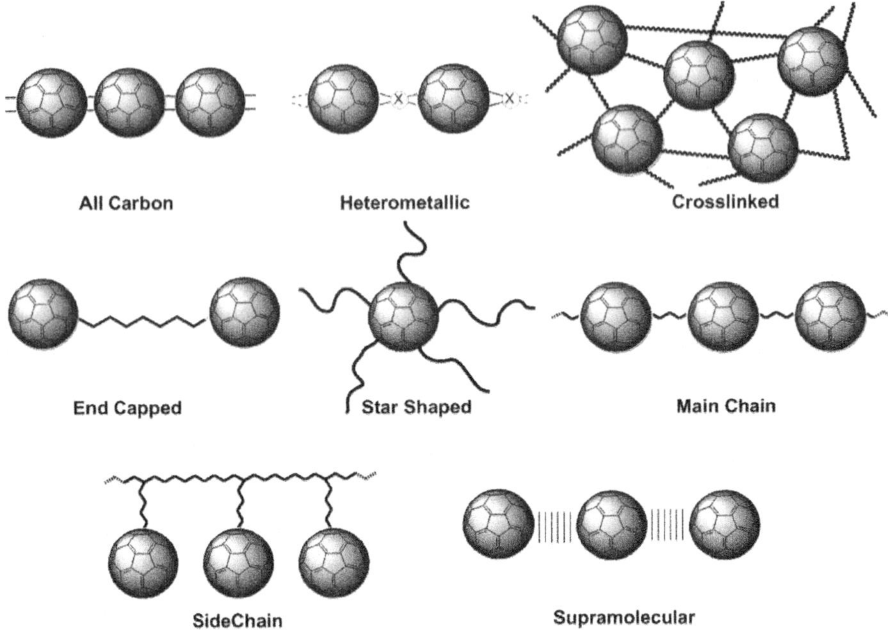

Fig. 6 Structural classification of the different families of C_{60}-polymers

Analogously, heterometallic polymers, a family of heteroatom-containing polymers, in which elements other than carbon are present in their structures, are obtained by means of charge-transfer polymerization mediated by metals [76] and also by electro-reduction in the case of fullerene epoxide [77, 78].

Another family of polyfullerenes that needs little chemical control is the cross-linked set. Their synthesis usually proceeds from tridimensional random and quick reactions involving several of the 30 equivalent double bonds of the fullerene cage. Nevertheless, some control over the addition reactions is required in order to avoid a dramatic intractability of the final products. Different synthetic strategies have been followed till now for their preparation including the reaction between fullerene (or a C_{60}-derivative) and a monomer or with the pending groups (or the end termini) of preformed polymers.

On the other hand, the incorporation of one or two fullerene units at the terminal positions of linear polymeric chains leads to the end-capped C_{60}-polymers. It is worth noting that the presence of C_{60} moieties strongly influences both the molecular and the bulk behavior of the parent polymers as a result of the modification of their hydrophobicity. Two different synthetic strategies have been employed to prepare this class of polymers: the capping of a polymer chain with C_{60} or the growth of a polymeric backbone from the surface of a fullerene moiety or a C_{60}-derivative, the so called "*graft to*" and "*graft from*" approaches. These protocols can also be easily used in the synthesis of the members of the star-shaped polymers, which are constituted by 2–12 long and flexible polymer chains covalently linked to a fullerene cage with topologies similar to that of sea-stars.

One of the less explored families of fullerene polymers is the main-chain one. Here, the C_{60} spheres are directly allocated in the polymer backbone forming a necklace-type structure. Unfortunately, the double addition on the C_{60} moieties results in the formation of complex regioisomeric mixtures (up to eight isomers) and also the formation of cross linking products by multiple additions can occur.

The preparation of these in-chain polymers can be carried out by direct reaction between the C_{60} unit and a suitable symmetrically difunctionalized monomer or by means of polycondensation between a fullerene bisadduct (or a mixture) and a difunctionalized monomer. In contrast, side-chain polymers, sometime called *on-chain* or "*charm-bracelet*," represent the most studied family of polyfullerenes and show the C_{60} pending from the main polymer chain. A century of studies on polymers has been exploited in the binding of C_{60} to all the "classic" families of polymers such as polystyrenes [79, 80], polyacrylates [81, 82], polyethers [83], polycarbonates [84], polysiloxanes [85], and polysaccharides [86] in the search for improved processability and enhanced properties, with a wide range of potential applications. In this family can also be included the "double-cable" polymers [87, 88], in which the π-conjugated semiconducting polymer (p-type cable) with electron-donating characteristics contains electron-accepting fullerene units covalently connected (n-type cable), with remarkable advantages for construction of photovoltaic devices. The synthesis of the members belonging to this family can be achieved by direct introduction of fullerene itself (or a C_{60}-derivative) into a preformed polymer, or by homo-/co-polymerization of a suitable C_{60}-derivative. Moreover, for the double-cable polymers, electropolymerization is also possible.

Finally, the most recent family of macromolecular fullerene is that composed by supramolecular polymers created through any type of self-assembly and via reversible interactions of one or more types of components. Interestingly, these reversible interactions can often allow supramolecular polymers to equilibrate thermally with their monomers or oligomers. There are several ways to obtain such supramolecular assemblies; among them several systems may be obtained by interactions between functionalized polymers and C_{60} derivatives or fullerene itself or through the assembly of self-complementary C_{60} derivatives. More recently, assemblies between ditopic concave guests and [60]fullerene by means of concave–convex complementary interactions have also been reported [89, 90].

3.2 Properties and Applications

Despite hundreds of examples of polyfullerenes having been reported over the last two decades, to date this class of smart material has not had a real application. However, the progress achieved year by year reveals new potential applications and improved properties with respect to early examples, or even with respect to the state of the art, as in the case of photovoltaic applications. In the present section, some of the most promising and recent applications for C_{60}-polymers will be shown. Since 1991 fullerene polymers have been studied as active materials in membranes both for gas separation [91, 92] and for proton exchange fuel cells [93], and also as active

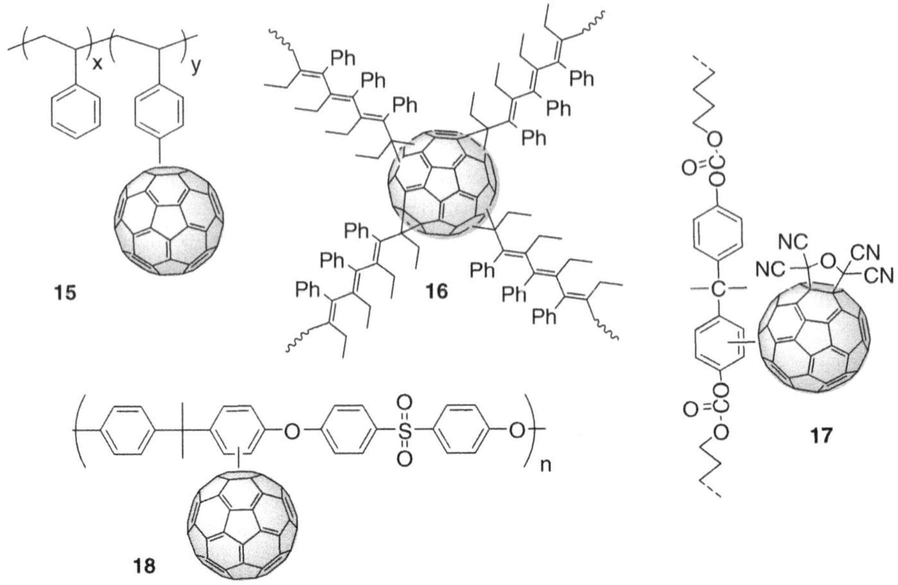

Fig. 7 Chemical structures of polyfullerenes with NLO properties

polymers in electroluminescent devices [77, 94, 95] and in non-volatile flash devices [96], among others. One of the most promising applications for polyfullerenes seems to be as an active layer in optical limiters. These are devices that strongly attenuate optical beams at high intensities while exhibiting higher transmittance at low intensities, which are used for protecting human eyes and optical sensors from intense laser beams. In this regard, C_{60} is an excellent optical limiter operating in a reverse saturable absorption mode [97], and its derivatives may be considered as potent broadband optical limiters due to the broad coverage of the characteristic ground- and excited-state absorptions over a wide wavelength range [98, 99]. However, the poor solubility and processibility of fullerene has limited its use in optical limiting devices and for this reason, in order to overcome such a drawback, C_{60} has been incorporated in polymeric matrices. In Fig. 7 some examples of fullerene-containing polymers used for optical limiting are collected.

In 1995 random solid fullerene-containing polystyrenes (PS) showed nonlinear optical (NLO) properties about five times greater than those of a C_{60} solution [100, 101]. Later on, random poly(methyl methacrylate) [102] and linear PS-containing fullerene **15** (Fig. 7) [103] were found to be optical limiters. In 2000 the star shaped C_{60} containing poly(1-phenyl-1-butyne) **16** turned out to be stable, and film-forming with a fullerene content of up to 9.1 wt%. NLO studies revealed that at 532 nm (Nd:YAG laser) **16** shows improved optical limiting performances compared to C_{60} in solution at a linear transmission of 43% [104]. In the same year the side chain polyfullerene **17**, prepared by Friedel–Crafts addition of C_{60} and of a C_{60}-derivative to a polycarbonate, also displayed good optical limiting properties

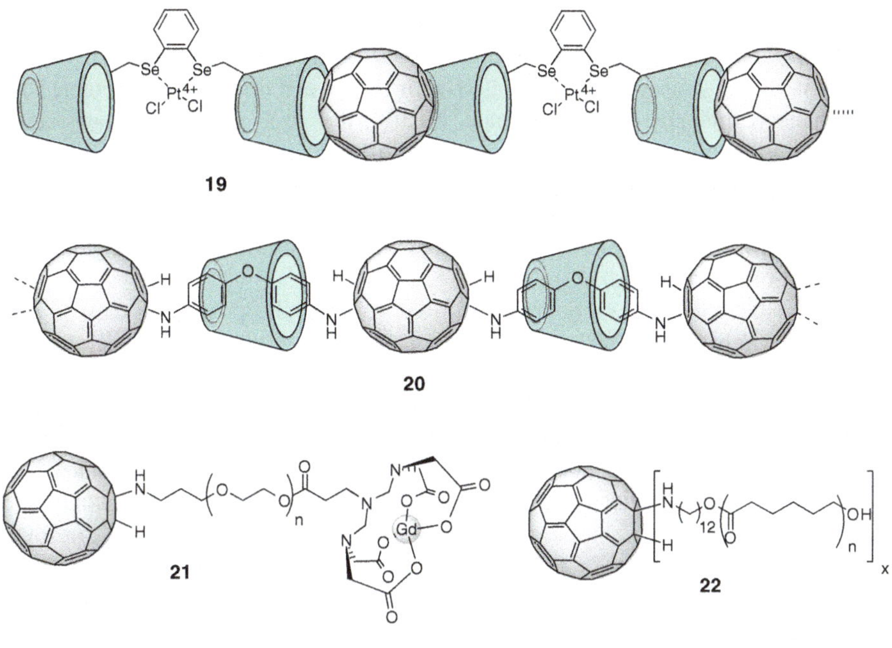

Fig. 8 Structures of DNA-cleaving and fullerene-polymers for PCT

[105]. Interestingly, the polycarbonate endowed with the fullerene derivative behaved better than that functionalized with fullerene itself. Very recently the same synthetic protocol has been used in order to prepare fullerene-functionalized polysulfones **18** [106]. These materials show not only a very high thermal stability and glass transition temperatures depending on the C_{60} content, but also optical limiting properties.

On the other hand, it is well known that fullerene may act as a scavenger. When excited in the ultraviolet region (340–400 nm) it generates reactive oxygen species (ROS) acting as an effective photosensitizer, also useful in the visible-light cleavage of DNA in the photodynamic cancer therapy (PCT) [107]. However, in order to be successfully employed C_{60} needs to be transformed in a water-soluble derivative and, among other things, its incorporation in hydrophilic polymers proved to be an excellent opportunity. In this regard, different polyfullerenes have been tested as DNA cleavers such as the supramolecular polymer **19** [108] formed by fullerene units complexed within the upper rims of cyclodextrin dimers (Fig. 8). Cyclodextrins were also employed in order to afford water solubility to the main chain polymer **20**, formed through nucleophilic polyaddition reaction between C_{60} and the β-cyclodextrin-bis(*p*-aminophenyl) ether [109]. Once again, under visible light conditions **20** proved to be a highly efficient DNA-cleaving agent for oligonucleotides. More recently a photosensitizer with magnetic resonance imaging (MRI) activity has been achieved by linking polyethylene glycol to fullerene at one end and the diethylenetriaminepentaacetic acid Gd^{3+} complex at the other

(**21**, Fig. 8) [110]. Intravenous injection of **21** into tumor bearing mice followed by irradiation showed a significant anti-tumor PCT effect which depended on the timing of light exposure that correlated with tumor accumulation as detected by the enhanced intensity of MRI signals. Finally, the photoactivity of star-shaped poly(ε-caprolactone)-C_{60} **22** has been successfully proven in the benchmark transformation of 9,10-anthracene dipropionic acid into its endoperoxide [111]. In fact, large amounts of $^1O_2$ have been obtained upon irradiation of **22** with visible light. These kinds of polymers are of interest especially because they are biodegradable, biocompatible, and non-toxic to living organisms.

Among the possible practical applications of fullerene-polymers, their use as electron acceptors in the active layer of organic photovoltaic devices seems to be one of the most realistic. Despite the disappointing results, in terms of efficiency, associated with the "double-cable" polymers, investigations of which were time consuming, the very latest few years are witnessing a new momentum in the use of polyfullerenes. This is probably due to improved technologies and methodologies that are allowing scientists to achieve efficiencies comparable to those obtained in bulk heterojunction cells based on molecular C_{60}-derivatives. The first improvement in organic solar cells performances has been obtained by using a new approach in which the glycidol ester of [6,6]-phenyl C_{61}butyric acid has been first prepolymerized in the presence of a Lewis acid as the initiator (**23**, Fig. 9) [112]. Second, after spin coating the prepolymer in a blend with P3HT (poly-3-hexyl thiophene), the ring-opening polymerization has been completed by heating the photovoltaic device which showed 2.0% conversion energy efficiency, probably due to the morphological stabilization of the active layers architecture. Soon after, a new approach was reported in which the amphiphilic diblock-polymer **24** carries both the C_{60} units and P3HT fragments acting as a compatibilizer between PCBM and P3HT in the active layer [113]. When added to a blend of PCBM: P3HT at 17 wt% the photovoltaic device prepared displayed an efficiency of 2.8%, along with enhanced stability of the devices against destructive thermal phase segregation. This improvement has been accounted by for the higher control in the blend morphology of the active layer due to the presence of fragments of P3HT in the polymer backbone which act as compatibilizer between PCBM and P3HT. Analogously, rod-coil block copolymer **25** has been used at various concentrations as surfactant/compatibilizer for the active layer of bulk-heterojunction solar cells in blends with PCBM [114]. This approach resulted in 35% increase of the photocurrent efficiency, increasing from 2.6% to 3.5% when the copolymer was used at 5 wt%. Such enhancement has been ascribed to the improvement in the bicontinuous interpenetrating network due to the compatibilizing action of the copolymer, as also evidenced by AFM studies.

Finally, a revolutionary approach has recently been described in which the cross-linked C_{60}-polymer **26** is generated in situ allowing the subsequent deposition of the active layer to avoid interfacial erosion [115]. The inverted solar cell ITO/ZnO/**26**/P3HT:PCBM/PEDOT:PSS/Ag showed an outstanding device performance with a PCE of 4.4% and an improved cell lifetime with no need for encapsulation. The strength of this new approach is its wide and general application. In fact, changing

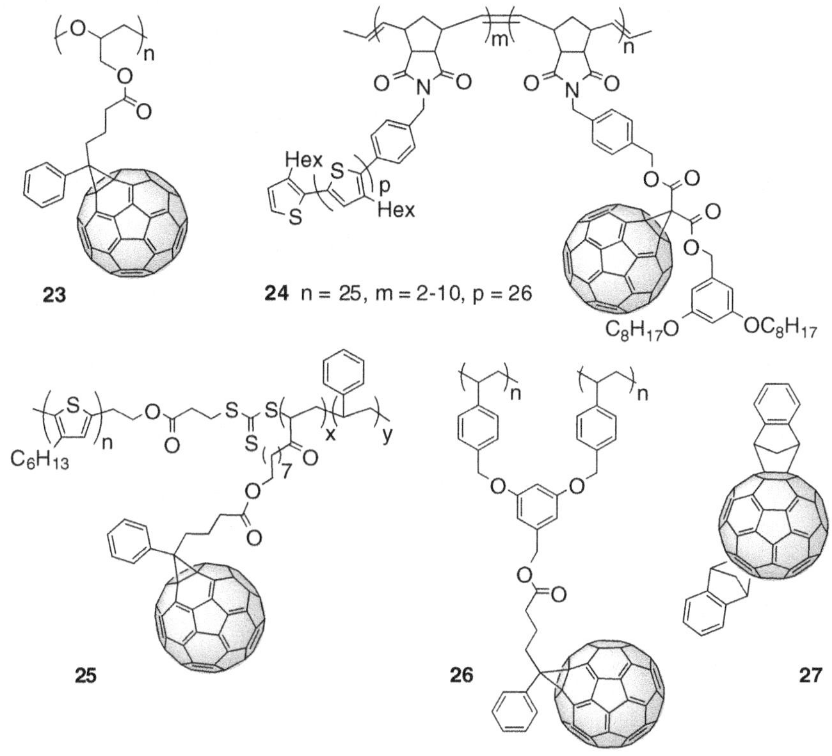

Fig. 9 Fullerene-containing polymers used in organic photovoltaic devices

PCBM with fullerene bis-adduct **27** as acceptor in an inverted bulk heterojunction cell with architecture ITO/ZnO/**26**/P3HT:**27**/PEDOT:PSS/Ag gave rise to the impressive value of 6.22% of efficiency, which retains 87% of the magnitude of its original PCE value after being exposed to ambient conditions for 21 days [116].

4 Supramolecular Chemistry of Fullerenes

The examples of covalent modification of fullerenes we have seen so far imply the saturation of at least one of the double bonds of the polyenic structure. This might be beneficial or detrimental for the application in mind. Alternatively, the chemist can choose to interact with the fullerenes or their derivatives in a supramolecular fashion, making use of weak noncovalent interactions. In the following we will very briefly review some of the most important advances in H-bonded fullerene assemblies as well as the host–guest chemistry of fullerenes, organized according to the main type of noncovalent interaction present in the associates.

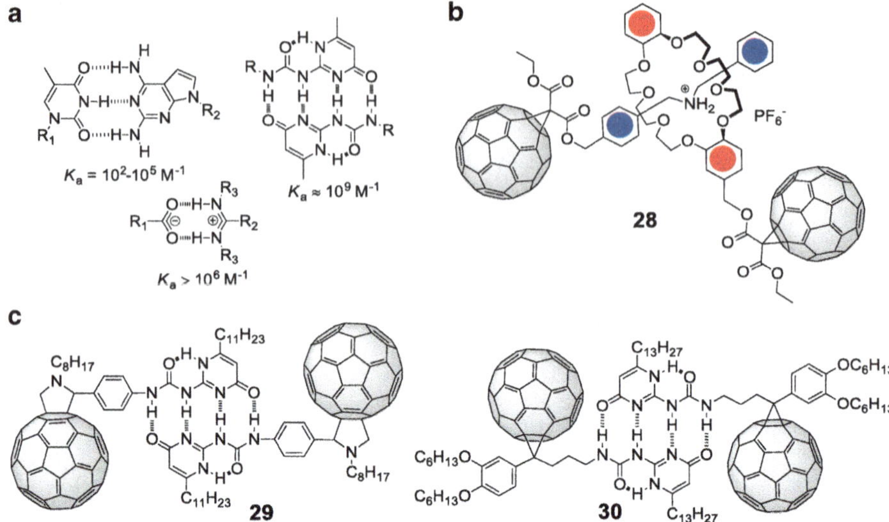

Fig. 10 (**a**) Illustration of different H-bonded supramolecular architectures. (**b**) First C_{60}-based supramolecular dimer. (**c**) Chemical structures of supramolecular dimers of C_{60}

4.1 H-Bonded Fullerene Assemblies

H-bonds are, perhaps, the best studied non-covalent interactions. Although they are weak interactions, with binding energies in the range of ~ 5 kcal/mol, hydrogen bonds are also selective and directional [117]. When molecules interact by forming two or more hydrogen bonds, secondary electrostatic interactions can give rise to dramatic differences in the stability of the supramolecular complexes. The combination of different non-covalent interactions, such as ionic, π–π interactions, etc., with hydrogen bonds allows one to modulate the affinity between the interacting molecules, giving rise to a wide spread of supramolecular architectures (Fig. 10).

The importance of hydrogen bonds in determining the geometry and, overall, the function of biomolecules such as DNA, RNA, proteins, tobacco mosaic virus, and so forth is well known. Another natural example comes from the photosynthetic apparatus, in which a highly ordered supramolecular array of electron-donors (chlorophylls) and electron-acceptors (quinones) harvests and converts sunlight into chemical potential energy through cascades of short-range electron transfer steps [118]. Inspired by the natural photosynthetic event, intramolecular photoinduced electron transfer processes have been thoroughly studied in different covalent and non-covalent systems formed by donor and acceptor electroactive moieties for their implementation in molecular electronic devices [119, 120]. In this context, fullerene C_{60} has probably been the most studied electroactive entity owing to its unique electron-acceptor properties and low reorganization energy in electron transfer processes.

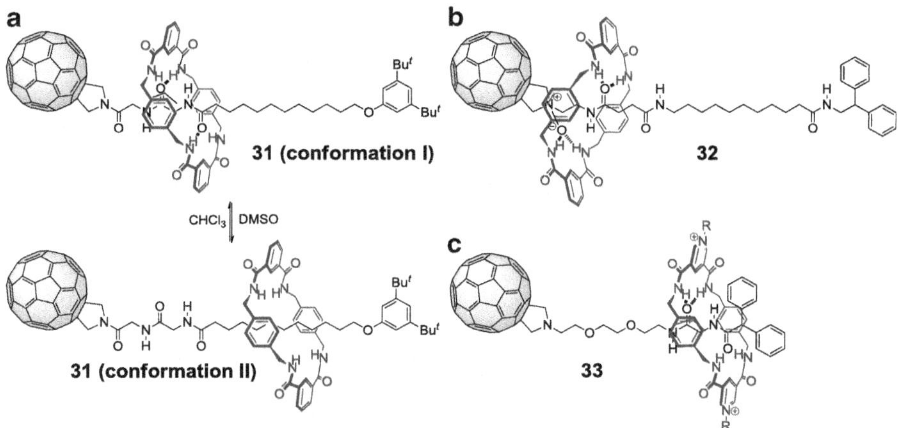

Fig. 11 (**a**) Solvent switchable rotaxane. (**b**) Stabilization of an N-oxide fulleropyrrolidine by encapsulation in the rotaxane. (**c**) Electrochemically driven molecular shuttle

The first example of a supramolecular H-bonded architecture involving fullerene was a dimer with a pseudorotaxane-type structure (**28**, Fig. 10) reported by Diederich et al. [121]. For this dimer a K_a value of ~970 M^{-1} was obtained. More robust supramolecular dimers of fullerene were obtained by employing a quadruple array of H-bonds based on 2-ureido-4-pyrimidone (UP) (**29** and **30**) [122, 123]. In these cases the DDAA H-bonds disposition gives rise to complementary pairs with affinities in the range of $K_a \sim 10^6\ M^{-1}$.

Leigh-type clipping reactions have allowed the preparation of several fullerene containing rotaxanes. The first bistable rotaxane containing a fullerene is depicted in Fig. 11 (**31**) [124]. Depending on the polarity of the solvent, the macrocycle resides preferentially over the glycylglicyne unit or over the alkyl chain. A slight modification of the structure of the thread allows the macrocycle to shuttle in the opposite direction, from the amide to the fullerene moiety [125].

This displacement of the macrocycle has been used to increase the stability of a fulleropyrrolidine N-oxide (**32**, Fig. 11) [126]. The formation of hydrogen bonds between the macrocycle and the N-oxide inhibits the deoxygenation reaction, thus enhancing the stability of the N-oxide derivative.

Remarkably, the shuttling of the macrocycle can also be stimulated electrochemically by the formation of the fullerene trianion [127]. This electrochemically driven molecular shuttling has recently been achieved at very low reduction potential by introducing positive charges in the macrocycle (**33**, Fig. 11). In the latter case, only one electron is needed to induce the operation of the shuttle [128].

The first reported donor–acceptor supramolecular dyad based on C_{60} is a pseudorotaxane formed between the dibenzylammonium salt of a Bingel-type fullerene and the crown ether of a zinc phthalocyanine (**34**) [129]. This complex has an association constant of ~ $1.4 \times 10^4\ M^{-1}$ and, what is more important, an

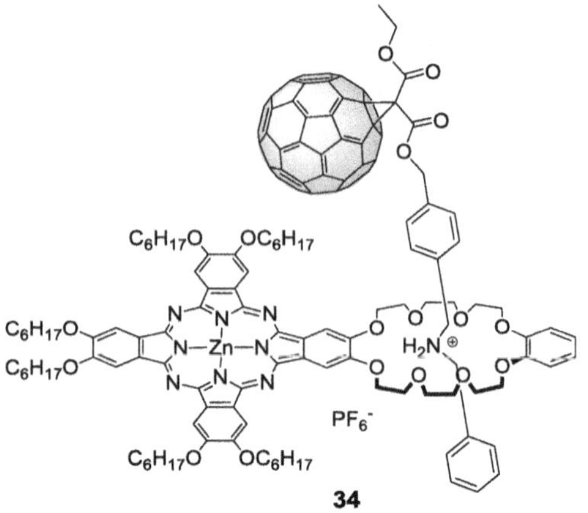

Fig. 12 First fullerene containing supramolecular donor–acceptor dyad

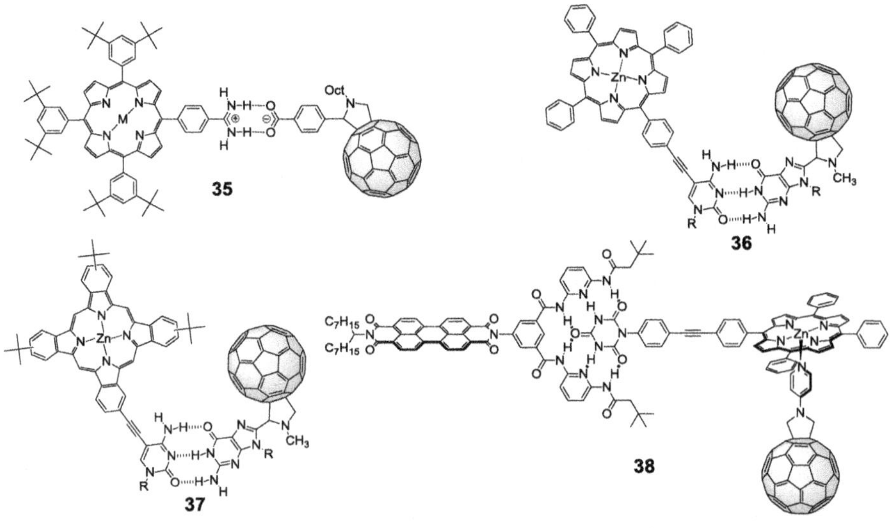

Fig. 13 Donor–acceptor dyads with different H-bonding motifs

intracomplex photoinduced electron transfer process leads to a radical pair species ($C_{60}^{\bullet -}$–$ZnPc^{\bullet +}$) with a remarkable lifetime in the region of microseconds (Fig. 12).

Since that, a plethora of donor–acceptor dyads in which the acceptor unit is a fullerene C_{60} have been studied. These studies have revealed that a strong electronic communication is found through the supramolecular ensembles. That is the case for dyad **35** (Fig. 13) [130], formed by a two point amidinium–carboxylate bonding motif which gives rise to an extraordinarily high affinity, with a K_a value up to $10^7\ M^{-1}$ in toluene. The strong electronic coupling (36 cm^{-1}) facilitates the formation of a long-lived radical pair with a lifetime of ~1 μs in THF.

Dyads **36** and **37** employ a three point guanosine–cytidine couple to form fullerene based hybrids with a porphyrin or phthalocyanine respectively. While

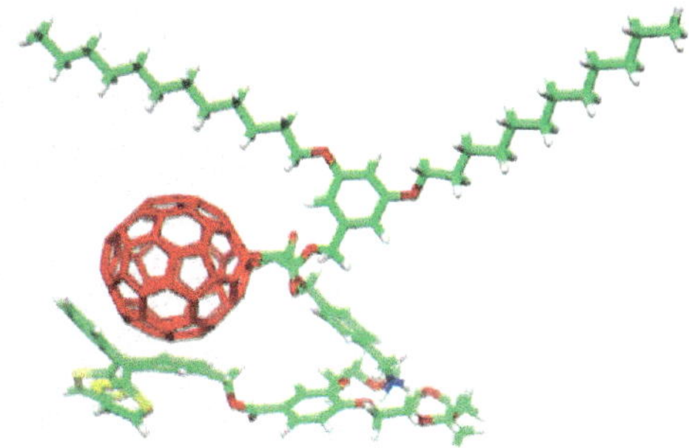

Fig. 14 Molecular model of exTTF-crown ether/fullerene ammonium salt complex

for the first dyad the formation of a charge separated state after irradiation with a long lifetime of ~2.02 μs was observed, the lifetime of the radical pair in the case of the zinc phthalocyanine dyad was only 3 ns [131, 132]. This difference has been ascribable to a pronounced coupling between the ZnPc and C_{60} in **37**, which is reflected in a large binding constant of $1.7 \times 10^7\ M^{-1}$ (vs a K_a of $5.1 \times 10^4\ M^{-1}$ found for **36**).

The six-point Hamilton array has also been used to form highly stable dyads when interacting with a cyanuric acid moiety, with association constants usually in the range of 10^4–$10^5\ M^{-1}$ for monotopic receptors [133] and even higher when a ditopic Hamilton receptor is employed [134]. In recent work, Hirsch et al. have employed this Hamilton receptor together with metal complexation to control the step-by-step assembly of the different components in triad **38** [135]. In this triad the perylenediimide (PDI) moiety acts as a light harvester unit and, after selective photoexcitation, an energy transfer to the porphyrin unit takes place, which has been corroborated by steady-state and time-resolved measurements. The energy transfer is followed by an electron transfer event from the porphyrin to the axially complexated fullerene moiety driving to the formation of a radical ion pair with a lifetime of 3.8 ns. This lifetime is longer than that of the corresponding dyad from the porphyrin and the fullerene.

Tetrathiafulvalene (TTF) and π-extended tetrathiafulvalene (exTTF) have also been used as the electron donor counterpart in D–A nanohybrids with fullerenes in supramolecular assemblies. In pseudorotaxane-type structures between exTTF and C_{60}, it has been found that the interaction between the donor moiety of exTTF and the acceptor unit of fullerene is stronger when a flexible spacer allows the involvement of the intramolecular interaction between the convex fullerene surface and the concave face of exTTF (Fig. 14) [136]. In this case, a binding constant of $1.58 \times 10^6\ M^{-1}$ in chlorobenzene was observed, more than two orders of magnitude higher than the K_a values obtained for related systems in which only the crown ether-ammonium salt motif can interact. For this dyad a remarkable anodic shift of ~ 100 mV of the oxidation potential of exTTF reflects the strong donor–acceptor interaction. Transient absorption spectroscopy showed the formation of a radical ion pair with a lifetime of 9.3 ps.

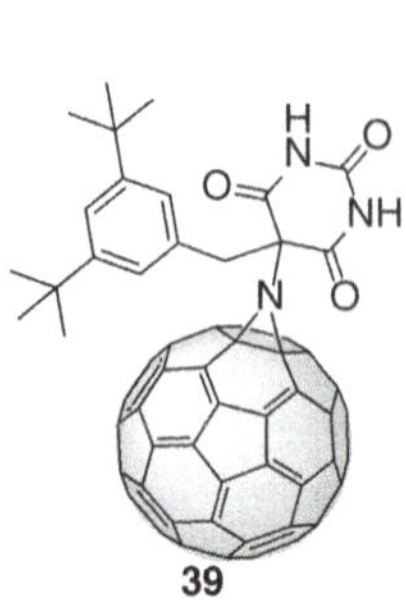

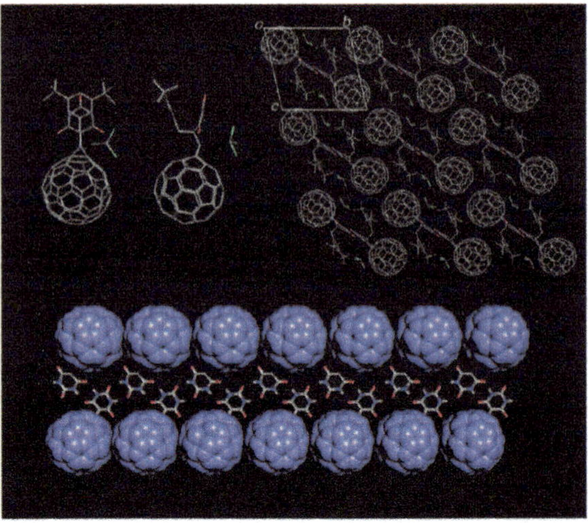

Fig. 15 Chemical structure and solid-state crystal structure of hydrogen-bonding barbiturate fullerene **39** highlighting the close van der Waals contacts between fullerenes in the H-bonding ribbon. (Reprinted with permission from [137]. Copyright 2010 American Chemical Society.)

An appealing topic in molecular electronics is the use of supramolecular interactions to control the assembly of the electron donors and acceptors in order to obtain highly ordered supramolecular entities with specific functions. In this context, Bassani et al. reported the hierarchical self-assembly of a barbituric acid appended fullerene and a thiophene oligomer substituted with a melamine moiety to build a photovoltaic device. They found that the photocurrent is 2.5-fold greater in this device than in analogous ones constructed with fullerene C_{60} and the oligomer without the H-bonding units, which is attributable to higher order at the molecular level [137].

Recently, trying to take advantage of the ability of supramolecular assembly to control the electronic interactions between the fullerene units, OFET devices were constructed with derivative **39**, which combines the solubilizing 3,4-ditertbutylbenzene group with the barbituric acid motif [138]. The fabricated OFET devices showed a mobility approximately two orders of magnitude lower than the devices constructed with pristine fullerene, owing to the anisotropy of the electrical conductivity of the crystals of **39** (Fig. 15).

4.2 On Concave–Convex Interactions

We have just seen some examples of supramolecular associates of fullerene derivatives based on hydrogen bonding. All of these rely on the previous covalent modification of the fullerene to introduce adequate chemical groups. Pristine fullerenes, on the other hand, are unfunctionalized, approximately spherical polyenes. From the point of view of their supramolecular chemistry this means

Fig. 16 Scheme depicting the concave–convex interaction vs planar π–π interaction

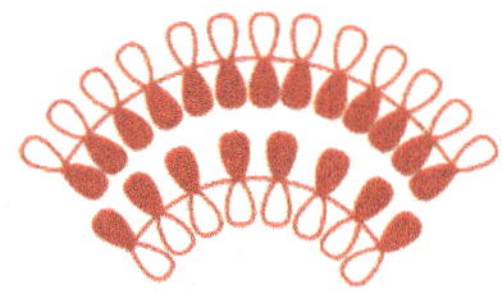
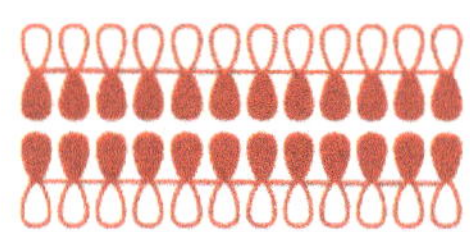

that very weak, non-directional dispersion interactions (mainly van der Waals and π–π) will account for the vast majority of the binding energy in host–guest complexes [139]. Since these forces depend directly on surface area, the shape complementarity between host and guest becomes critical. In this sense, distorted concave recognition motifs seem ideally suited for the association of the convex fullerenes. The importance of this shape complementarity is beautifully illustrated by the distortion from planarity observed in the solid state structure of some porphyrin–fullerene supramolecular complexes reported by Aida. In the associates the porphyrins adopt a non-planar concave conformation to maximize the positive interactions with the C_{60} guest, even at the expense of some degree of conjugation [140]. Besides the optimization of dispersion interactions due to shape complementarity between concave hosts and convex fullerene guests, in 2006 Kawase and Kurata suggested that there might be an additional positive effect arising from the unsymmetrical nature of the π orbitals of the contorted molecules with respect to the convex (outer in the case of the fullerenes) and concave (inner in the case of the fullerene) sides [141]. They termed this "concave–convex interaction" (Fig. 16).

Despite this, most of the examples of receptors for fullerene reported to date rely on planar recognition motifs. Curved molecules are geometrically tensioned structures, with bond angles away from the preferred ones and consequently are not always an easy synthetic target. However, some concave molecules are synthetically accessible and have been employed as hosts for fullerenes. We will now overview some prominent examples.

Corannulene consists of five benzene rings fused into a central five-member ring. As could be expected considering its size – nearly identical to that of C_{60} and thus too small to associate with it – chemical derivatization is necessary to enlarge the cavity of corannulene and observe binding. Following this strategy, several monotopic receptors have been reported [142].

A ditopic receptor, "buckycatcher" **40**, was synthesized by Sygula et al. joining together two units of coronene via a rigid aromatic spacer [143], forming a tweezers-like host [144]. It forms stable complexes with C_{60} (log $K_a = 3.9$, d_8-toluene, room temperature),[1] in which the fullerene is included between the two coronene units, as demonstrated through X-ray diffraction studies on cocrystals of **40** and C_{60} (Fig. 17).

Another very interesting family of curved aromatic hosts for fullerenes are the cyclic [*n*]*para*-phenyl acetylenes (CPPAs, **41–44** in Fig. 18) reported by Kawase

[1] All through the article we will report binding constants as logarithms, and without an error interval, for simplicity. The reader can refer to the original publications for these data.

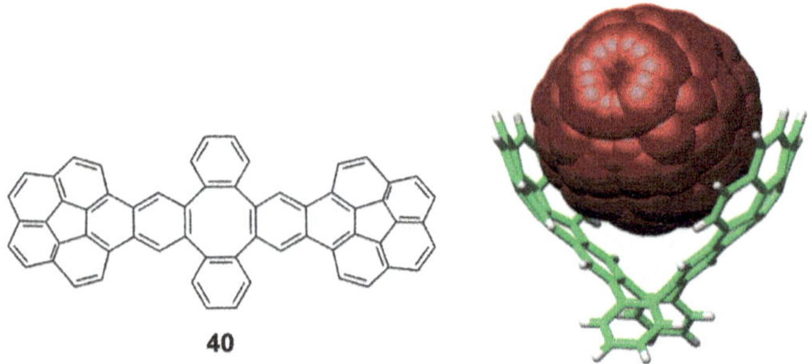

Fig. 17 Chemical structure of the "buckycatcher" and solid state structure of its complex with C_{60}. Note that the fullerene unit is disordered in the crystal. Reprinted with permission from [144]

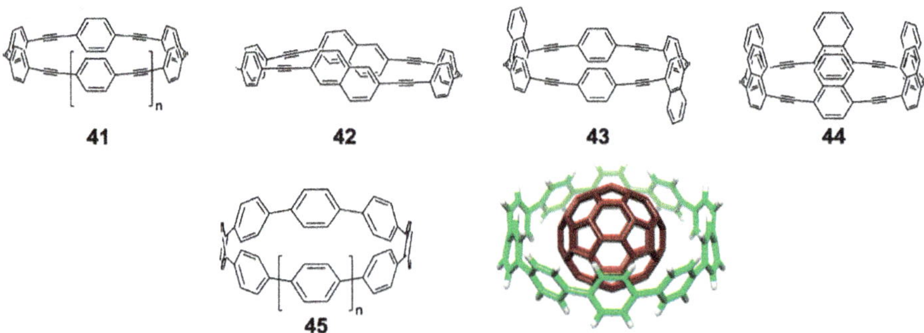

Fig. 18 Chemical structure of some CPPAs and general structure of CPPs. The solid state structure of the [10]CPP·C_{60} associate is also shown

et al. in 1996 [145]. The diameter of the cavity of **41** is 1.33 nm as found in its solid state structure, a close to perfect fit for C_{60}. Consequently, the ability of **41** to associate with [60] fullerene was investigated through UV–vis titrations, the analysis of which afforded a binding constant of log $K_a = 4.2$ in benzene at room temperature [146]. In order to increase the depth of the cavity, and in turn increase interactions between host and fullerene, macrocycles **42–44** (Fig. 18) were synthesized [147]. In these carbon nanorings at least two of the *p*-phenylene moieties are substituted by naphthalene units. This structural optimization bore fruit, as all of the newly synthesized hosts form complexes of remarkable stability with both C_{60} and C_{70}. In fact, the binding constants in benzene were log $K_a > 5$ as estimated through fluorescence quenching experiments.

Cycloparaphenylenes (CPPs, **45** in Fig. 18), in which the phenyl units are connected directly, have also been synthesized [148], and their association with fullerenes investigated. In particular, [10]CPP presents a concave cavity of 1.34 nm in diameter, ideally suited to associate with C_{60}. Indeed, it does so with a remarkable association constant of log $K_a = 6.4$ in toluene at room temperature [149]. These experiments have been followed up by the group of Jasti, who have reported

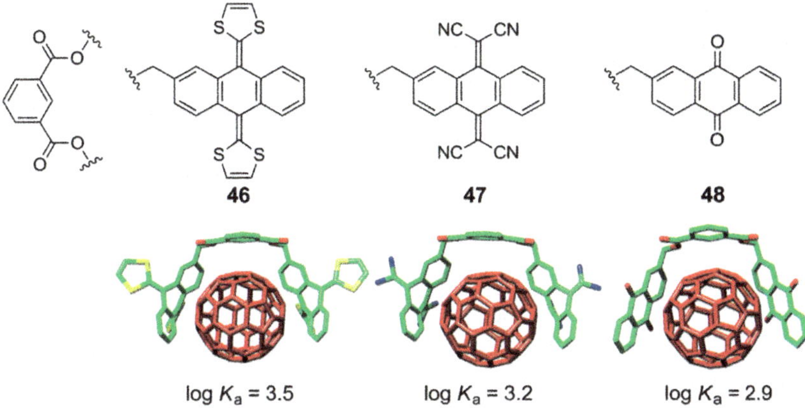

Fig. 19 Chemical structure of receptors **46–48** and molecular models of their complexes calculated at the BH&H/6-31 G** level

the synthesis of gram quantities of [8]- and [10]cycloparaphenylenes, and the crystal structure of the [10]CPP·C_{60} complex [150].

We have reported extensively on the ability of the curved electron donor exTTF to serve as a recognition motif for fullerenes. The geometric and electronic complementarity between the concave aromatic face of exTTF and the convex surface of the C_{60} was exploited to build very simple tweezers-like receptors (**46**, in Fig. 19) which associate with C_{60} with respectable binding constants in the order of log K_a = 3–4, in several solvents at room temperature [151, 152]. Thanks to their synthetic accessibility we could access a collection of hosts in which we could investigate the specific contribution of concave–convex interactions to the molecular recognition of C_{60} (**46–48** in Fig. 19) [153]. All three receptors bear the same number of aromatic rings and are approximately equal in size, so the contribution of π–π and dispersion interactions can be considered equivalent. Hosts **46** and **47** both feature concave recognition motifs, and differ only in their electronic character: while the exTTF moiety in **46** is an electron donor, the TCAQ units in **47** are electron acceptors. Meanwhile, the anthraquinone fragments in **48** have a similar electronic character to **47**, but are completely flat. Unsurprisingly, **46** shows the highest association constant, due to the combination of electronic and geometric complementarity towards C_{60}. Remarkably, there is a noticeable difference between the binding affinities of **47** and **48** that can be attributed to concave–convex interactions. As we noted in our original communication, whether concave–convex interactions should be treated as a new kind of intermolecular force or just a particular case of preorganization is not straightforward. In any case, it is worth considering them for the few curved guests of interest, mainly fullerenes and carbon nanotubes [154].

We have also investigated the binding abilities of π-extended derivatives of tetrathiafulvalene with a truxene core (truxTTF), which feature up to three dithiole units connected covalently to the conjugated aromatic core. To accommodate the dithioles, the truxene moiety breaks down its planar structure and adopts an all-*cis* sphere-like geometry with the three dithiole rings protruding. The concave shape adopted by the truxene core perfectly mirrors the convex surface of fullerenes, indicating that van der Waals and concave–convex π–π interactions between them should be maximized. Indeed, the association of trux-TTF and fullerenes in solution was investigated by ^{1}H NMR titrations with C_{60} and C_{70} as guests affording binding constants of log $K_a = 3.1$ and 3.9 for C_{60} and C_{70} in $CDCl_3/CS_2$, respectively [155].

4.3 Tweezers and Macrocycles for the Molecular Recognition of Fullerenes

As we have seen in some of the examples described above, tweezer-like hosts have been a particularly popular design for the construction of receptors for fullerenes. This is so because tweezers are usually synthetically accessible, since it requires only connecting two recognizing units symmetrically via a spacer.

One of the earliest examples of molecular tweezers for C_{60} was reported by the group of Fukazawa in 1998 [156]. They connected two units of calix[5]arene through a variety of rigid spacers, and obtained the best results for host **49**, which showed a binding constant of log $K_a = 4.9$ in toluene at room temperature, a world record in complex stability at the time.

The positive interaction between porphyrins and fullerenes has often been exploited to construct this kind of receptor [157]. The first example of such porphyrin tweezers for fullerenes was reported by Boyd, Reed and co-workers over a decade ago [158]. They connected two porphyrin units appended with pyridine ligands through coordination of palladium. The structure of the tweezers **50** is shown in Fig. 20. The binding constant of this receptor towards C_{60} was estimated to be log $K_a = 3.7$ in toluene-d_8 at room temperature.

After this first example the same authors reported very similar receptors, in which the coordination link between the porphyrin units was substituted with covalent bonding through the amides of either isophthalic or terephthalic acid [159]. This resulted in a decrease in the association constant. Later, in collaboration with the group of Armaroli, they replaced the benzenedicarboxamide spacers with several calixarenes, reaching binding constants as high as log $K_a = 5.4$ and 6.4 for C_{60} and C_{70} respectively, both in toluene at room temperature [160].

A more sophisticated example of molecular tweezers for C_{60} (**51** in Fig. 21), with a mechanism to turn "on" and "off" their ability to bind the fullerene guest, was reported by Shinkai and co-workers [161]. As synthesized, each of the appending pyridines coordinates to the porphyrin metal, keeping the two porphyrins on opposite sides and preventing association of C_{60}. When an external palladium center is added, the conformation changes to bring both porphyrins to the same side of the molecule, allowing for association with fullerene, as shown in Fig. 21. In the

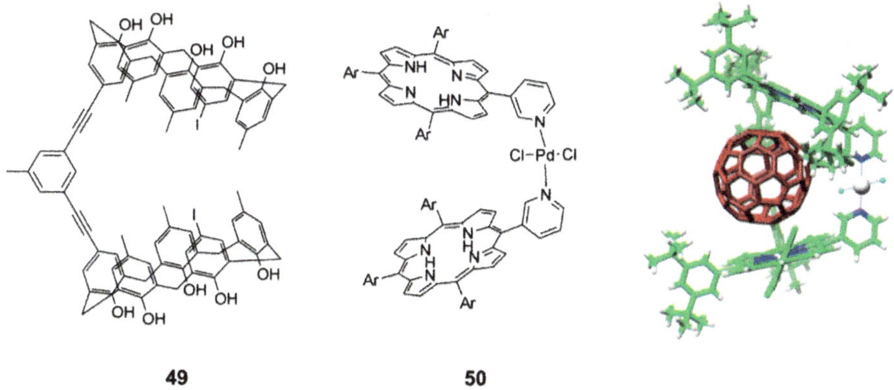

Fig. 20 Chemical structure of the biscalix[5]arene host **49** [157], the "jaws" receptor **50** [158], and X-ray crystal structure of the **50**·C_{60} complex

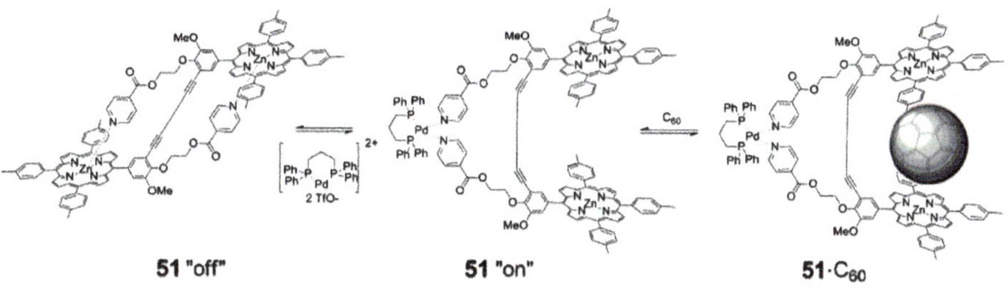

Fig. 21 Structure of the switchable receptor **51** and scheme showing its switching between the "off" and "on" states and its binding to C_{60}

"on" state, the binding constant was estimated to be log $K_a = 3.7$ in toluene/ CH_2Cl_2 (50:1) at room temperature.

A distinctive advantage of the simplicity of the tweezers-like design is that it can easily be adapted to construct more elaborate supramolecular assemblies at relatively low synthetic cost. For instance, in our group, based on the exTTF tweezers **46**, we have built both linear [162] and hyperbranched [163] supramolecular polymers [164], covalent dendrimers capable of associating several units of C_{60} [165], and have extended the design to associate with and solubilize carbon nanotubes in aqueous solution.

On the negative side, tweezers are not well preorganized unless very rigid spacers are used to link the two recognizing units. This results in relatively modest association constants, in the case of fullerenes typically in the order of 10^3–10^4 M^{-1}. To obtain complexes of higher stability, a tried and tested strategy is to move from tweezers to macrocycles. Macrocycles are much better preorganized, and have better expressed binding sites, increasing the binding constants significantly. In return, they are more challenging synthetic targets, and the cavity of the host needs to be fine-tuned to match the size of the guest, or the preorganization will

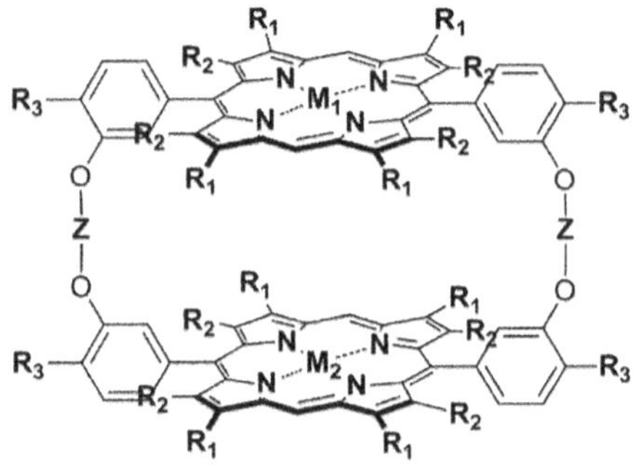

	Z	R_1	R_2	R_3	M_1	M_2
52a	$-(CH_2)_6-$	Hex	Hex	H	Zn(II)	Zn(II)
52b	$-CH_2-C{\equiv}C-C{\equiv}C-CH_2-$	Hex	Hex	H	Zn(II)	Zn(II)
52c	$-(CH_2)_6-$	Me	Hex	H	Ir(III)	Ir(III)
52d	$-(CH_2)_6-$	H	H	*t*Bu	Me, H	Rh(III)

Fig. 22 Structure of the macrocyclic bisporphyrin receptors reported by Aida

actually be detrimental to the binding event [166]. We will now see a few examples of macrocyclic hosts for fullerenes to illustrate these points.

In 1999, Aida, Saigo and co-workers published the first example of a bisporphyrin macrocyclic host for C_{60} (**52a** in Fig. 22) [167], initiating what would become one of the largest and most successful families of receptors for fullerenes [140]. Through multiple structural variations they have been able to establish some clear structure–binding affinity relationships. For instance, it has become apparent that not only the length but also the flexibility of the links between the porphyrins is critical to the association event. The synthetic precursors to the alkane spacers are the corresponding alkynes, **52b**. The macrocycles with these rigid spacers do not show *any* sign of binding towards C_{60}, despite theoretically having a cavity of the right size. In contrast, the macrocycles with flexible alkane spacers show remarkably high binding constants, including the world-record in complex stability, **52c**, with an incredibly high log $K_a = 8.1$ toward C_{60} in 1,2-dichlorobenzene (*o*-DCB) at room temperature [168]. This is a dramatic example of the price one has to pay for increased preorganization: it is often a make or break situation.

In the last few years, examples of macrocyclic hosts for fullerenes including more than one porphyrin recognition motif have been reported. Anderson's group has described a rigid cyclic porphyrin trimer, **53**, which associates with C_{60} with log $K_a = 6.2$ [169]. The cavity of **53** is actually better suited to host higher

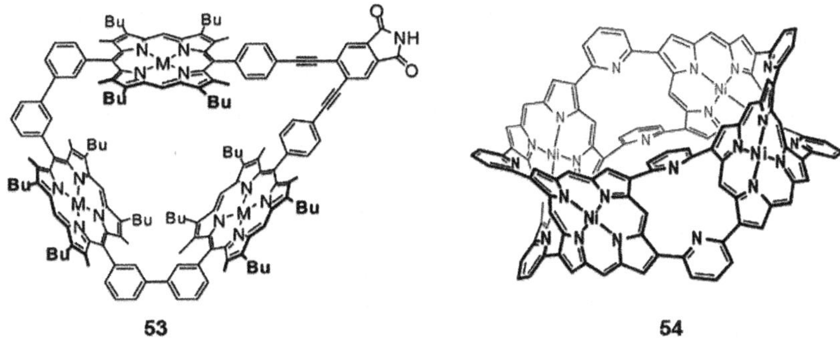

Fig. 23 Structures of the cyclic porphyrin trimer **53** and of nanobarrel **54**

fullerenes, so it shows a binding constant of log $K_a = 8.2$ for C_{70} and log $K_a > 9$ for C_{86}, all in toluene at room temperature.

Osuka and co-workers have gone one step further, linking four porphyrin units to form what they called a "nanobarrel" (**54** in Fig. 23) [170]. The solid state structure of **54** shows a rigid concave cavity of adequate size to associate with C_{60}. Despite this, the authors reported a binding constant of log $K_a = 5.7$ in toluene at room temperature, perhaps not as large as could be anticipated. This might be due to the use of Ni porphyrins, since previous studies have shown a decrease in the binding constant of approximately one order of magnitude from zinc to nickel in macrocyclic porphyrin dimers [171]. Alternatively it might be that the very rigid structure of **54** is not sufficiently flexible to optimize the C_{60}-porphyrin distances, again showing the adverse effects of an excess of preorganization.

Our group has also gone the distance from tweezers to macrocycles. We have recently synthesized a family of nine macrocyclic exTTF-based receptors, in which we have conserved the basic features of the tweezers design (two exTTF units linked through an aromatic spacer) and added alkene-terminated alkyl spacers, to perform ring-closing metathesis [172, 173]. We produced systematic variations of both the aromatic and the alkyl spacers, as shown in Fig. 24. The structural variation strategy proved to be successful, as among the family of hosts we found some of the best purely organic hosts for both C_{60} and C_{70}. For instance, *p*-xylmac12 associates with C_{60} with log $K_a = 6.5$ in chlorobenzene and 7.5 in benzonitrile, both at room temperature. Perhaps more interestingly, the synthesis of such a complete family of macrocyclic receptors showed that even very small variations in structure can lead to huge changes in binding abilities. For example, there is a difference of three orders of magnitude between the association constant of *p*-xylmac12 towards C_{60} and that of *p*-xylmac10 (log $K_a = 3.5$ under the same experimental conditions). Besides changes in the stability constants, we even found variations in the stoichiometry of the associates. The smaller members of the family, *p*-xylmac10 and *m*-xylmac10, associate with C_{70} forming both 1:1 and 2:1 host:guest complexes, while naphmac10, with only a slightly bigger cavity, forms exclusively 1:1 associates.

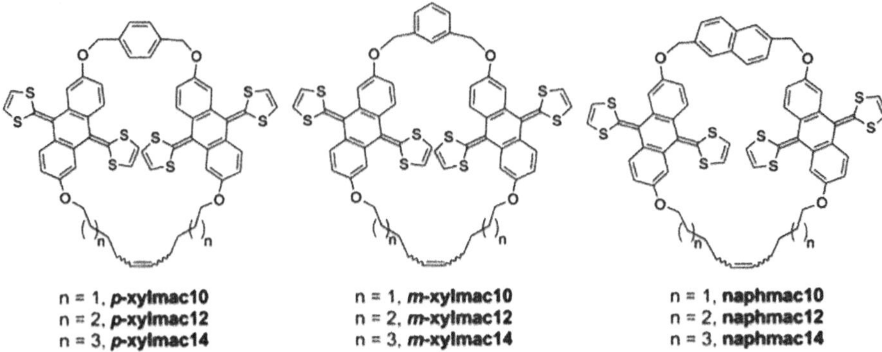

Fig. 24 Chemical structure of the exTTF-based macrocyclic hosts. Reprinted with permission from [173]

There are many other examples of tweezer-like and macrocyclic hosts for fullerenes, which we will not discuss in this chapter. The interested reader can refer to more comprehensive reviews [144, 166]. In this very brief account we intended to outline general considerations regarding the design principles for the construction of hosts for fullerenes, illustrating them with a few selected examples.

5 Endohedral Fullerenes: Improving Size, Shape, and Electronic Properties

Almost immediately after the discovery of the fullerene family, scientists wondered about the possibility of introducing different atoms and molecules into the carbon cages and speculated about the new properties that the hybrid materials could exhibit. This expectation, far from decreasing, is actually growing with the application of these materials in different fields such as medicine or molecular electronics [174]. The properties of endohedral fullerenes can easily be modified depending upon the species entrapped and the fullerene cages. In this sense, endohedral fullerenes are classified in two principal categories:

1. *Metallofullerenes*, which are those that contain elemental metals or their combined forms [175]. More specifically, this family includes *classical metallofullerenes*, $M@C_{2n}$ or $M_2@C_{2n}$, with $60 \leq 2n \leq 88$ [176], *metallic carbides*, $M_2C_2@C_{2n}$ and $M_3C_2@C_{2n}$, where $68 \leq 2n \leq 92$ [177], *metallic nitrides*, $M_3N@C_{2n}$, with $68 \leq 2n \leq 96$ [178], and, more recently, *metallic oxides*, $M_4O_2@C_{80}$ [179]. These fullerenes are typically produced by laser-vaporization or arc discharge techniques of graphite-metal oxides composite materials in the atmosphere of certain gases [180].

2. *Fullerenes encapsulating small molecules*, such as noble gas atoms (helium, neon, argon, krypton, and even xenon have been introduced inside fullerene cages), which are obtained by treating the fullerene powder under forced conditions (650°C

Fig. 25 Drawing for the cationic metallofullerene $Li^{+}@C_{60}$

and 3,000 atm of noble gas), although the occupation level of the guest is as low as 0.1–1% [181]. In a second approach, molecules as water have been introduced in C_{60} and C_{70} cages by using organic reactions in the so-called "molecular surgery" [182]. This strategy consists in a series of steps which involve making an incision in the fullerene cage to form an opening on the surface, inserting the desired molecule through the opening, and finally closing the hole to reproduce the fullerene cage while retaining the guest species.

The case of $H_2O@C_{60}$ is quite remarkable, since single-crystal X-ray analysis of the complex $[H_2O@C_{60}\cdot(NiOEP)_2]$ (OEP = octaethylporphyrin) reveals that, in contrast to many metallofullerenes where the metal often adopts an off-center location and does not move freely, the O atom is located at the center of C_{60}, with the O–H bonds pointing towards the Ni atoms. In addition, $H_2O@C_{60}$ and the empty fullerene can be separated quite easily by HPLC on a pyrenylated stationary phase, in stark contrast to the case of noble gas atoms or H_2 endohedrals. This easy access to a non-hydrogen-bonded H_2O molecule inside the apolar fullerene cage allows the investigation of the properties of the isolated H_2O molecule as well as the modification of the exohedral chemical reactivity of a unique *wet* C_{60}.

From the two categories of endohedral fullerenes mentioned above, the electronic properties of metallofullerenes are particularly promising considering that they are featured by a charge transfer from the encapsulated metal atoms to the carbon cage, forming a non-dissociating salt that consists of metal cation(s) encapsulated in a fulleride anion [183]. This electron-transfer was regarded to stabilize not only the encapsulated species but also the fullerene cage that can sometimes be otherwise unstable in the empty form. A striking example of this ionic model was the isolation of the cationic endohedral metallofullerene $Li^{+}@C_{60}$ (Fig. 25) [184], which can only be stabilized significantly in ambient conditions when it co-exists with an appropriate counteranion. For example, the crystal structure of the salt $[Li^{+}@C_{60}](PF_6)^{-}$ exhibits a strong interaction between Li^{+}, residing inside the C_{60} cage, and PF_6^{-} on the outside, the interaction being shown to occur through the six-membered rings [185].

In the following sections we are going to concentrate on mono- and divalent metallofullerenes and metallic nitrides, since these metallofullerenes have been extensively investigated and their chemical reactivity reasonably explored.

5.1 Metallofullerenes and TNT Endofullerenes

Smalley and co-workers demonstrated in 1991 that a family of lanthanum containing fullerenes were produced under a modified Krätschmer–Huffman reactor and that extraction with toluene yielded mostly La@C_{82} (Fig. 26), which was the first endohedral fullerene to be isolated [186]. La@C_{82} has an electronic state best described as $[La]^{3+}[C_{82}]^{3-}$ with an open-shell electronic structure that is a consequence of a three-electron transfer from lanthanum to C_{82} [187]. The resulting electron spin of La@C_{82} imposes a unique chemical reactivity comparable to radical species, inducing magnetism on the molecular scale, or an enhanced electron conductivity [188, 189].

Since the isolation and characterization of the first mono-endohedral metallofullerene, La@C_{82}, many other classical M@C_{2n} (M = Sc, Y, La, Ce, Gd, etc.) have been obtained, with a C_{2v}-C_{82} most abundant cage [176]. In all these fullerene cages filled with a single metal, the metal is not in the center of the cage but tends to coordinate with the cage carbons, being situated under a hexagonal ring along the C_2 axis. As a result, the distribution of charge density is highly anisotropic over the surface, with electrophiles and nucleophiles selectively attacking the two different regions [190].

In the case of the endohedral fullerenes containing two metal, the M_2@C_{80} (M = La, Ce, etc.) cage is typically obtained, with the two isomers I_h and D_{5h} as the most abundant [176]. In the case of these endohedral metallofullerenes, not only is the metal–cage interaction important but also the metal–metal interaction is crucial for the positioning and moving of the metal atoms. In the case of M_2@I_h-C_{80} structures, it has been demonstrated how the metal atoms circulate three-dimensionally [191], in contrast to M_2@D_{5h}-C_{80} species, where the metallic atoms circulate two-dimensionally along a band of ten contiguous hexagons inside the D_{5h}-C_{80} cage [192].

However, for a long period of time the development of the chemistry of endohedral metallofullerenes was impeded by the relatively low yields in which they were produced. An important breakthrough in this chemistry occurred in 1999, when Dorn and co-workers reported the production of trimetallic nitride clusters with high yields [193]. In the trimetallic nitride template (TNT) method, packed graphite rods (metal oxide/carbon/catalyst) are burned in the presence of a dynamic flow of He/N_2 and afforded macroscopic quantities of materials such as Sc_3N@C_{80}, with yields that exceed those of the third most abundant (next to C_{60} and C_{70}) empty cage C_{84}, produced under normal conditions.

The isolation of Sc_3N@C_{80} in macroscopic quantities has facilitated the study of its physical structure and chemical reactivity [180]. From the seven possible constitutional isomers for C_{80} satisfying the isolated pentagon rule (IPR), interestingly only the two least stable empty isomers with I_h and D_{5h} symmetries are the ones that predominate when they are filled with metallic nitride clusters, the I_h isomer certainly being the most abundant [194]. When considering the electronic structure of I_h-C_{80} it is possible to rationalize this observation. It is characterized by

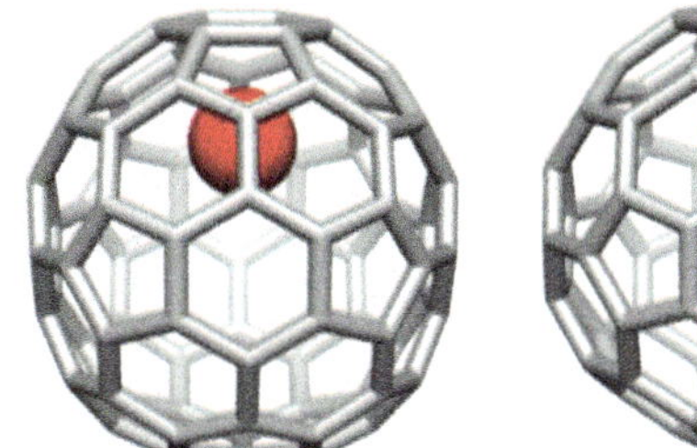

Fig. 26 Molecular structures of La@C_{2v}-C_{82}, La_2@I_h-C_{80}, and Gd_3N@I_h-C_{80}

the presence of an energetically low-lying fourfold degenerate LUMO, which enables this fullerene to accept up to six electrons, and the electronic distribution may be represented by the ionic model: $(Sc_3N)^{6+}$@$(C_{80})^{6-}$ [195].

Considering the TNT method, endohedrals of the type M_3N@I_h-C_{80} have been prepared for very different metals (M = Sc, Y, La, Ce, Nd, Gd, Tb, Dy, Ho, Er, etc.) and, in general, the metal cluster rotates freely inside the fullerene, with the M_3N unit adopting a planar geometry, except for the case of Gd_3N@I_h-C_{80} (Fig. 26), where the nitride ion is out of the plane of the three gadolinium ions [196].

Upon increasing the size of the encapsulated metal, the yield of cluster fullerenes usually decreases, and a distribution of molecules is obtained with larger metal ions favoring larger cages. For example, with gadolinium six cages are formed: Gd_3N@C_{78}, Gd_3N@C_{80}, Gd_3N@C_{82}, Gd_3N@C_{84}, Gd_3N@C_{86}, and Gd_3N@C_{88} [197], whereas for the larger lanthanum only the formation of three very large cages, La_3N@C_{88}, La_3N@C_{92}, and La_3N@C_{96}, is observed [198]. In addition, Echegoyen and co-workers found a remarkable influence of the cage structure on the electrochemistry of the Gd_3N@C_{2n} family [197]. The cage size does not seem to affect significantly the reduction potential of these compounds, which displayed very similar first reduction potentials, but the oxidation potentials shift from +0.58 V vs Fc^+/Fc in Gd_3N@C_{80} to +0.06 V vs Fc^+/Fc in Gd_3N@C_{88}, which suggests that the HOMO of the TNT-endofullerenes is probably cage-centered.

5.2 *Chemical Reactivity of Endohedral Fullerenes*

The chemical functionalization of endohedral metallofullerenes is essential to generate materials easy to process for multiple potential applications. Initial experiments on the functionalization of endohedral fullerenes demonstrated a high reactivity and the formation of multiple adducts or regioisomeric mixtures [199]. However, a remarkable regioselectivity has been observed in a few cases depending on the encapsulated cluster, metal species, carbon cage size or symmetry.

More specifically, the I_h isomer of the C_{80} carbon cage presents high symmetry, with only two possible [1,2] addition sites: bonds between two hexagonal rings

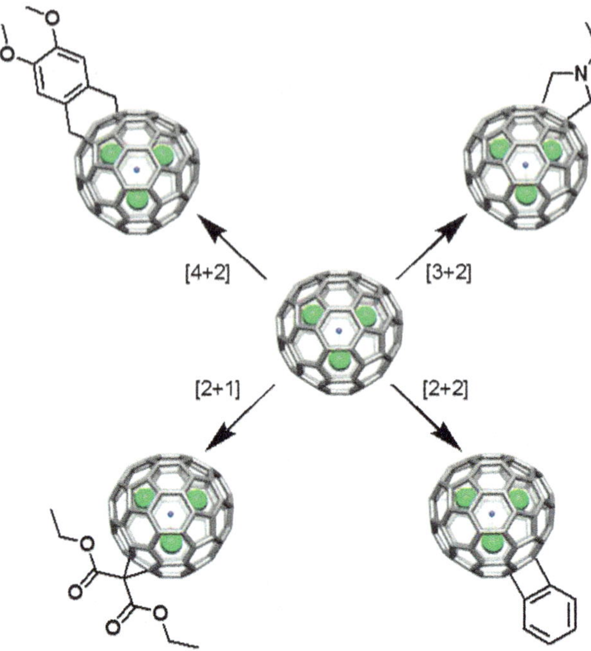

Fig. 27 Representative examples of cycloaddition monoadducts obtained by chemical reaction with Sc_3N-I_h@C_{80}

([6,6]-junctions) and those between pentagonal and hexagonal rings ([5,6]-junctions) [200]. Several reactions have already been reported considering these C_{80} cages: carbene additions [201], radical trifluoromethylations [202], or free radical additions [203], even though cycloaddition reactions are the most effective way to generate covalent derivatives of endohedral metallofullerenes to date.

[4+2]Cycloadditions. The Diels–Alder monoadduct resulting from heating an excess of 6,7-dimethoxyisochroman-3-one and Sc_3N@C_{80} in trichlorobenzene (Fig. 27) was the first isolated organic derivative of a TNT endofullerene [204]. NMR experiments served to identify the regioisomer obtained as the one localized in a [5,6] ring junction of the I_h isomer, which was later corroborated by solving the X-ray structure of the Diels–Alder monoadduct [205]. TNT endofullerenes have a much lower reactivity than empty fullerenes and classical endohedral metallofullerenes and Dorn and co-workers used this selective reactivity to purify TNT endohedral metallofullerenes directly from as-prepared soots in a single facile step in Diels–Alder reactions with a cyclopentadiene-functionalized resin [206].

[3+2]Cycloadditions. So far these have been the processes more intensively investigated, in particular 1,3-dipolar cycloaddition reactions of azomethyne ylides to yield fulleropyrrolidines (Fig. 27) [200, 207], disilirane additions [208], and more recently cycloadditions with epoxides such as tetracyanoethylene oxide [209], or azides to form azafulleroids after nitrogen extrusion [210].

Disilirane additions were used to control the motion of the encapsulated Ce atoms inside the C_{80} cage of $Ce_2@C_{80}$. In this compound, the free random motion of two Ce atoms is regulated under a hexagonal ring on the equator by the electron donation from the silyl groups, exohedrically introduced, to the C_{80} cage [211]. The motion of La atoms encapsulated inside fullerenes has also been controlled exohedrically, by the addition positions, in pyrrolidine adducts of $La_2@C_{80}$. Two different compounds are obtained: in the [5,6]-adduct the two La atoms rotate rather freely, whereas in the [6,6]-adduct the metallic atoms are in fixed positions inside the cage [212].

The regioselectivity of 1,3-dipolar cycloaddition reactions of in situ generated azomethine ylides can also be controlled by the trimetallic nitride cluster. In this sense, Echegoyen et al. demonstrated that when the fullerene inner metal cluster was Sc_3N, the only product detected was the adduct at a [5,6]-junction (product of thermodynamic control). On the other hand, the 1,3-dipolar cycloaddition reaction occurred initially at a [6,6]-junction for the $Y_3N@C_{80}$, and underwent rearrangement to the thermodynamically more stable $Y_3N@C_{80}$ [5,6]-monoadducts upon heating [213]. These experimental results, as well as the computational studies by Poblet and Echegoyen, seem to indicate that, after thermalization of the kinetically favored product, a pirouette-kind of mechanism gives rise to the [5,6]-monoadduct that is thermodynamically preferred [214]. The rate of this rearrangement depends on the internal cluster and on the pyrrolidine addend.

Pyrrolidine adducts have been used as the organic addend to connect electron-acceptor $Sc_3N@C_{80}$, $Y_3N@C_{80}$, or $La_2@C_{80}$ units to powerful donors such as ferrocene [215] or π-extended tetrathiafulvalene derivatives (exTTFs) [216] in the preparation of electron–donor–acceptor (D–A) systems that, upon photoexcitation, yield radical ion pair states with remarkable lifetimes.

[2+2]Cycloadditions. The reaction of benzyne, generated from isoamyl nitrite and anthranilic acid, with $Sc_3N@I_h$-C_{80} was successfully carried out recently to afford both [5,6]- and [6,6]-monoadducts (Fig. 27) [217]. Interestingly, when the reaction was carried out with 2-amino-4,5-diisopropoxybenzoic acid instead of anthranilic acid and under an aerobic atmosphere, in addition to the expected [2+2] benzyne adducts, oxygenation of the [5,6] regioisomer produces an intriguing third product, which is an open-cage metallofullerene. Under an inert atmosphere, the reaction gave only the expected [2+2] adducts [218].

[2+1]Cycloadditions. The investigation of the Bingel reaction on endohedral metallofullerenes demonstrates the remarkable regioselective control exerted by the encapsulated species. The [2+1] cycloaddition of bromodiethylmalonate (the Bingel–Hirsch reaction) in the presence of the non-nucleophilic base 1,8-diazabicyclo[5.4.0]-undec-7-ene (DBU) produced extremely stable derivatives with $Y_3N@C_{80}$ and $Er_3N@C_{80}$ [219], while $Sc_3N@C_{80}$ did not react under the same experimental conditions. Under these conditions, the cyclopropanation of $Sc_3N@C_{78}$ with diethyl bromomalonate produced only one monoadduct and one dominant symmetric bisadduct. The high regioselectivity in the second addition is supported by the highest LUMO surface electron density value for the reactive bond, which corresponds to the kinetically preferred site for a second nucleophilic attack [220].

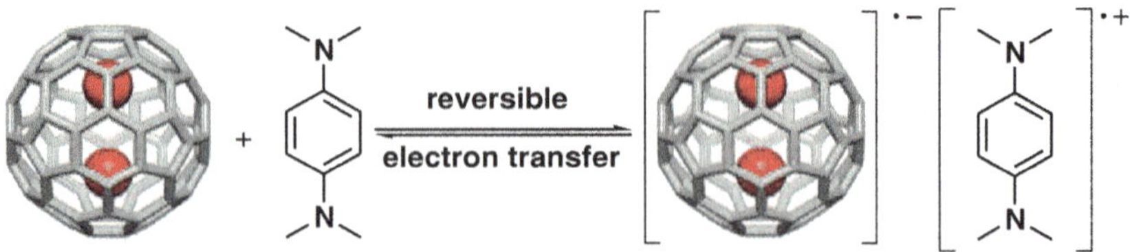

Fig. 28 Intermolecular complexation and electron transfer behavior between $La_2@C_{80}$ and the organic donor molecules TMPD

Another useful [2+1] cycloaddition process is the addition of diazocompounds, generated in situ from tosylhydrazones in basic media, to fullerenes. This reaction has been used for the preparation of relevant endohedral metallofullerene adducts in the context of organic photovoltaics, such as endohedral metallofullerene ($Sc_3N@C_{80}$, $Lu_3N@C_{80}$, or $Ce_2@C_{80}$) PCBM-based derivatives [221] or the first D–A dyad where an endohedral metallofullerene, $Lu_3N@C_{80}$, acts as an electron donor upon photoexcitation. In particular, when connected to perylenebisimide units [222].

Electrosynthetic routes have also been recently explored for the preparation of endohedral metallofullerene adducts that might not be accessible by conventional synthetic routes. In this sense, the different nucleophilicity observed by TNT dianions [223] and trianions [224] seems to be particularly relevant for the preparation of $Sc_3@C_{80}$ derivatives.

Surprisingly, the non-covalent functionalization of endohedral fullerenes has been scarcely investigated, only a few examples about the complexation with calixarenes and crown or thiacrown ethers are known, and size matching was found to be critical to the stability of the resulting complexes [176]. The most promising example is the formation of stable radical ion pairs of *N*-substituted *p*-phenylenediamine with $La@C_{82}$ and $La_2@C_{80}$ [188, 225]. Such spin-site exchange processes are reversible in solution and are readily controllable by changing the temperature and the solvent (Fig. 28).

6 Fullerenes for Organic Electronics

As mentioned above, a huge number of fullerene derivatives have been prepared, many of which have been tested in so-called organic molecular electronics. Their remarkable electron-accepting ability and low reorganization energy combined with solubility in organic solvents and outstanding photophysical properties have made fullerenes a most appealing system to be used in the preparation of electronic devices such as organic photovoltaics and the study of molecular wires, which are discussed in detail in the following sections.

6.1 Fullerenes for Organic Photovoltaics

Global dependence on fossil fuels is a key issue with important consequences in the world today. A reasonable alternative to overcome this need is the use of renewable energy sources, like solar energy, which could, in principle, fulfil our energy requirements with environmentally clean procedures and low prices. Actually, the energy received from the Sun, calculated to be 120,000 TW (5% ultraviolet, 43% visible, and 52% infrared), surpasses by several thousandfold that consumed by the planet. [226].

Photovoltaic (PV) solar cells are currently a hot topic in science and since the original silicon-based device was prepared by Chapin in 1954 exhibiting an efficiency around 6% [227], different semiconducting materials (inorganic, organic, molecular, polymeric, hybrids, quantum dots, etc.) have been used for transforming sunlight into chemical energy. Among them, photo- and electro-active organic materials are promising due to key advantages such as the possibility of processing directly from solution, thus affording lighter, cheaper, and flexible all-organic PV devices. The most widely used configuration of polymer solar cells is based on the use of a fullerene derivative as the acceptor component. Indeed, fullerenes have been demonstrated to be the ideal acceptor because of their singular electronic and geometrical properties and the ability of their chemically functionalized derivatives to form a bicontinuous phase network with π-conjugated polymers acting as electron conducting (n type) material (Fig. 29).

A great variety of chemically modified fullerenes were initially synthesized for blending with semiconducting polymers and to prepare photovoltaic devices. These fullerene derivatives were covalently linked to different chemical species such as electron acceptors, electron donors, π-conjugated oligomers, etc. [119] (Fig. 30). However, in general, the obtained blends resulted in PV devices exhibiting low energy conversion efficiencies [228].

The best known and most widely used fullerene derivative as acceptor for PV devices is [6,6]-phenyl-C_{61} butyric acid methyl ester (PCBM, **55**) [229]. Since its first reported application in solar cells [230], it has been by far the most widely used fullerene, being considered as a benchmark material for testing new devices. This initial report inspired the synthesis of many other PCBM analogues [231] in an attempt to increase the efficiencies of the cells by improving the stability or PV parameters such as the open circuit voltage (V_{oc}) by raising the LUMO energies of the fullerene acceptor.

In this regard, only small shifts (<100 meV) of the LUMO level have been obtained by attaching a single substituent on the fullerene sphere, even by using electron-donating groups. In contrast, significantly higher V_{oc} values have been achieved through the polyaddition of organic addends to the fullerene cage (~100 mV raising the LUMO per saturated double bond). Recently, an externally verified power-conversion efficiency of 4.5% has been reported by Hummelen et al. employing a regioisomeric mixture of PCBM bisadducts as a result of an enhanced open-circuit voltage, while maintaining a high short-circuit current (J_{sc}) and fill factor (FF) values [232].

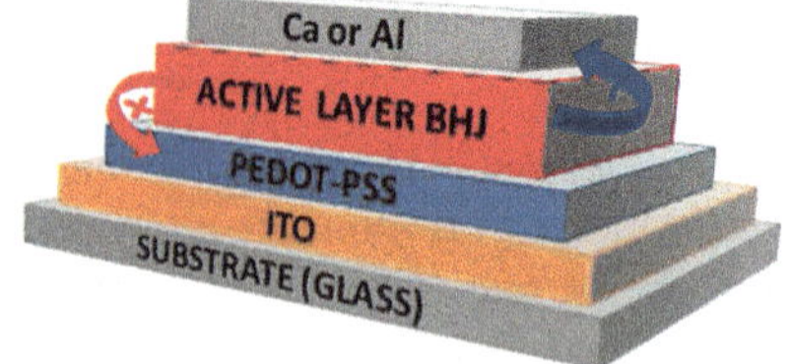

Fig. 29 Typical sandwich-like architecture for a (OPV) solar cell. Polymers and fullerene derivatives are bended on the active layer

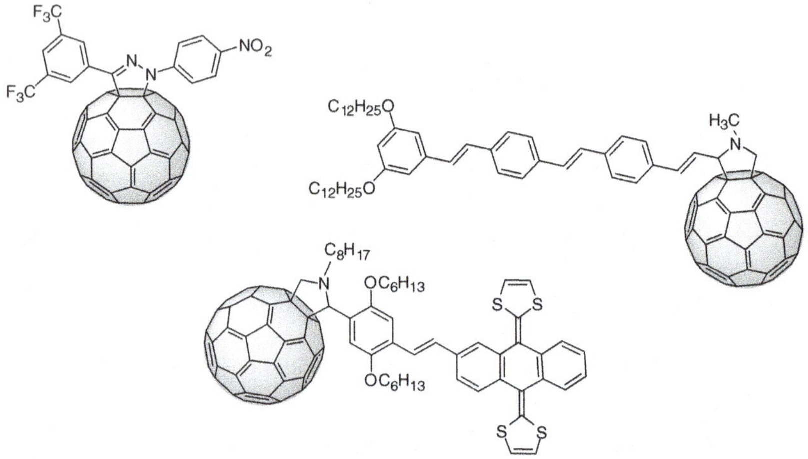

Fig. 30 Some examples of modified fullerenes bearing different organic addends used to prepare photovoltaic devices

The cyclopropanation reaction to superior fullerenes to form PCBM analogues is more complex than for C_{60}. Indeed, the less symmetry and the presence of more than one reactive double bond are often responsible for the formation of regioisomeric mixtures. Nevertheless, the loss of symmetry of C_{70} induces a stronger absorption, even in the visible region. As a result, $PC_{71}BM$ [233] is considered a suitable candidate for more efficient polymer solar devices. Moreover, such devices displayed the highest verified efficiency determined so far in a BHJ solar cell, with an internal quantum efficiency approaching 100% [234]. Analogously, $PC_{84}BM$ [235] has been obtained as a mixture of three major isomers. The stronger electron affinity and the diminished solubility gave rise, however, to poor power conversion efficiencies.

Although the PCBMs are the acceptors that guarantee best performances, it does not mean that they are necessarily the optimum fullerene derivatives. Therefore, a variety of other fullerene derivatives [236] have been synthesized in order to improve the device efficiency or to achieve a better understanding of the dependence of the cell parameters on the structure of the acceptor.

One of the most promising modified fullerene prepared so far is the diphenylmethanofullerene (DPM12, **56**) prepared by Martín et al. [237] endowed with two

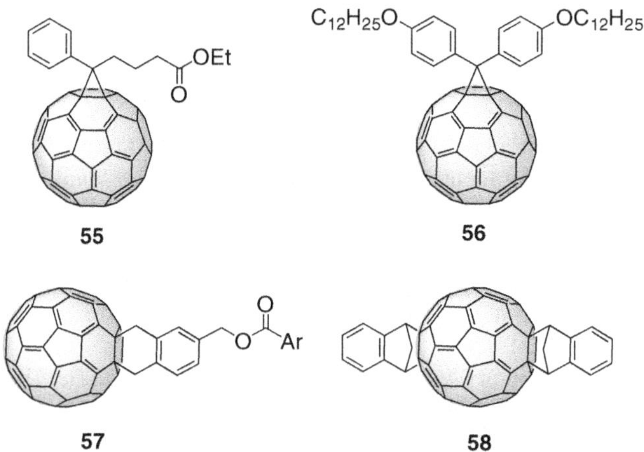

Fig. 31 PCBM (**55**), DPM-12 (**56**) and other modified fullerenes used as successful acceptors for photovoltaic devices

alkyl chains which improve dramatically the solubility of the acceptor in the blend to reach efficiencies in the region of 3% (Fig. 31). Although the LUMO energy level for DPM12 is the same as that for PCBM, an increase in the V_{oc} of 100 mV for the DPM12 over PCBM has been observed [238]. This is currently an important issue for improving the design of future fullerene-based acceptors.

Other fullerene derivatives such as the dihydronaphthylfullerene benzyl alcohol benzoic ester (**57**) described by Fréchet et al. [239] have also shown their ability to produce efficient PV devices (PCE up to 4.5%).

Hou and Li reported the preparation of a bis-adduct fullerene derivative **58** formed by two indene units covalently connected to the C_{60} sphere [240]. Interestingly, the presence of two aryl groups improves the visible absorption compared to the parent PCBM, as well as its solubility (>90 mg/mL in chloroform) and the LUMO energy level, which is 0.17 eV higher than PCBM. Photovoltaic devices formed with P3HT as semiconducting polymer revealed PCE values of 5.44% under illumination of AM1.5, 100 mW/cm^2, thus surpassing PCBM which afforded an efficiency of 3.88% under the same experimental conditions.

Although some of the fullerene derivatives prepared to date exhibit outstanding performances in PV devices, the synthesis of new fullerene derivatives with stronger visible and NIR absorption and higher LUMO energy levels than PCBM is currently a challenge in organic photovoltaics.

6.2 Fullerenes for Molecular Wires

Donor–bridge–acceptor (DBA) systems, in which the bridge mediates the transport of charge between the donor and the acceptor, provide good models to study the electron transfer processes at the molecular level. In these systems the rate of charge transfer (CT) is a combination of a strongly distance-dependent tunneling mechanism (or superexchange) and a weakly distance-dependent incoherent transport (or hopping).

The attenuation factor, β, is the parameter usually employed to describe the quality of a system as molecular wire, so that the lower the β value the longer the distance that the charge can be moved efficiently. The β parameter is characteristic for the decay of CT rate constant, k_{ET}, as a function of distance, R_{DA}:

$$k_{ET} = k_0 \, e^{-\beta R_{DA}}$$

Typical β values can vary from 1.0–1.4 $Å^{-1}$ for proteins to 0.001–0.06 $Å^{-1}$ for highly conjugated bridges [241]. Other parameters exert an impact on the rate of charge transfer, in particular the underlying driving forces ($-\Delta G°$), the corresponding reorganization energies (λ), and the electronic couplings (V) between the donor and acceptor moieties. Therefore β value depends not only on the bridge but rather on the wire system as a whole, that is, the DBA system, whether the D and A termini are molecular units or metallic contacts.

The unique electronic properties of fullerenes have prompted the study of the molecular wire behavior of different molecular bridges connected to fullerene C_{60} as the electron acceptor and different electron donor fragments [242, 243]. The study of these systems with various bridge lengths allows the measurement of distance-dependent charge separation (CS) and charge recombination (CR) rates and, therefore, the determination of the corresponding β values.

It is well documented that molecular wires that exhibit *para*-conjugation facilitate charges to be transferred over larger distances and hence show lower β values related to non-conjugated bridges. Very fast electron transfer for both CS and CR has been reported for compound **59**, with two acetylene bonds connecting the C_{60} unit with the β-position of ZnP ($k_{CS} > 1 \times 10^{11}$ s^{-1}, $k_{CR} = 2.5 \times 10^{10}$ s^{-1}) [244]. In contrast, for compounds **60**, in which the polyacetylene bridge is connected through a phenyl ring in the meso position of the porphyrin, CS and CR are in the regions of $7.5 \pm 2.4 \times 10^9$ s^{-1} and $1.6 \pm 0.2 \times 10^6$ s^{-1}, with a β value of 0.06 $Å^{-1}$ in PhCN [245]. This difference in the rate of electron transfer is due, at least in part, to more favorable orbital interactions between ZnP and C_{60} in **59** and suggests that the phenyl ring retards the electron transfer processes (Fig. 32).

When comparing these systems with a ZnP–*p*-phenylenebutadiynilenes–C_{60} series (**61**), the k_{CS} and k_{CR} values decrease drastically as the distances increase from 22 to 40 Å (k_{CS} range from 1.0×10^{10} s^{-1} to 1.1×10^8 s^{-1} and k_{CR} range from 2.1×10^6 s^{-1} to 4.6×10^5 s^{-1}) giving a quite large attenuation factor of 0.25 $Å^{-1}$ in PhCN [246]. This β value suggests that the phenyl ring inserted in the polyalkyne bridge acts as a resistor for electron transfer, giving rise to a less effective superexchange mechanism.

For the β-substituted ZnP–oligophenyleneethynylene–C_{60} systems **62** (ZnP–oPPE–C_{60}), a strong increase in the lifetime of the charge separated state

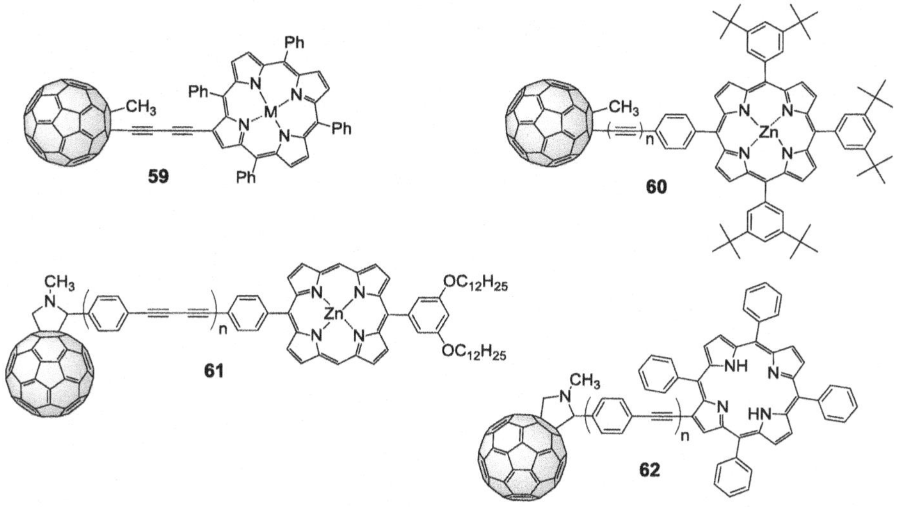

Fig. 32 Chemical structures of some ZnP/C_{60} conjugates

with increasing distance between the donor and acceptor moieties is observed, which implies a through-bond mechanism where the bridge plays an important role. A damping factor of 0.11 ± 0.05 Å^{-1} has been determined for these compounds [247]. This value is lower than the β value reported for similar systems bearing exTTF as donor moiety (exTTF-oPPE-C_{60}), for which an attenuation factor of 0.21 Å^{-1} was determined. This difference has been accounted for by a much more uniform distribution of the local electron affinity in ZnP–oPPE–C_{60} **62** due to higher electron density of ZnP in comparison to exTTF. These results underline the dependence of the β value and hence the wire-like behavior on the whole DBA system and not only on the linker.

An extraordinarily small attenuation factor of 0.01 ± 0.005 Å^{-1} was determined for compounds in which the donor exTTF and the acceptor C_{60} are connected through a *p*-phenylenevinylene oligomer **63** (exTTF–oPPV–C_{60}) [248, 249]. This low β value has been explained attending to the *para*-conjugation of the bridge with the donor exTTF and the homogeneous distribution of the local electron affinity throughout the whole bridge. A remarkable value of ~5.5 cm^{-1} for the coupling constant (V) was determined for these systems, unusually strong over a distance of 40 Å from the electron donor to the electron acceptor. Analogous systems in which the donor moiety has been substituted by a porphyrin **66** (ZnP–oPPV–C_{60}) give rise to a slightly higher β value (0.03 ± 0.005 Å^{-1}) due probably to a loss of conjugation between the donor ZnP and the linker [250] (Fig. 33).

The systematic study of the wire-like behavior in systems in which ZnP/C_{60} conjugates are connected through a *para*-cyclophane–oPPV (pCp–oPPV) bridge has shown a β value of 0.039 ± 0.001 Å^{-1}. The inhomogeneous and weaker

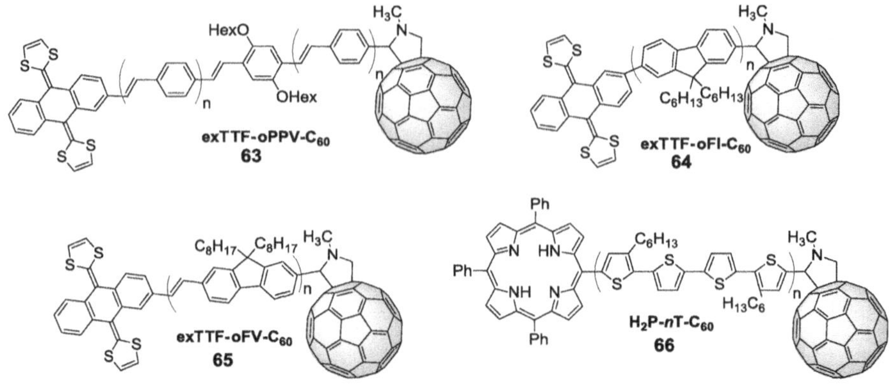

Fig. 33 Oligophenylenevinylene, oligofluorene, oligofluorenevinylene, and polythiophene donor/acceptor conjugates

π-conjugation of pCp–oPPV related to oPPV accounts for the approximately four times greater damping factor [251].

The comparison of the charge transfer characteristics of oligofluorenes (oFl) **64** vs oligofluorenevinylenes (oFV) **65** shows that it is possible to tune the electronic properties by slight structural alterations of the compounds. Specifically, β values of 0.075 $Å^{-1}$ are obtained for the oFV wires, lower than the value determined for oFl (0.09 $Å^{-1}$) [252]. The vinylene groups improve the π-conjugation of the wire in such a way that charge injection into the bridge is favored, facilitating CS and CR processes [253].

When oligothiophenes (nT) are employed as bridges, distances up to 55.7 Å have been reached between D and A. For these H_2P–nT–C_{60} systems, a dependency of the β value on solvent polarity was observed (0.11 $Å^{-1}$ in *o*-dichlorobenzene and 0.03 $Å^{-1}$ in benzonitrile) [254, 255]. Due to the electron-donating ability of the polythiophene oligomers, these bridges participate more actively in the CS/CR processes, which take place almost via the electron-hopping mechanism.

Recently, supramolecular ZnP/C_{60} hybrids connected by a Hamilton receptor and employing different spacers as oPPE, oPPV, *p*-ethynylene, or fluorene have been systematically studied [256]. These studies demonstrate that the electronic communication between the donor and acceptor moieties is governed by the charge transfer properties of the conjugated spacers. Thus, spacers with low β values facilitate the charge transfer along the supramolecular bridge. The study of the dependence of the rate constants of CS and CR on the D–A distances allowed the calculation of the attenuation factor for the first time in a hydrogen-bonding-mediated electron transfer, rendering a value of 0.11 $Å^{-1}$, which is similar to that obtained for other covalent systems (Fig. 34).

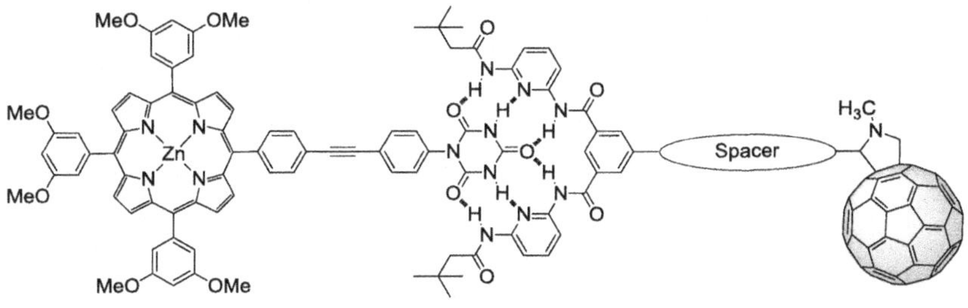

Fig. 34 Supramolecular ZnP/C_{60} molecular wire

7 The Future: Non-IPR Fullerenes

As stated above, C_{60} is by far the most abundant and common of all fullerenes. The question is why, among the many possible cages that can be formed with carbon atoms, the one containing 60 atoms is favored? Furthermore, since all fullerenes Cn are constituted by hexagons ($n \geq 20$ with the exception of $n = 22$) and pentagons (12 for all fullerene cages, which are responsible for the curved geometry), why among the 1,812 possible isomers for 60 carbon atoms was only the icosahedral symmetry I_h-C_{60} molecule (soccer-ball shape) formed?

These intriguing questions were answered by Kroto, who proposed that the local strain increases with the number of bonds shared by two pentagons (pentalene units), thus affording less-stable molecules. This rule was coined as the "isolated pentagon rule" (IPR), stating that all pentagons must be surrounded by hexagons, thus forming the corannulene moiety [16]. The resonance destabilization that results from the adjacent pentagons (8π electrons which do not satisfy the Hückel rule) and reduction of the π-orbital overlapping because of cage curvature explains the lower stability of non-IPR fullerenes [257]. A head-to-tail exclusion rule has also been proposed to explain the higher stability of fullerenes obeying the IPR rule [258].

For a precise number of carbon atoms forming a cage, the number of non-IPR fullerene isomers is very much larger than the IPR ones. Furthermore, in addition to doubly fused pentagons found in non-IPR fullerenes, triple directly fused pentagons and more recently triple sequentially fused pentagons have been reported [259]. Therefore there is great interest in the search for the huge number of expected non-IPR fullerenes whose chemical reactivity and properties should be different from those known for IPR fullerenes [260].

In order to achieve non-IPR fullerenes, two different strategies have been developed to increase their stability, namely through endohedral and exohedral derivatization [261]. In both approaches the key issue to stabilize non-IPR fullerenes focuses on how to release or decrease the strains of fused pentagons.

The first endohedral strategy involves encaging a metal cluster inside the fullerene cage. The bending strain on fused pentagons is significantly decreased because of the strong interactions between the fused pentagons and the metal cluster.

Endohedral fullerenes have been known from the earliest times of fullerenes (see above). Although theoretical studies predicted in the early 1990s that elusive non-IPR fullerenes could be stabilized by the presence of clusters encapsulated in the fullerene cage, the first non-IPR fullerenes, namely $Sc_2@C_{66}$ [262] and $Sc_3N@C_{68}$ [263], were obtained in 2000. It is important to note that carbon cages in endohedral fullerenes are different from those observed in empty fullerenes. Therefore, favorable electronic interactions between the encapsulated species with the carbon cage are required for the stabilization of the resulting endohedral fullerene.

This method brings about the productions of other different non-IPR metallofullerenes, such as trimetallic nitride fullerenes (i.e., $Tb_3N@C_{84}$ [264], $Sc_3N@C_{70}$ [265], and $Y_3N@C_{78}$ [266] etc.), metal carbide endofullerene $Sc_2C_2@C_{68}$ [267], and metal cyanide endofullerene $Sc_3NC@C_{78}$ [268].

A simple and qualitative rule to predict the stability of a given endohedral is based on the calculated HOMO–LUMO gap for the resulting "ionic" endofullerene. This energy gap can be roughly calculated from the (LUMO-3)–(LUMO-4) gap determined for the neutral cage, thus predicting the most stable IPR and non-IPR endofullerenes [269].

Remarkably, non-IPR endohedral metallofullerenes (fullerenes containing one or more metal atoms in the inner cavity) show a strong coordination of the metal atoms to the fused pentagons, similar to that observed for a variety of organometallic species in which the concave face of the pentalene unit is coordinated to the metal atom [270]. In contrast, IPR endofullerenes show a motion for the encapsulated metals or clusters in the inner cavity. In some cases involving endofullerenes endowed with only one metal atom, the metal generally coordinates with the cage and motion is difficult.

On the other hand, the second strategy based on exohedral derivatization has afforded a variety of non-IPR derivatives based on the remarkable reactivity of the fused pentagons, thus changing the carbon bond hybridization and releasing the bending strains. The first small fullerene $^{\#271}C_{50}$ (the Fowler–Manolopoulos nomenclature to differentiate isomers is specified by symmetry and/or by spiral algorithm) was trapped and stabilized by chlorine atoms as $^{\#271}C_{50}Cl_{10}$ in 2004 [271]. Since then the fullerene cage has been exohedrically functionalized by introducing hydrogen or chloride atoms on the cage surface, to produce a variety of non-IPR fullerene derivatives, such as $C_{64}H_4$ [272], $C_{71}H_2$ [273], $C_{78}Cl_8$ [274], etc.

This stabilization of the resulting non-IPR fullerene derivatives has been accounted for by the "strain-relief principle" resulting from the rehybridization from sp^2 to sp^3 carbon atoms, as well as by the "local aromaticity principle," which involves maintaining the local aromaticity of the un-derivatized sp^2 carbon skeleton that remains after the derivatization process. Based on both principles, it has been possible to predict the stability of a variety of exohedrically functionalized non-IPR fullerenes.

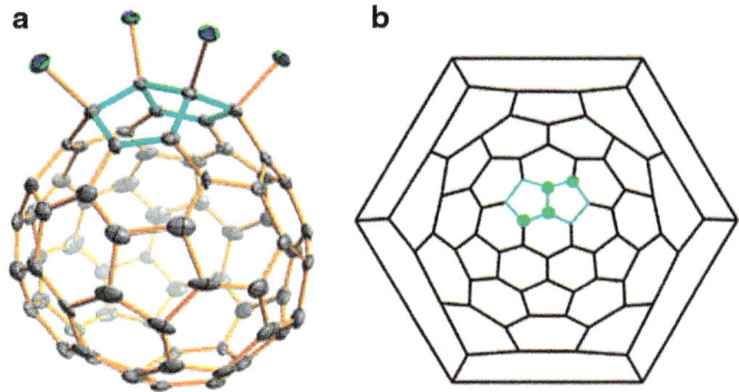

Fig. 35 X-ray structure of non-IPR $^{\#11188}C_{72}Cl_4$. (**a**) Fused pentagons are shown in *blue*. (**b**) Schlegel diagram showing the position of the four chlorine atoms. (Reprinted with permission from [279].)

Particular attention has been devoted to the structure of fullerene C_{72}, a so-called "missing fullerene" because empty C_{72}, IPR or non-IPR isomer, had never been isolated. However, along with further studies, recently several C_{72}-based species have been successfully prepared and characterized, namely $La_2@D_2(10611)$-C_{72} [275], $Ce_2@D_2(10611)$-C_{72} [276], $La@C_2(10612)$-C_{72} [277], and $Sc_2S@Cs$ (10528)-C_{72} [278].

Experimental results have recently reported, for the first time, the higher stability of a non-IPR fullerene compared to its related IPR isomer for $C_{72}Cl_4$ [279, 280] (Fig. 35). These new results violate the "universal" IPR rule for fullerenes, but confirm the valuable "strain-release" and "local aromaticity" principles that have been so useful to predict the stability of a wide variety of fullerene derivatives. The IPR rule is therefore valid for pristine fullerenes, whereas for fullerene derivatives additional factors emerge that could eventually force a non-IPR cage to be the most stable one. These new results pave the way to the advent of a huge family of, so far, unknown non-IPR fullerenes whose number would be almost infinite. These thermodynamically less stable carbon cages should exhibit significant differences compared to those IPR obeying fullerenes, thus enhancing the properties and applications of these molecular allotropes of carbon. The scientific community should be ready for the advent of the unprecedented non-IPR fullerenes, the fullerenes for the near future!

8 Summary, Conclusions, Outlook

In the above sections we have presented some of the important achievements in the past few years with buckyballs from the viewpoint of a synthetic chemist and with an eye on the most remarkable properties of these molecules of interest for practical purposes, namely in organic electronics.

Fullerenes form a broad family of carbon nanostructures which, in principle, must obey the IPR rule as a criteria of stability. The so-called hollow fullerenes undergo a very rich covalent chemistry affording a wide variety of fullerene derivatives. In this review we have mainly focussed on those reactions carried out involving metals as catalysts, which have significantly enhanced the scope of derivatives prepared so far. Furthermore, in our group we reported the detailed retro-cycloaddition reaction of well-known fullerenopyrrolidines in a quantitative manner, thus providing a new protection–deprotection protocol in fullerene chemistry. This interesting reaction was later extended to other fullerene cycloadducts.

An important topic is combining the intriguing properties of fullerenes with those of highly versatile polymers, affording the new interdisciplinary field of "fullerene polymers." This field has been reviewed with the most outstanding and recent advances and applications according to the rational structural classification given in the text.

The convex surface of fullerenes is an ideal scenario to interact with the concave surface of different organic molecules. Therefore the supramolecular chemistry of fullerenes has been presented with special emphasis on H-bonding interactions and π,π concave–convex interactions in which our group has been mainly engaged. This singular approach has been complemented with the most significant examples in the search for fullerene receptors from other authors.

Just a few years have been enough to show that the carbon family is much larger than initially thought and many other forms of carbon, sometimes encapsulating atoms, molecules, or clusters in their inner cavity (endohedral fullerenes), have been produced. Although the number of endohedral fullerenes is becoming larger and larger and new chemical species are placed at the inner cavity of fullerenes, the most frequent and studied endofullerenes have been presented according to their properties and chemical reactivity, thus complementing the related hollow fullerenes.

The applications of fullerenes is an open question which has been mainly focussed on the use of fullerenes for organic electronics, namely with their use in organic photovoltaics which, no doubt, represents the most realistic application of fullerenes, as well as their use in the study of molecular wires. Both topics give an idea of the interest of these carbon allotropes in the emerging fields of nanoscience and nanotechnology.

Finally, we have stressed the future of fullerenes on those which are nowadays considered an almost scientific curiosity, the non-IPR fullerenes. Since they are thermodynamically less stable than the corresponding fullerenes obeying the IPR rule, their syntheses as endo- or exohedrically functionalized species is currently a synthetic challenge and some of the most outstanding examples have been presented. Since the potential number is huge compared with those known so far, the history and development of fullerenes might just be in its infancy. The future will tell us how these and other fullerene species still to come will impact new technologies based on these new carbon nanoforms.

References

1. Kroto HW, Heath JR, O'Brien SC et al (1985) C_{60}: buckminsterfullerene. Nature 318:162–163
2. Cami J, Bernard-Salas J, Peeters E et al (2010) Detection of C_{60} and C_{70} in a young planetary nebula. Science 329:1180–1182
3. Iijima S (1991) Helical microtubules of graphitic carbon. Nature 354:56–58
4. Iijima S, Ichihashi T (1993) Single-shell carbon nanotubes of 1-nm diameter. Nature 363:603–605
5. Bethune DS, Kiang CH, de Vries MS et al (1993) Cobalt-catalyzed growth of carbon nanotubes with single-atomic-layer walls. Nature 363:605–607
6. Novoselov KS, Geim AK, Morozov SV et al (2004) Electric field effect in atomically thin carbon films. Science 306:666–669
7. Delgado JL, Herranz MA, Martín N (2008) The nanoforms of carbon. J Mater Chem 18:1417–1426
8. Akasaka T, Nagase S (2002) Endofullerenes: a new family of carbon cluster. Kluwer, Dordrecht
9. Kroto HW (1997) Symmetry, space, stars, and C_{60}. Angew Chem Int Ed 36:1578–1593
10. Smalley RE (1997) Discovering the fullerrenes. Angew Chem Int Ed 36:1594–1601
11. Curl RF (1997) Dawn of the fullerenes: conjecture and experiment. Angew Chem Int Ed 36:1566–1576
12. Martín N (2006) New challenges in fullerene chemistry. Chem Commun 2093–2104
13. Krätschmer W, Lamb LD, Fostiropoulos K et al (1990) Solid C_{60}: a new form of carbon. Nature 347:354–358
14. Jones DEH (1966) Hollow molecules. New Sci 32:245
15. Chuvilin A, Kaiser U, Bichoutskaia E et al (2010) Direct transformation of graphene to fullerene. Nat Chem 2:450–453
16. Kroto HW (1987) The stability of the fullerenes C_n, with $n = 24, 28, 32, 36, 50, 60$ and 70. Nature 329:529–531
17. Haddon RC (1992) Electronic structure, conductivity, and superconductivity of alkali metal doped C_{60}. Acc Chem Res 25:127–133
18. Hirsch A, Chen Z, Jiao H (2000) Spherical aromaticity in I_h symmetrical fullerenes: the 2 $(N + 1)^2$ rule. Angew Chem Int Ed 39:3915–3917
19. Guldi DM, Martín N (eds) (2002) Fullerenes: from synthesis to optoelectronic properties. Kluwer Academic, Dordrecht
20. Hirsch A, Brettreich M (2005) Fullerenes, chemistry and reactions. Wiley-VCH, Weinheim
21. Langa F, Nierengarten JF (eds) (2012) Fullerenes: principles and applications. Royal Society of Chemistry, Cambridge
22. Haddon RC, Brus LE, Raghavachari K (1986) Electronic structure and bonding in icosahedral carbon cluster (C_{60}). Chem Phys Lett 125:459–464
23. Xie Q, Perez-Cordero E, Echegoyen L (1992) Electrochemical detection of C_{60} and C_{70}: enhanced stability of fullerides in solution. J Am Chem Soc 114:3978–3980
24. Martin N, Altable M, Filippone S et al (2006) Thermal [2+2] intramolecular cycloadditions of fuller-1,6-enynes. Angew Chem Int Ed 45:1439–1442
25. Altable M, Filippone S, Martin-Domenech A et al (2006) Intramolecular ene reaction of 1,6-fullerenynes: a new synthesis of allenes. Org Lett 8:5959–5962
26. Li H, Risko C, Seo JH et al (2011) Fullerene–carbene Lewis acid–base adducts. J Am Chem Soc 133:12410–12413
27. Cozzi F, Powell WH, Thilgen C (2005) Numbering of fullerenes. Pure Appl Chem 77:843–923
28. Komatsu K, Murata Y, Takimoto N et al (1994) Synthesis and properties of the first acetylene derivatives of C_{60}. J Org Chem 59:6101–6102

29. Nagashima H, Terasaki H, Kimura E et al (1994) Silylmethylations of C_{60} with Grignard reagents: selective synthesis of $HC_{60}CH_2SiMe_2Y$ and $C_{60}(CH_2SiMe_2Y)_2$ with selection of solvents. J Org Chem 59:1246–1248
30. Hirsch A, Soi A, Karfunhel HR (1992) Titration of C_{60}: a method for the synthesis of organofullerenes. Angew Chem Int Ed 31:766–768
31. Sawamura M, Iikura H, Nakamura E (1996) The first pentahaptofullerene metal complexes. J Am Chem Soc 118:12850–12851
32. Matsuo Y, Nakamura E (2008) Selective multiaddition of organocopper reagents to fullerenes. Chem Rev 108:3016–3028
33. Martin N, Altable M, Filippone S et al. (2004) Highly efficient Pauson–Khand reaction with C_{60}: regioselective synthesis of unprecedented *cis*-1 biscycloadducts. Chem Commun 1338–1339
34. Martín N, Altable M, Filippone S et al (2005) Regioselective intramolecular Pauson–Khand reactions of C_{60}: an electrochemical study and theoretical underpinning. Chemistry 11:2716–2729
35. Nambo M, Noyori R, Itami K (2007) Rh-catalyzed arylation and alkenylation of C_{60} using organoboron compounds. J Am Chem Soc 129:8080–8081
36. Nambo M, Segawa Y, Wakamiya A et al (2011) Selective introduction of organic groups to C_{60} and C_{70} using organoboron compounds and rhodium catalyst: a new synthetic approach to organo(hydro)fullerenes. Chem Asian J 6:590–598
37. Lu S, Jin T, Bao M et al (2011) Cobalt-catalyzed hydroalkylation of [60]fullerene with active alkyl bromides: selective synthesis of monoalkylated fullerenes. J Am Chem Soc 133:12842–12848
38. Xiao Z, Matsuo Y, Nakamura E (2010) Copper-catalyzed formal [4+2] annulation between alkyne and fullerene bromide. J Am Chem Soc 132:12234–12236
39. Zhu B, Wang G-W (2009) Palladium-catalyzed heteroannulation of [60]fullerene with anilides via C–H bond activation. Org Lett 11:4334–4337
40. Thilgen C, Gosse I, Diederich F (2003) Chirality in fullerene chemistry. Top Stereochem 23:1–124
41. Thilgen C, Diederich F (2006) Structural aspects of fullerene chemistry: a journey through fullerene chirality. Chem Rev 106:5049–5135
42. Nishimura T (2004) Macromolecular helicity induction on a poly(phenylacetylene) with C_2-symmetric chiral [60]fullerene-bisadducts. J Am Chem Soc 126:11711–11717
43. Friedman SH, Ganapathi PS, Rubin Y et al (1998) Optimizing the binding of fullerene inhibitors of the HIV-1 protease through predicted increases in hydrophobic desolvation. J Med Chem 41:2424–2429
44. Hizume Y, Tashiro K, Charvet R et al (2010) Chiroselective assembly of a chiral porphyrin–fullerene dyad: photoconductive nanofiber with a top-class ambipolar charge-carrier mobility. J Am Chem Soc 132:6628–6629
45. Filippone S, Maroto EE, Martín-Domenech A et al (2009) An efficient approach to chiral fullerene derivatives by catalytic enantioselective 1,3-dipolar cycloadditions. Nat Chem 1:578–582
46. Maroto EE, Filippone S, Martin-Domenech A et al (2012) Switching the stereoselectivity: (fullero)pyrrolidines “a la carte”. J Am Chem Soc 134:12936–12938
47. Maroto EE, de Cózar A, Filippone S et al (2011) Hierarchical selectivity in fullerenes: site-, regio-, diastereo-, and enantiocontrol of the 1,3-dipolar cycloaddition to C70. Angew Chem Int Ed 50:6060–6064
48. Sawai K, Takano Y, Izquierdo M et al (2011) Enantioselective synthesis of endohedral metallofullerenes. J Am Chem Soc 133:17746–17752
49. Bosi S, Da Ros T, Spalluto G et al (2003) Fullerene derivatives: an attractive tool for biological applications. Eur J Med Chem 38:913–923
50. Prato M, Martín N (eds) (2002) Special issue: Functionalised fullerenes. J Mater Chem 12:1931–2159

51. Manoharan M, de Proft F, Geerlings P (2000) Aromaticity interplay between quinodimethanes and C60 in Diels–Alder reactions: insights from a theoretical study. J Org Chem 65:6132–6137
52. Kräutler B, Maynollo J (1995) A highly symmetric sixfold cycloaddition product of fullerene C_{60}. Angew Chem Int Ed Engl 34:87–88
53. Herranz MA, Martín N, Ramey J et al (2002) Thermally reversible C_{60}-based donor–acceptor ensembles. Chem Commun 2002:2968–2969
54. Bingel C (1993) Cyclopropanierung von fullerenen. Chem Ber 126:1957–1959
55. Kessinger R, Crassous J, Herrmann A et al (1998) Preparation of enantiomerically pure C_{76} with a general electrochemical method for the removal of di(alkoxycarbonyl)methano bridges from methanofullerenes: the retro-Bingel reaction. Angew Chem Int Ed 37:1919–1922
56. Kessinger R, Fender NS, Echegoyen LE et al (2000) Selective electrolytic removal of bis (alkoxycarbonyl)methano addends from C_{60} bis-adducts and electrochemical stability of C_{70} derivatives. Chemistry 6:2184–2192
57. Moonen NNP, Thilgen C, Echegoyen L et al (2000) The chemical retro-Bingel reaction: selective removal of bis(alkoxycarbonyl)methano addends from C_{60} and C_{70} with amalgamated magnesium. Chem Commun 5:335–336
58. Prato M, Maggini M (1998) Fulleropyrrolidines: a family of full-fledged fullerene derivatives. Acc Chem Res 31:519–526
59. Martín N, Altable M, Filippone S et al (2006) Retro-cycloaddition reaction of pyrrolidinofullerenes. Angew Chem Int Ed 45:110–114
60. Brunetti FG, Herrero MA, Muñoz JM et al (2007) Reversible microwave-assisted cycloaddition of aziridines to carbon nanotubes. J Am Chem Soc 129:14580–14581
61. Guryanov I, Montellano Lopez A, Carraro M et al (2009) Metal-free, retro-cycloaddition of fulleropyrrolidines in ionic liquids under microwave irradiation. Chem Commun 3940–3942
62. Filippone S, Izquierdo Barroso M, Martín-Domenech A et al (2008) On the mechanism of the thermal retrocycloaddition of pyrrolidinofullerenes (retro-Prato reaction). Chemistry 14:5198–5206
63. Lukoyanova O, Cardona CM, Altable M et al (2006) Selective electrochemical retro-cycloaddition reaction of pyrrolidinofullerenes. Angew Chem Int Ed 45:7430–7433
64. Martín N, Altable M, Filippone S et al (2007) Highly efficient retro-cycloaddition reaction of isoxazolino[4,5:1,2][60]- and -[70]fullerenes. J Org Chem 72:3840–3846
65. Delgado JL, Oswald F, Cardinali F et al (2008) On the thermal stability of [60]fullerene cycloadducts: retro-cycloaddition reaction of 2-pyrazolino[4,5:1,2][60]-fullerenes. J Org Chem 73:3184–3188
66. Olah GA, Bucsi I, Lambert C et al (1991) Polyarenefullerenes, $C_{60}(H\text{-}Ar)_n$, obtained by acid-catalyzed fullerenation of aromatics. J Am Chem Soc 113:9387–9388
67. Giacalone F, Martín N (2006) Fullerene polymers: synthesis and properties. Chem Rev 106:5136–5190
68. Giacalone F, Martín N (eds) (2009) Fullerene polymers: synthesis, properties and applications. Wiley VCH, Weinheim
69. Giacalone F, Martín N (2010) New concepts and applications in the macromolecular chemistry of fullerenes. Adv Mater 22:4220–4248
70. Special issue on polymeric fullerenes (1997) Appl Phys A: Mater Sci Process 64:223–330
71. Sundqvist B (1999) Fullerenes under high pressures. Adv Phys 48:1
72. Rao AM, Zhou P, Wang K-A et al (1993) Photo-induced polymerization of solid C_{60} films. Science 259:955–957
73. Iwasa Y, Arima T, Fleming RM et al (1994) New phases of C_{60} synthesized at high-pressure. Science 264:1570–1572
74. Takahashi N, Dock H, Matsuzawa N et al (1993) Plasma-polymerized C_{60}/C_{70} mixture films: electric conductivity and structure. J Appl Phys 74:5790–5798

75. Nuñez-Regueiro M, Marques L, Hodeau JL et al (1995) Polymerized fullerite structures. Phys Rev Lett 74:278–281
76. Rao AM, Eklund PC, Venkateswaran UD et al (1997) Properties of C_{60} polymerized under high pressure and temperature. Appl Phys A: Mater Sci Process 64:231–2239
77. Fedurco M, Costa DA, Balch AL et al (1995) Electrochemical synthesis of a redox-active polymer based on buckminsterfullerene epoxide. Angew Chem Int Ed Engl 34:194–196
78. Winkler K, Costa DA, Balch AL et al (1995) A study of fullerene epoxide electroreduction and electropolymerization processes. J Phys Chem 99:17431–17436
79. Liu B, Bunker CE, Sun T-P (1996) Preparation and characterization of soluble pendant [60] fullerene-polystyrene polymers. Chem Commun 1241–1242
80. Stalmach U, de Boer B, Videlot C et al (2000) Semiconducting diblock copolymers synthesized by means of controlled radical polymerization techniques. J Am Chem Soc 122:5464–5472
81. Zheng JW, Goh SH, Lee SY (2000) Miscibility of C_{60}-containing poly(methyl methacrylate)/poly(vinylidene fluoride) blends. J Appl Polym Sci 75:1393–1396
82. Wang C, Tao Z, Yang W et al (2001) Synthesis and photoconductivity study of C_{60}-containing styrene/acrylamide copolymers. Macromol Rapid Commun 22:98–103
83. Gutiérrez-Nava M, Masson P, Nierengarten J-F (2003) Synthesis of copolymers alternating oligophenylenevinylene subunits and fullerene moieties. Tetrahedron Lett 44:4487–4490
84. Vitalini D, Mineo P, Iudicelli V et al (2000) Preparation of functionalized copolymers by thermal processes: porphyrination and fullerenation of a commercial polycarbonate. Macromolecules 33:7300–7309
85. Kraus A, Müllen K (1999) [60]Fullerene-containing poly(dimethylsiloxane)s: easy access to soluble polymers with high fullerene content. Macromolecules 32:4214–4219
86. Ungurenasu C, Pienteala M (2007) Syntheses and characterization of water-soluble C_{60}–curdlan sulfates for biological applications. J Polym Sci Part A: Polym Chem 45:3124–3128
87. Cravino A, Sariciftci NS (2002) Double-cable polymers for fullerene based organic optoelectronic applications. J Mater Chem 12:1931–1943
88. Cravino A, Sariciftci NS (2003) Organic electronics: molecules as bipolar conductors. Nat Mater 2:360–361
89. Kawase T (2012) Receptors for pristine fullerenes based on concave-convex π-π interactions. In: Martín N, Nierengarten J-F (eds) Supramolecular chemistry of fullerenes and carbon nanotubes. Wiley-VCH, Weinheim, pp 55–78 (Chap. 3)
90. Martín N, Nierengarten J-F (2012) Supramolecular chemistry of fullerenes and carbon nanotubes. Wiley-VCH, Weinheim
91. Sterescu DM, Stamatialis DF, Mendes E, Wibbenhorst M, Wessling M (2006) Fullerene-modified poly(2,6-dimethyl-1,4-phenylene oxide) gas separation membranes: why binding is better than dispersing. Macromolecules 39:9234–9242
92. Vinogradova LV, Polotskaya GA, Shevtsova AA et al (2009) Gas-separating properties of membranes based on star-shaped fullerene (C_{60})-containing polystyrenes. Polym Sci Ser A 51:209–215
93. Wang H, DeSousa R, Gasa J et al (2007) Fabrication of new fullerene composite materials and their application in proton exchange membrane fuel cells. J Membr Sci 289:277–283
94. Chen X, Gholamkhass B, Han X et al (2007) Polythiophene-*graft*-styrene and polythiophene-*graft*-(styrene-*graft*-C_{60}) copolymers. Macromol Rapid Commun 28:1792–1797
95. Nanjo M, Cyr PW, Liu K et al (2008) Donor–acceptor C_{60}-containing polyferrocenylsilanes: synthesis, characterization, and applications in photodiode devices. Adv Funct Mater 18:470–477
96. Ling Q-D, Lim S-L, Song Y et al (2007) Nonvolatile polymer memory device based on bistable electrical switching in a thin film of poly(*N*-vinylcarbazole) with covalently bonded C_{60}. Langmuir 23:312–319

97. Tutt LW, Kost A (1992) Optical limiting performance of C_{60} and C_{70} solutions. Nature 356:225–226
98. Cha M, Sariciftci NS, Heeger AJ et al (1995) Enhanced nonlinear absorption and optical limiting in semiconducting polymer/methanofullerene charge transfer films. Appl Phys Lett 67:3850–3852
99. Maggini M, Scorrano G, Prato M et al (1995) C_{60} derivatives embedded in sol–gel silica films. Adv Mater 7:404–406
100. Bunker CE, Lawson GE, Sun YP (1995) Fullerene-styrene random copolymers. Novel optical properties. Macromolecules 28:3744–3746
101. Kojima Y, Matsuoka T, Takahashi H et al (1995) Optical limiting property of polystyrene-bound C_{60}. Macromolecules 28:8868–8869
102. Lu Z, Goh SH, Lee SY et al (1999) Synthesis, characterization and nonlinear optical properties of copolymers of benzylaminofullerene with methyl methacrylate or ethyl methacrylate. Polymer 40:2863–2867
103. Sun YP, Riggs JE (1997) Non-linear absorptions in pendant [60]fullerene–polystyrene polymers. J Chem Soc Faraday Trans 93:1965–1969
104. Tang BZ, Xu HY, Lam JWY et al (2000) C_{60}-containing poly(1-phenyl-1-alkynes): synthesis, light emission, and optical limiting. Chem Mater 12:1446–1449
105. Li FY, Li YL, Guo ZX et al (2000) Synthesis and optical limiting properties of polycarbonates containing fullerene derivative. J Phys Chem Solids 61:1101–1103
106. Celli A, Marchese P, Vannini M et al (2011) Synthesis of novel fullerene-functionalized polysulfones for optical limiting applications. React Funct Polym 71:641–647
107. Mroz P, Tegos GP, Gali H et al (2007) Photodynamic therapy with fullerenes. Photochem Photobiol Sci 6:1139–1149
108. Liu Y, Wang H, Liang P et al (2004) Water-soluble supramolecular fullerene assembly mediated by metallobridged β-cyclodextrins. Angew Chem Int Ed 43:2690–2694
109. Samal S, Choi B-J, Geckeler KE (2001) DNA-cleavage by fullerene-based synzymes. Macromol Biosci 1:329–331
110. Liu J, Ohta S, Sonoda A et al (2007) Preparation of PEG-conjugated fullerene containing Gd^{3+} ions for photodynamic therapy. J Control Release 117:104–110
111. Stoilova O, Jérôme C, Detrembleur C et al (2007) C_{60}-containing nanostructured polymeric materials with potential biomedical applications. Polymer 48:1835–1843
112. Drees M, Hoppe H, Winder C et al (2005) Stabilization of the nanomorphology of polymer–fullerene "bulk heterojunction" blends using a novel polymerizable fullerene derivative. J Mater Chem 15:5158–5163
113. Sivula K, Ball ZT, Watanabe N et al (2006) Amphiphilic diblock copolymer compatibilizers and their effect on the morphology and performance of polythiophene:fullerene solar cells. Adv Mater 18:206–210
114. Yang C, Lee JK, Heeger AJ et al (2009) Well-defined donor–acceptor rod–coil diblock copolymers based on P3HT containing C_{60}: the morphology and role as a surfactant in bulk-heterojunction solar cells. J Mater Chem 19:5416–5423
115. Hsieh C-H, Cheng Y-J, Li P-J et al (2010) Highly efficient and stable inverted polymer solar cells integrated with a cross-linked fullerene material as an interlayer. J Am Chem Soc 132:4887–4893
116. Cheng Y-J, Hsieh C-H, He Y et al (2010) Combination of indene-C_{60} bis-adduct and cross-linked fullerene interlayer leading to highly efficient inverted polymer solar cells. J Am Chem Soc 132:17381–17383
117. Jeffery GA (1997) An introduction to hydrogen bonding. Oxford University Press, Oxford
118. Collins AF, Critchley C (2005) Artificial photosynthesis: from basic biology to industrial applications. Wiley, Weinheim
119. Delgado JL, Bouit PA, Filippone S et al (2010) Organic photovoltaics: a chemical approach. Chem Commun 46:4853–4865

120. Pinzón JR, Villalta-Cerdas A, Echegoyen L (2012) Fullerenes, carbon nanotubes, and graphene for molecular electronics. Top Curr Chem 312:127–174
121. Diederich F, Echegoyen L, Gómez-López M et al (1999) The self-assembly of fullerene-containing [2]pseudorotaxanes: formation of a supramolecular C_{60} dimer. J Chem Soc Perkin Trans 2:1577–1586
122. Rispens MT, Sánchez L, Knol J et al (2001) Supramolecular organization of fullerenes by quadruple hydrogen bonding. Chem Commun 161–162
123. González JJ, González S, Priego E et al (2001) A new approach to supramolecular C_{60}-dimers based in quadruple hydrogen bonding. Chem Commun 163–164
124. Da Ros T, Guldi DM, Morales AF et al (2003) Hydrogen bond-assembled fullerene molecular shuttle. Org Lett 5:689–691
125. Mateo-Alonso A, Fioravanti G, Marcaccio M et al (2006) Reverse shuttling in a fullerene-stoppered rotaxane. Org Lett 8:5173–5176
126. Mateo-Alonso A, Brough P, Prato M (2007) Stabilization of fulleropyrrolidine N-oxides through intrarotaxane hydrogen bonding. Chem Commun 1412–1414
127. Mateo-Alonso A, Fioravanti G, Marcaccio M et al (2007) An electrochemically driven molecular shuttle controlled and monitored by C_{60}. Chem Commun 1945–1947
128. Scarel F, Valenti G, Gaikwad S et al (2012) A molecular shuttle driven by fullerene radical-anion recognition. Chemistry 44:14063–14068
129. Guldi DM, Ramey J, Martínez-Díaz MV et al (2002) Reversible zinc phthalocyanine fullerene ensembles. Chem Commun 2774–2775
130. Sánchez L, Sierra M, Martín N et al (2006) Exceptionally strong electronic communication through hydrogen bonds in porphyrin–C_{60} pairs. Angew Chem Int Ed 45:4637–4641
131. Sessler JL, Jayawickramarajah J, Gouloumis A et al (2005) Synthesis and photophysics of a porphyrin-fullerene dyad assembled through Watson–Crick hydrogen bonding. Chem Commun 1892–1894
132. Torres T, Gouloumis A, Sánchez-García D et al (2007) Photophysical characterization of a cytidine-guanosine tethered phthalocyanine-fullerene dyad. Chem Commun 292–294
133. Wessendorf F, Gnichwitz J-F, Sarova GH et al (2007) Implementation of a Hamilton-receptor-based hydrogen-bonding motif toward a new electron donor-acceptor prototype: electron versus energy transfer. J Am Chem Soc 129:16057–16071
134. Maurer K, Grimm B, Wessendorf F et al (2010) Self-assembling depsipeptide dendrimers and dendritic fullerenes with new *cis*- and *trans*-symmetric Hamilton receptor functionalized Zn–porphyrins: synthesis, photophysical properties and cooperativity phenomena. Eur J Org Chem 5010–5029
135. Grimm B, Schornbaum J, Jasch H et al (2012) Step-by-step self-assembled hybrids that feature control over energy and charge transfer. Proc Natl Acad Sci U S A 109:15565–15571
136. Santos J, Grimm B, Illescas BM et al (2008) Cooperativity between π-π and H-bonding interactions – a supramolecular complex formed by C_{60} and exTTF. Chem Commun 5993–5995
137. Huang C-H, McClenaghan ND, Kuhn A et al (2005) Enhanced photovoltaic response in hydrogen-bonded all-organic devices. Org Lett 7:3409–3412
138. Chu C-C, Raffy G, Ray D et al (2010) Self-assembly of supramolecular fullerene ribbons via hydrogen-bonding interactions and their impact on fullerene electronic interactions and charge carrier mobility. J Am Chem Soc 132:12717–12723
139. Pérez EM, Martín N (2008) Curves ahead: molecular receptors for fullerenes based on concave-convex complementarity. Chem Soc Rev 37:1512–1519
140. Tashiro K, Aida T (2007) Metalloporphyrin hosts for supramolecular chemistry of fullerenes. Chem Soc Rev 36:189–197
141. Kawase T, Kurata H (2006) Ball-, bowl-, and belt-shaped conjugated systems and their complexing abilities: exploration of the concave-convex π–π interaction. Chem Rev 106:5250–5273

142. Mizyed S, Georghiou PE, Bancu M et al (2001) Embracing C_{60} with multiarmed geodesic partners. J Am Chem Soc 123:12770–12774
143. Sygula A, Fronczek FR, Sygula R et al (2007) A double concave hydrocarbon buckycatcher. J Am Chem Soc 129:3842–3843
144. Pérez EM, Martín N (2010) Molecular tweezers for fullerenes. Pure Appl Chem 82:523–533
145. Kawase T, Darabi HR, Oda M (1996) Cyclic [6]- and [8]paraphenylacetylenes. Angew Chem Int Ed 35:2664–2666
146. Kawase T, Tanaka K, Fujiwara N et al (2003) Complexation of a carbon nanoring with fullerenes. Angew Chem Int Ed 42:1624–1628
147. Kawase T, Tanaka K, Seirai Y et al (2003) Complexation of carbon nanorings with fullerenes: supramolecular dynamics and structural tuning for a fullerene sensor. Angew Chem Int Ed 42:5597–5600
148. Omachi H, Segawa Y, Itami K (2012) Synthesis of cycloparaphenylenes and related carbon nanorings: a step toward the controlled synthesis of carbon nanotubes. Acc Chem Res 45:1378–1389
149. Iwamoto T, Watanabe Y, Sadahiro T et al (2011) Size-selective encapsulation of C_{60} by [10] cycloparaphenylene: formation of the shortest fullerene-peapod. Angew Chem Int Ed 50:8342–8344
150. Xia J, Bacon JW, Jasti R (2012) Gram-scale synthesis and crystal structures of [8]- and [10] CPP, and the solid-state structure of C_{60}·[10]CPP. Chem Sci 3:3018–3021
151. Pérez EM, Sánchez L, Fernández G et al (2006) exTTF as a building block for fullerene receptors. Unexpected solvent-dependent positive homotropic cooperativity. J Am Chem Soc 128:7172–7173
152. Gayathri SS, Wielopolski M, Pérez EM et al (2009) Discrete supramolecular donor-acceptor complexes. Angew Chem Int Ed 48:815–819
153. Pérez EM, Capodilupo AL, Fernández G et al (2008) Weighting non-covalent forces in the molecular recognition of C_{60}. Relevance of concave-convex complementarity. Chem Commun 4567–4569
154. Pérez EM, Martín N (2012) Chiral recognition of carbon nanoforms. Org Biomol Chem 10:3577–3583
155. Pérez EM, Sierra M, Sánchez L et al (2007) Concave tetrathiafulvalene-type donors as supramolecular partners for fullerenes. Angew Chem Int Ed 46:1847–1851
156. Haino T, Yanase M, Fukazawa Y (1998) Fullerenes enclosed in bridged calix[5]arenes. Angew Chem Int Ed 37:997–998
157. Uno H, Furukawa M, Fujimoto A et al (2011) Porphyrin molecular tweezers for fullerenes. J Porphyr Phthalocyanins 15:951–963
158. Sun D, Tham FS, Reed CA et al (2000) Porphyrin-fullerene host-guest chemistry. J Am Chem Soc 122:10704–10705
159. Sun D, Tham FS, Reed CA et al (2002) Supramolecular fullerene-porphyrin chemistry. Fullerene complexation by metalated "jaws porphyrin" hosts. J Am Chem Soc 124:6604–6612
160. Hosseini A, Taylor S, Accorsi G et al (2006) Calix[4]arene-linked bisporphyrin hosts for fullerenes: binding strength, solvation effects, and porphyrin-fullerene charge transfer bands. J Am Chem Soc 128:15903–15913
161. Ayabe M, Ikeda A, Shinkai S et al (2002) A novel [60]fullerene receptor with a Pd(II)-switched bisporphyrin cleft. Chem Commun 1032–1033
162. Fernández G, Pérez EM, Sánchez L et al (2008) Self-organization of electroactive materials: a head-to-tail donor-acceptor supramolecular polymer. Angew Chem Int Ed 47:1094–1097
163. Fernández G, Pérez EM, Sánchez L et al (2008) An electroactive dynamically polydisperse supramolecular dendrimer. J Am Chem Soc 130:2410–2411
164. Santos J, Pérez EM, Illescas BM et al (2011) Linear and hyperbranched electron-acceptor supramolecular oligomers. Chem Asian J 6:1848–1853

165. Fernández G, Sánchez L, Pérez EM et al (2008) Large exTTF-based dendrimers. Self-assembly and peripheral cooperative multiencapsulation of C_{60}. J Am Chem Soc 130:10674–10683
166. Canevet D, Pérez EM, Martín N (2011) Wraparound hosts for fullerenes: tailored macrocycles and cages. Angew Chem Int Ed 50:9248–9259
167. Tashiro K, Aida T, Zheng J-Y et al (1999) A cyclic dimer of metalloporphyrin forms a highly stable inclusion complex with C_{60}. J Am Chem Soc 121:9477–9478
168. Yanagisawa M, Tashiro K, Yamasaki M et al (2007) Hosting fullerenes by dynamic bond formation with an iridium porphyrin cyclic dimer: a "chemical friction" for rotary guest motions. J Am Chem Soc 129:11912–11913
169. Gil-Ramírez G, Karlen SD, Shundo A et al (2010) A cyclic porphyrin trimer as a receptor for fullerenes. Org Lett 12:3544–3547
170. Song J, Aratani N, Shinokubo H et al (2010) A porphyrin nanobarrel that encapsulates C_{60}. J Am Chem Soc 132:16356–16357
171. Zheng J-Y, Tashiro K et al (2001) Cyclic dimers of metalloporphyrins as tunable hosts for fullerenes: a remarkable effect of rhodium(III). Angew Chem Int Ed 40:1857–1861
172. Isla H, Gallego M, Pérez EM et al (2010) A bis-exTTF macrocyclic receptor that associates C_{60} with micromolar affinity. J Am Chem Soc 132:1772–1773
173. Canevet D, Gallego M, Isla H et al (2011) Macrocyclic hosts for fullerenes: extreme changes in binding abilities with small structural variations. J Am Chem Soc 133:3184–3190
174. Akasaka T, Wudl F, Nagase S (2010) Chemistry of nanocarbons. Wiley-VCH, Chichester
175. Yamada M, Akasaka T, Nagase S (2010) Endohedral metal atoms in pristine and functionalized fullerene cages. Acc Chem Res 43:92–102
176. Lu X, Akasaka T, Nagase S (2012) Chemistry of endohedral metallofullerenes: the role of metals. Chem Commun 47:5942–5957
177. Rodríguez-Fortea A, Balch AL, Poblet JM (2011) Endohedral metallofullerenes: a unique host-guest association. Chem Soc Rev 40:3551–3563
178. Dunsch L, Yang S (2007) Metal nitride cluster fullerenes: their current state and future prospects. Small 3:1298–1320
179. Stevenson S, Mackey MA, Stuart MA et al (2008) A distorted tetrahedral metal oxide cluster inside an icosahedral carbon cage. Synthesis, isolation, and structural characterization of $Sc_4(mu_3\text{-}O)_2@I_h\text{-}C_{80}$. J Am Chem Soc 130:11844–11845
180. Chaur MN, Melin F, Ortiz AL et al (2009) Chemical, electrochemical, and structural properties of endohedral metallofullerenes. Angew Chem Int Ed 48:7514–7538
181. Saunders M, Jiménez-Vázquez HA, Cross RJ et al (1993) Stable compounds of helium and neon. $He@C_{60}$ and $Ne@C_{60}$. Science 259:1428–1430
182. Kurotobi K, Murata Y (2011) A single molecule of water encapsulated in fullerene C_{60}. Science 333:613–616
183. Campanera JM, Bo C, Olmstead MM et al (2002) Bonding within the endohedral fullerenes $Sc_3N@C_{78}$ and $Sc_3N@C_{80}$ as determined by density functional calculations and reexamination of the crystal structure of $\{Sc_3N@C_{78}\}\cdot Co(OEP)\}\cdot 1.5(C_6H_6)\cdot 0.3(CHCl_3)$. J Phys Chem A 106:12356–12364
184. Aoyagi S, Nishibori E, Sawa H et al (2010) A layered ionic crystal of polar $Li@C_{60}$ superatoms. Nat Chem 2:678–683
185. Aoyagi S, Sado Y, Nishibori E et al (2012) Rock-salt-type crystal of thermally contracted C_{60} with encapsulated lithium cation. Angew Chem Int Ed 51:3377–3381
186. Chai Y, Guo T, Jin C et al (1991) Fullerenes with metals inside. J Phys Chem 95:7564–7568
187. Nagase S, Kobayashi K (1994) The ionization energies and electron affinities of endohedral metallofullerenes $MC_{82}(M = Sc, Y, La)$: density functional calculations. J Chem Soc Chem Commun 1837–1838
188. Tsuchiya T, Sato K, Kurihara H et al (2006) Spin-site exchange system constructed from endohedral metallofullerenes and organic donors. J Am Chem Soc 128:14418–14419

189. Sato S, Seki S, Honsho Y et al (2011) Semi-metallic single-component crystal of soluble La@C_{82} derivative with high electron mobility. J Am Chem Soc 133:2766–2771
190. Feng L, Tsuchiya T, Wakahara T et al (2006) Synthesis and characterization of a bisadduct of La@C_{82}. J Am Chem Soc 128:5990–5991
191. Wakahara T, Yamada M, Takahashi S et al (2007) Two-dimensional hopping motion of encapsulated La atoms in silylated La_2@C_{80}. Chem Commun 2680–2682
192. Yamada M, Mizorogi N, Tsuchiya T et al (2009) Synthesis and characterization of the D_{5h} isomer of the endohedral dimetallofullerene Ce_2@C_{80}: two-dimensional circulation of encapsulated metal atoms inside a fullerene cage. Chemistry 15:9486–9493
193. Stevenson S, Rice G, Glass T et al (1999) Small-bandgap endohedral metallofullerenes in high yield and purity. Nature 401:55–57
194. Popov AA, Dunsch L (2007) Structure, stability, and cluster-cage interactions in nitride clusterfullerenes M_3N@C_{2n} (M = Sc, Y; 2n = 68–98): a density functional theory study. J Am Chem Soc 129:11835–11849
195. Rodríguez-Fortea A, Alegret N, Balch AL et al (2010) The maximum pentagon separation rule provides a guideline for the structures of endohedral metallofullerenes. Nat Chem 2:955–961
196. Stevenson S, Phillips JP, Reid JE et al (2004) Pyramidalization of Gd_3N inside a C_{80} cage. The synthesis and structure of Gd_3N@C_{80}. Chem Commun 2814–2815
197. Chaur MN, Melin F, Elliott B et al (2007) Gd_3N@C_{2n} (n = 40, 42, and 44): remarkably low HOMO-LUMO gap and unusual electrochemical reversibility of Gd_3N@C_{88}. J Am Chem Soc 129:14826–14829
198. Chaur MN, Melin F, Ashby J et al (2008) Lanthanum nitride endohedral fullerenes La_3N@C_{2n} (43< or =n< or =55): preferential formation of La_3N@C_{96}. Chemistry 14:8213–8219
199. Cao B, Wakahara T, Maeda Y et al (2004) Lanthanum endohedral metallofulleropyrrolidines: synthesis, isolation, and EPR characterization. Chemistry 10:716–720
200. Cardona CM, Kitaygorodskiy A, Echegoyen L (2005) Trimetallic nitride endohedral metallofullerenes: reactivity dictated by the encapsulated metal cluster. J Am Chem Soc 127:10448–10453
201. Yamada M, Someya C, Wakahara T et al (2008) Metal atoms collinear with the spiro carbon of 6,6-open adducts, M_2@C_{80}(Ad) (M = La and Ce, Ad = adamantylidene). J Am Chem Soc 130:1171–1176
202. Shustova NB, Popov AA, Mackey MA et al (2007) Radical trifluoromethylation of Sc_3N@C_{80}. J Am Chem Soc 129:11676–11677
203. Shu C, Cai T, Xu L et al (2007) Manganese(III)-catalyzed free radical reactions on trimetallic nitride endohedral metallofullerenes. J Am Chem Soc 129:15710–15717
204. Iezzi EB, Duchamp JC, Harich K (2002) A symmetric derivative of the trimetallic nitride endohedral metallofullerene, Sc_3N@C_{80}. J Am Chem Soc 124:524–525
205. Lee HM, Olmstead MM, Iezzi E et al (2002) Crystallographic characterization and structural analysis of the first organic functionalization product of the endohedral fullerene Sc_3N@C_{80}. J Am Chem Soc 124:3494–3495
206. Ge Z, Duchamp JC, Cai T et al (2005) Purification of endohedral trimetallic nitride fullerenes in a single, facile step. J Am Chem Soc 127:16292–16298
207. Cai T, Ge Z, Iezzi EB et al (2005) Synthesis and characterization of the first trimetallic nitride templated pyrrolidino endohedral metallofullerenes. Chem Commun 3594–3596
208. Wakahara T, Iiduka Y, Ikenaga O et al (2006) Characterization of the bis-silylated endofullerene Sc_3N@C_{80}. J Am Chem Soc 128:9919–9925
209. Yamada M, Minowa M, Sato S et al (2011) Regioselective cycloaddition of La_2@I_h-C_{80} with tetracyanoethylene oxide: formation of an endohedral dimetallofullerene adduct featuring enhanced electron-accepting character. J Am Chem Soc 33:3796–3799
210. Liu T-X, Wei T, Zhu S-E et al (2012) Azide addition to an endohedral metallofullerene: formation of azafulleroids of Sc_3N@I_h-C_{80}. J Am Chem Soc 134:11956–11959

211. Yamada M, Nakahodo T, Wakahara T et al (2005) Positional control of encapsulated atoms inside a fullerene cage by exohedral addition. J Am Chem Soc 127:14570–14571
212. Yamada M, Wakahara T, Nakahodo T et al (2006) Synthesis and structural characterization of endohedral pyrrolidinodimetallofullerene: $La_2@C_{80}(CH_2)_2NTrt$. J Am Chem Soc 128:1402–1403
213. Cardona CM, Elliott B, Echegoyen L (2006) Unexpected chemical and electrochemical properties of $M_3N@C_{80}$ (M = Sc, Y, Er). J Am Chem Soc 128:6480–6485
214. Rodríguez-Fortea A, Campanera JM, Cardona CM et al (2006) Dancing on a fullerene surface: isomerization of $Y_3N@$(N-ethylpyrrolidino-C_{80}) from the 6,6 to the 5,6 regioisomers. Angew Chem Int Ed 45:8176–8180
215. Pinzón JR, Plonska-Brzezinska ME, Cardona CM et al (2008) $Sc_3N@C_{80}$-ferrocene electron-donor/acceptor conjugates as promising materials for photovoltaic applications. Angew Chem Int Ed 47:4173–4176
216. Takano Y, Herranz MA, Martín N et al (2010) Donor-acceptor conjugates of lanthanum endohedral metallofullerene and π-extended tetrathiafulvalene. J Am Chem Soc 132:8048–8055
217. Li FF, Pinzón JR, Mercado BQ et al (2011) [2+2]Cycloaddition reaction to $Sc_3N@I_h$-C_{80}. The formation of very stable [5,6]- and [6,6]-adducts. J Am Chem Soc 133:1563–1571
218. Wang GW, Liu TX, Jiao M et al (2011) The cycloaddition reaction of I_h-$Sc_3N@C_{80}$ with 2-amino-4,5-diisopropoxybenzoic acid and isoamyl nitrite to produce an open-cage metallofullerene. Angew Chem Int Ed 50:4658–4662
219. Lukoyanova O, Cardona CM, Rivera J et al (2007) Open rather than closed malonate methano-fullerene derivatives. The formation of methanofulleroid adducts of $Y_3N@C_{80}$. J Am Chem Soc 129:10423–10430
220. Cai T, Xu L, Shu C et al (2008) Selective formation of a symmetric Sc3N@C78 bisadduct: adduct docking controlled by an internal trimetallic nitride cluster. J Am Chem Soc 130:2136–2137
221. Rudolf M, Wolfrum S, Guldi DM et al (2012) Endohedral metallofullerenes–filled fullerene derivatives towards multifunctional reaction center mimics. Chemistry 8:5136–48
222. Feng L, Rudolf M, Wolfrum S et al (2012) A paradigmatic change: linking fullerenes to electron acceptors. J Am Chem Soc 34:12190–12197
223. Li FF, Rodríguez-Fortea A, Poblet JM et al (2011) Reactivity of metallic nitride endohedral metallofullerene anions: electrochemical synthesis of a $Lu_3N@I_h$-C_{80} derivative. J Am Chem Soc 133:2760–2765
224. Li FF, Rodríguez-Fortea A, Peng P et al (2012) Electrosynthesis of a $Sc_3N@I_h$-C_{80} methano derivative from trianionic $Sc_3N@Ih$-C_{80}. J Am Chem Soc 134:480–7487
225. Tsuchiya T, Wielopolski M, Sakuma N et al (2011) Stable radical anions inside fullerene cages: formation of reversible electron transfer systems. J Am Chem Soc 133:13280–13283
226. Armaroli N, Balzani V (2007) The future of energy supply: challenges and opportunities. Angew Chem Int Ed 46:52–66
227. Chapin DM, Fuller CS, Pearson GL (1954) A new silicon *p-n* junction photocell for converting solar radiation into electrical power. J Appl Chem 25:676–678
228. Rispens MT, Hummelen JC (2002) Fullerenes: from synthesis to optoelectronic properties. In: Guldi DM, Martín N (eds) Photovoltaic applications. Kluwer Academic, Dordrech, pp 387–435 (Chap. 12)
229. Hummelen JC, Knight BW, LePeq F et al (1995) Preparation and characterization of fulleroid and methanofullerene derivatives. J Org Chem 60:532–538
230. Yu G, Gao J, Hummelen JC et al (1995) Polymer photovoltaic cells: enhanced efficiencies via a network of internal donor-acceptor heterojunctions. Science 270:1789–1791
231. Zhang Y, Yip HL, Acton O et al (2009) A simple and effective way of achieving highly efficient and thermally stable bulk-heterojunction polymer solar cells using amorphous fullerene derivatives as electron acceptor. Chem Mater 21:2598–2600

232. Lenes L, Wetzelaer GJAH, Kooistra FB et al (2008) Fullerene bisadducts for enhanced open-circuit voltages and efficiencies in polymer solar cells. Adv Mater 20:2116–2119
233. Wienk MM, Kroon JM, Verhees WJH et al (2003) Efficient methano[70]fullerene/MDMO-PPV bulk heterojunction photovoltaic cells. Angew Chem Int Ed 42:3371–3375
234. Park SH, Roy A, Beaupré S et al (2009) Bulk heterojunction solar cells with internal quantum efficiency approaching 100%. Nat Photonics 3:297–302
235. Kooistra FB, Mihailetchi VD, Popescu LM et al (2006) New C_{84} derivative and its application in a bulk heterojunction solar cell. Chem Mater 18:3068–3073
236. Li CZ, Yip HL, Jen AKY (2012) Functional fullerenes for organic photovoltaics. J Mater Chem 22:4161–4177
237. Riedel I, von Hauff E, Parisi J et al (2005) Diphenylmethanofullerenes: new and efficient acceptors in bulk-heterojunction solar cells. Adv Funct Mater 15:1979–1987
238. Riedel I, Martín N, Giacalone F et al (2004) Polymer solar cells with novel fullerene-based acceptor. Thin Solid Films 451:43–47
239. Backer S, Sivula K, Kavulak DF et al (2007) High efficiency organic photovoltaics incorporating a new family of soluble fullerene derivatives. Chem Mater 19:2927–2929
240. He Y, Chen HY, Hou J et al (2010) Indene–C_{60} bisadduct: a new acceptor for high-performance polymer solar cells. J Am Chem Soc 132:1377–1382
241. Weiss EA, Wasielewski MR, Ratner MA (2005) Molecules as wires: molecule-assisted movement of charge and energy. Top Curr Chem 257:103–133
242. Guldi DM, Illescas BM, Atienza CM et al (2009) Fullerene for organic electronics. Chem Soc Rev 38:1587–1597
243. Ito O, Yamanaka K (2009) Roles of molecular wires between fullerenes and electron donors in photoinduced electron transfer. Bull Chem Soc Jpn 82:316–332
244. Vail SA, Schuster DI, Guldi DM et al (2006) Energy and electron transfer in beta-alkynyl-linked porphyrin-[60]fullerene dyads. J Phys Chem B 110:14155–14166
245. Vail SA, Krawczuk PJ, Guldi DM et al (2005) Energy and electron transfer in polyacetylene-linked zinc–porphyrin–[60]fullerene molecular wires. Chemistry 11:3375–3388
246. Tashiro K, Sato A, Yuzawa T et al (2006) Long-range photoinduced electron transfer mediated by oligo-p-phenylenebutadiynylene conjugated bridges. Chem Lett 35:518–519
247. Lembo A, Tagliatesta P, Guldi DM et al (2009) Porphyrin-β-oligo-ethynylenephenylene-[60] fullerene triads: synthesis and electrochemical and photophysical characterization of the new porphyrin-oligo-PPE-[60]fullerene systems. J Phys Chem A 113:1779–1793
248. Giacalone F, Segura JL, Martín N et al (2004) Exceptionally small attenuation factors in molecular wires. J Am Chem Soc 126:5340–5341
249. Giacalone F, Segura JL, Martín N et al (2005) Probing molecular wires: synthesis, structural, and electronic study of donor-acceptor assemblies exhibiting long-range electron transfer. Chemistry 11:4819–4834
250. de la Torre G, Giacalone F, Segura JL et al (2005) Electronic communication through π-conjugated wires in covalently linked porphyrin/C_{60} ensembles. Chemistry 11:12671280
251. Molina-Ontoria A, Wielopolski M, Gebhardt J (2011) [2,2′]Paracyclophane-based π-conjugaed molecular wires reveal molecular-junction behavior. J Am Chem Soc 133:2370–2373
252. Atienza-Castellanos C, Wielopolski M, Guldi DM et al (2007) Determination of the attenuation factor in fluorene-based molecular wires. Chem Commun 5164–5166
253. Wielopolski M, Santos J, Illescas BM et al (2011) Vinyl spacers – tuning electron transfer through fluorene-based molecular wires. Energy Environ Sci 4:765–771
254. Ikemoto J, Takimiya K, Aso Y et al (2002) Porphyrin–oligothiophene–fullerene triads as an efficient intramolecular electron-transfer system. Org Lett 4:309–311
255. Nakamura T, Fujitsuka M, Araki Y et al (2004) Photoinduced electron transfer in porphyrin-oligothiophene-fullerene linked triads by excitation of a porphyrin moiety. J Phys Chem B 108:10700–10710
256. Wessendorf F, Grimm B, Guldi DM et al (2010) Pairing fullerenes and porphyrins: supramolecular wires that exhibit charge transfer activity. J Am Chem Soc 132:10786–10795

257. Schmalz TG, Seitz WA, Klein DJ et al (1986) C_{60} carbon cages. Chem Phys Lett 130:203–207
258. Schein S, Friedrich TA (2008) A geometric constraint, the head-to-tail exclusion rule, may be the basis for the isolated-pentagon rule in fullerenes with more than 60 vertices. Proc Natl Acad Sci U S A 105:19142–19147
259. Tan T-Z, Li J, Zhu F et al (2010) Chlorofullerenes featuring triple sequentially fused pentagons. Nat Chem 2:269–273
260. Martín N (2011) Fullerene $C_{72}C_{14}$: the exception that proves the rule? Angew Chem Int Ed 50:5431–5433
261. Tan Y-Z, Xie S-Y, Huang R-B et al (2009) The stabilization of fused-pentagon fullerene molecules. Nat Chem 1:450–460
262. Wang CR, Kai T, Tomiyama T et al (2000) Materials science – C_{66} fullerene encaging a scandium dimer. Nature 408:426–427
263. Stevenson S, Fowler PW, Heine T et al (2000) Materials science: a stable non-classical metallofullerene family. Nature 408:427–428
264. Beavers CM, Zuo TM, Duchamp JC et al (2006) $Tb_3N@C_{84}$: an improbable, egg-shaped endohedral fullerene that violates the isolated pentagon rule. J Am Chem Soc 128:11352–11353
265. Yang SF, Popov AA, Dunsch L (2007) Violating the isolated pentagon rule (IPR): the endohedral non-IPR C_{70} cage of $Sc_3N@C_{70}$. Angew Chem Int Ed 46:1256–1259
266. Ma YH, Wang TS, Wu JY et al (2011) Size effect of endohedral cluster on fullerene cage: preparation and structural studies of $Y_3N@C_{78}$-C_2. Nanoscale 3:4955–4957
267. Shi ZQ, Wu X, Wang CR et al (2006) Isolation and characterization of $Sc_2C_2@C_{68}$: a metal-carbide endofullerene with a non-IPR carbon cage. Angew Chem Int Ed 45:2107–2111
268. Wu JY, Wang TS, Ma YH et al (2011) Synthesis, isolation, characterization, and theoretical studies of $Sc_3NC@C_{78}$-C_2. J Phys Chem C 115:23755–23759
269. Campanera JM, Bo C, Poblet JM (2005) General rule for the stabilization of fullerene cages encapsulating trimetallic nitride templates. Angew Chem Int Ed 44:7230–7233
270. Summerscales OT, Cloke FGN (2006) The organometallic chemistry of pentalene. Coord Chem Rev 250:1122–1140
271. Xie SY, Gao F, Lu X et al (2004) Capturing the labile fullerene[50] as $C_{50}Cl_{10}$. Science 304:699–699
272. Wang CR, Shi ZQ, Wan LJ et al (2006) C64H4: production, isolation, and structural characterizations of a stable unconventional fulleride. J Am Chem Soc 128:6605–6610
273. Li B, Shu CY, Lu X et al (2010) Addition of carbene to the equator of C(70) to produce the most stable C(71)H(2) isomer: 2 aH-2(12)a-homo(C(70)-D(5 h(6)))[5,6]fullerene. Angew Chem Int Ed 49:962–966
274. Tan YZ, Li J, Zhou T, Feng YQ et al (2010) Pentagon-fused hollow fullerene in C_{78} family retrieved by chlorination. J Am Chem Soc 132:12648–12652
275. Kato H, Taninaka A, Sugai T et al (2003) Structure of a missing-caged metallofullerene: $La_2@C_{72}$. J Am Chem Soc 125:7782–7783
276. Yamada M, Wakahara T, Tsuchiya T et al (2008) Spectroscopic and theoretical study of endohedral dimetallofullerene having a non-IPR fullerene cage: $Ce_2@C_{72}$. J Phys Chem A 112:7627–7631
277. Wakahara T, Nikawa H, Kikuchi T et al (2006) $La@C_{72}$ having a non-IPR carbon cage. J Am Chem Soc 128:14228–14229
278. Chen N, Beavers CM, Mulet-Gas M et al (2012) $Sc_2S@C(s)(10528)$-C_{72}: a dimetallic sulfide endohedral fullerene with a non isolated pentagon rule cage. J Am Chem Soc 134:7851–7860
279. Tan Y-Z, Zhou T, Bao J, Shan G-J, Xie S-Y, Huang R-B, Zheng L-S (2010) $C_{72}Cl_4$: a pristine fullerene with favorable pentagon-adjacent structure. J Am Chem Soc 132:17102–17104
280. Ziegler K, Mueller A, Amsharov KY, Jansen M (2010) Disclosure of the elusive C_{2v}-C_{72} carbon cage. J Am Chem Soc 132:17099–17101

Top Curr Chem (2014) 350: 65–110
DOI: 10.1007/128_2012_403

Published online: 14 February 2013

Carbon Nanotubes: Synthesis, Structure, Functionalization, and Characterization

Valeria Anna Zamolo, Ester Vazquez, and Maurizio Prato

Abstract Carbon nanotubes have generated great expectations in the scientific arena, mainly due to their spectacular properties, which include a high aspect ratio, high strain resistance, and high strength, along with high conductivities. Nowadays, carbon nanotubes are produced by a variety of methods, each of them with advantages and disadvantages. Once produced, carbon nanotubes can be chemically modified, using a wide range of chemical reactions. Functionalization makes these long wires much easier to manipulate and dispersible in several solvents. In addition, the properties of carbon nanotubes can be combined with those of organic appendages. Finally, carbon nanotubes need to be carefully characterized, either as pristine or modified materials.

Keywords Carbon nanotubes · Characterization · Functionalization

Contents

V.A. Zamolo and M. Prato (✉)
Department of Chemical and Pharmaceutical Sciences, Center of Excellence for Nanostructured Materials (CENMAT) and INSTM, Unit of Trieste, University of Trieste, Piazzale Europa 1, 34127 Trieste, Italy
e-mail: prato@units.it

E. Vazquez
Departamento de Química Orgánica, Facultad de Ciencias Químicas-IRICA, Universidad de Castilla-La Mancha, 13071 Ciudad Real, Spain

Abbreviations

AmB	Amphotericin B
^{13}C-NMR	Carbon nuclear magnetic resonance
CNTs	Carbon nanotubes
CVD	Chemical vapor deposition
DBU	1,8-Diazabicyclo[5,4,0]undec-7-ene
DENs	Dendrimer encapsulated nanoparticles
DMAD	Dimethyl acetylenedicarboxylate
DMAP	4-Dimethylaminopyridine
DMF	*N,N*-Dimethylformamide
DOS	Density of states
EDC	1-Ethyl-3-[3-dimethylaminopropyl]carbodiimide hydrochloride
EGF	Epidermal growth factor
EtOH	Ethanol
FAD	Flavin adenine dinucleotide
FIR	Far infrared
HiPco	High-pressure carbon monoxide
^{1}H-NMR	Proton nuclear magnetic resonance
HOBt	Hydroxybenzotriazole
HR-SEM	High-resolution scanning electron microscopy
HR-TEM	High-resolution tunneling electron microscopy
i-PrOH	Iso-propanol
MRI	Magnetic resonance imaging
MWCNTs	Multiwalled carbon nanotubes
NIR	Near infrared
PABS	Poly(aminobenzene sulfonic acid)
PEG	Polyethylene glycol
PL-PEG	PEGylated phospholipids
PmPV	Poly(*m*-phenylvinylene)
PSS	Polystyrenesulfonate
PVD	Polyvinylpyrrolidone

RBM	Radial breathing modes
SDBS	Sodium dodecylbenzene sulfonate
SDS	Sodium dodecylsulfate
SEM	Scanning electron microscopy
STM	Scanning tunneling microscopy
SWCNTs	Single walled carbon nanotubes
TEM	Transmission electron microscopy
TGA	Thermogravimetric analysis
THF	Tetrahydrofuran

1 Introduction

Diamond and graphite are the two main allotropic forms of carbon. In diamond, each sp^3 hybridized carbon atom is arranged in the corresponding tetrahedral configuration, giving an extended and highly ordered three-dimensional network, where the motif is represented by the chair conformation of cyclohexane. Graphite is made by planar sheets where sp^2 carbon atoms are disposed following the benzene ring motif and giving a honeycomb arrangement.

A third carbon structure, fullerene or buckminsterfullerene, was discovered in 1985 and added to the family of carbon allotropes [1]. The best known fullerene, C_{60}, consists of sp^2 hybridized carbon atoms combined together in hexagons and pentagons to form a truncated icosahedral-shaped molecule that reminds a soccer ball.

The first evidence of carbon nanotubes dates back to 1952, when the Russian scientists Radushkevich and Lukyanovich published some transmission electron micrographs of carbon filaments with 50 nm diameters [2]. However, the first time that carbon nanotubes were significantly introduced to scientific imagery was only in 1990, at a carbon composite workshop, when Rick Smalley envisioned a fourth carbon tubular structure, formed by the elongation of a C_{60} molecule.

One year later, experimental evidence of nanotube existence and synthesis was reported by Iijima, who, by means of a high-resolution transmission electron microscope (HR-TEM), observed fullerene tubular needles in the soot produced by an arch discharge between two graphite electrodes [3]. The powder analyzed was made of multi-walled carbon nanotubes (MWCNTs), up to several tens of hollow concentric cylinders of graphene, capped on the edges by fullerene-type hemispheres (Fig. 1).

In 1993, Iijima, Ichihashi, and Bethune reported the existence of a similar single-walled construction (SWCNTs), whose structure was investigated and elucidated in order to confirm the fundamental properties of the tubes [4–8].

According to a widely known definition, it is sufficient to cut a strip of graphene from an infinite two-dimensional sheet and roll it up to form a nanotube. However, the deceptively simple idea of rolling up a graphene sheet brings about different

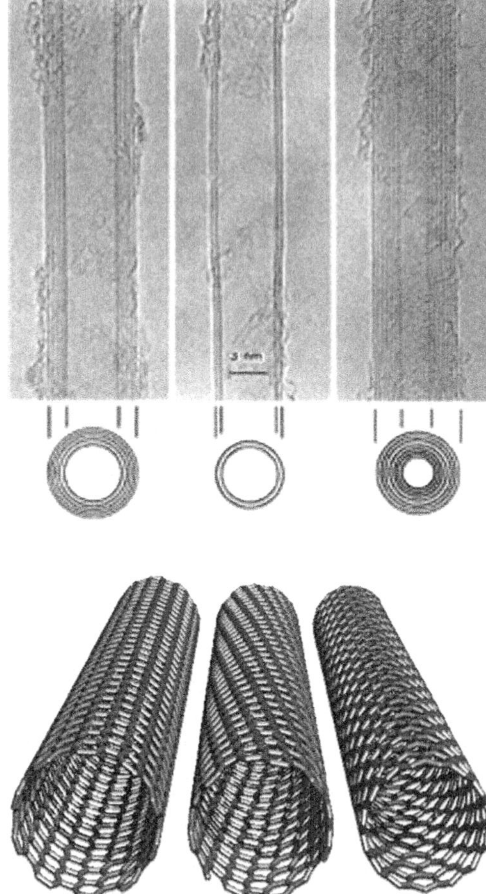

Fig. 1 Transmission electron micrographs of MWCNTs reported by Iijima in 1991. Reprinted with permission from [3]

Fig. 2 The way the graphene sheet is rolled up results in tubes with different morphologies and electronic behaviors. (**a**) Metallic "armchair" tube. (**b**) Semiconducting "chiral" tube. (**c**) Conducting "zig zag" tube. Reprinted with permission from [14]

types of tubes with very different properties. For instance, depending on the wrapping angle of the two-dimensional sheet, three types of SWCNTs can be defined: armchair, zigzag, and chiral (Fig. 2). The electronic behavior of the tubes is determined by their exact structure; thus the zigzag and chiral nanotubes can be either metallic or semiconducting while armchair nanotubes are always metallic [9]. In general, the available synthetic methods produce SWCNTs as a mixture of metallic and semiconducting material while MWCNTs are regarded as metallic. This is important, as semiconducting SWCNTs could be useful in applications that involve charge transfer processes, including sensors and photovoltaic devices, while metallic CNTs are preferred in electronic tools or as conductive filler in CNT-composites [10, 11].

Furthermore, nanotubes present intrinsic advantages, including high mechanical strength, structural flexibility, and high surface area, which make them potential candidates in the design of high-performance composite materials. Another key opportunity arises from the fact that CNTs can be easily internalized by cells and therefore can act like delivery vehicles for a variety of biological cargos, ranging from small molecules to biomacromolecules such as proteins and DNA/RNA. Furthermore, the intrinsic physical properties of CNTs can be utilized for multimodal imaging and therapy [12, 13].

This chapter is intended to introduce the reader to the world of CNTs, giving a snapshot of the most salient production and characterization techniques and putting particular emphasis on their properties and their chemical modification, which can lead to many interesting applications.

2 Production of SWCNTs and MWCNTs

There are several methods of making CNTs. Most probably carbon nanotubes had been around for a long time before their discovery as secondary products of various vapor deposition or carbon combustion processes, but electron microscopy was not advanced enough to identify them.

The first mass production of CNTs was obtained just by tuning the arc discharge synthesis method described by Bacon and Iijima in their pioneer article of 1991. In fact, CNTs were first produced in reasonable quantity just 1 year later by Ebbesen and Ajayan [15].

Other methods successfully explored were laser ablation, whose principle is similar to the arc discharge technique but may result in different purity of nanotubes, together with chemical vapor deposition (CVD), which nowadays is one of the most promising methods employed for large-scale production of carbon fibers and fullerenes [16].

Whichever is the method chosen for CNT production, common prerequisites are an active catalyst, a source of carbon, and suitable energy.

2.1 Arc Discharge

The arc discharge method provides both SWCNTs and MWCNTs on a large scale. In a typical arc discharge process, two graphite electrodes are used to produce a direct current electric arc discharge under inert gas atmosphere, such as Ar or He, at pressures between 100 and 1,000 Torr. The synthesis is carried out at low voltage (25–40 V) and high current (50–150 A) and the reaction time varies from 30 s to 10 min (Fig. 3).

The temperature in the inter-electrode zone is so high (4,000°C) that carbon sublimes from the positive electrode, which is consumed. The plasma formed between the electrodes acquires the shape of carbon nanotubes and fullerenes when it condenses at the cathode.

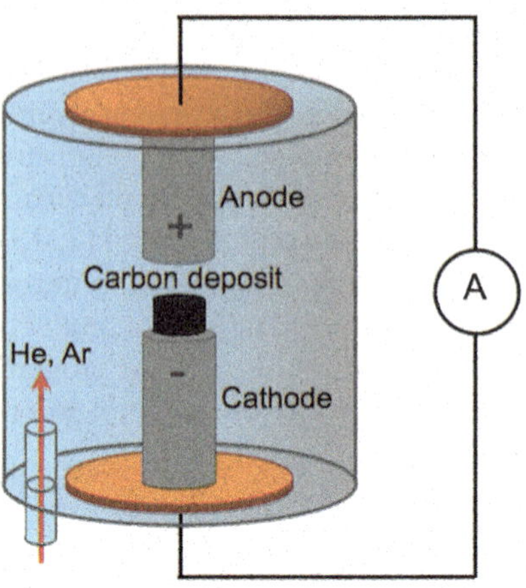

Fig. 3 Schematic representation of an arc discharge chamber

Iijima found that the use of a small amount of metal catalyst was critical for SWCNTs production. By drilling a hole in the anode and filling it with a small piece of iron he observed formation of SWCNTs with a broad diameter distribution between 0.7 and 1.65 nm. SWCNTs synthesis was also reported by Bethune, utilizing Co as catalyst instead of Fe [4, 5]. Actually several metal catalyst compositions have been found to produce SWCNTs, but the most widely used is a Y:Ni mixture able to yield up to 90% SWCNTs.

In general, MWCNTs obtained with this method present an inner diameter from 1 to 3 nm, a maximum length of 1 μm, and closed tips, while SWCNTs are smaller in diameter (maximum 1.4 nm) and longer (several microns), with closed tips as well.

Possible variations to the final product could be introduced by tailoring the Ar:He ratio (higher Ar content for smaller diameters) or generally by playing with the chemical and physical factors influencing the nucleation and growth of the nanotubes, such as the overall gas pressure, presence of hydrogen, or carbon vapor dispersion [17, 18].

One of the most important drawbacks of this technique lies in the extensive tube purification required before use. On the other hand, the method can produce large quantities of product at relatively low cost.

2.2 *Laser Ablation*

Arc discharge and laser ablation are similar methods, since they both exploit a graphitic source which is consumed at high temperatures in order to generate a plasma plume of vaporized carbon.

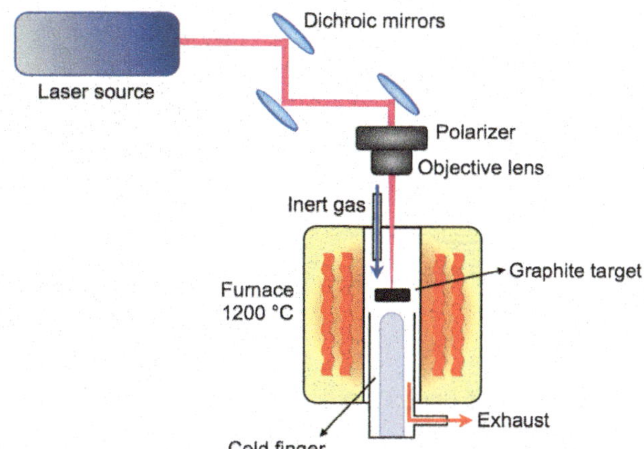

Fig. 4 Illustration of the components forming a laser ablation apparatus

The method, discovered by Guo et al. in 1995 and developed for the large-scale production of nanotubes by Smalley et al. in 1996, was conceived with the aim to improve the synthesis of SWCNTs in terms of quantity and purity [18, 19].

The laser ablation technique utilizes a laser beam focused onto a graphite target with small amounts of Ni and Co, which is located in a 1,200°C furnace with an inert atmosphere of Ar or He. The carbonaceous products formed after target vaporization are collected in a water-cooled chamber outside the furnace (Fig. 4).

Like the arc discharge method, this technique allows one to obtain both SWCNTs and MWCNTs depending on the composition of the graphitic source.

The greatest strength of this technique is to produce SWCNTs with higher yields (>70%) and purity (>90%) with respect to the arc discharge process.

2.3 *Chemical Vapor Deposition*

In 1993 a new candidate for nanotube synthesis emerged with the promise to achieve large-scale production and controlled direction of growth on a substrate [20].

The method is based on the catalytic decomposition of a gaseous carbon feedstock (usually a mixture of hydrocarbon gas and nitrogen) at high temperatures (550–1,200°C) from which precipitation of carbon follows after cooling to room temperature. The catalysis is mediated by transition metals (most frequently Fe, Co, and Ni) that can be introduced into the gas furnace in the form of free nanoparticles or deposited on a substrate, allowing the formation of forests of vertical aligned nanotubes for use in electronics [21] (Fig. 5).

The choice of catalyst nanoparticle size and substrate strongly influences the forest development, affecting respectively the diameter of the tubes and the speed

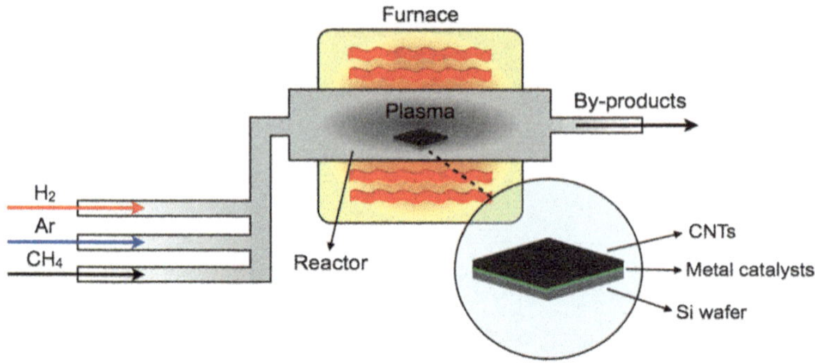

Fig. 5 Depiction of the CVD-based synthesis of carbon nanotubes

of growth (length per minute) [22]. It has been observed that nanotubes grow faster and better aligned on porous silicon substrates compared to flat silicon or alumina supports.

Depending on the parameter setting (temperature, operation pressure, active particle nature and size, supports, reaction time) the morphology of the product is widely tunable, giving selective SWCNT or MWCNT growth in powder or forest form. CVD can afford yields higher than 99% with remarkable purity and low amorphous carbon amounts.

For all these reasons, the majority of the methods currently employed for nanotube production are reinterpretations of the CVD technique. One of the most common is high-pressure carbon monoxide (HiPco) synthesis, renowned for the unique capacity to make SWCNTs on a kilogram per day scale [23].

2.3.1 HiPco Synthesis

This new synthetic approach uses a volatile organometallic precursor for the generation of the catalyst inside the reaction chamber. More precisely, a high-pressure carbon monoxide stream (30–50 atm) is introduced continuously into the furnace (900–1,100°C) together with $Fe(CO)_5$ or $Ni(CO)_4$. Under such conditions of temperature and pressure, carbon monoxide disproportionates, acting as a carbon feedstock, while the organometallic species decompose, giving the catalytic clusters on which nanotubes can nucleate and elongate (Fig. 6).

Average diameters of the nanotubes produced vary between 0.7 and 1.1 nm, depending on the pressure applied in the reaction chamber and the catalyst nature.

The synthesis product yield is up to 97% with purity exceeding 70%. This method is particularly suitable for large-scale synthesis since the reaction can be carried out continuously and allows the reuse of carbon monoxide [24].

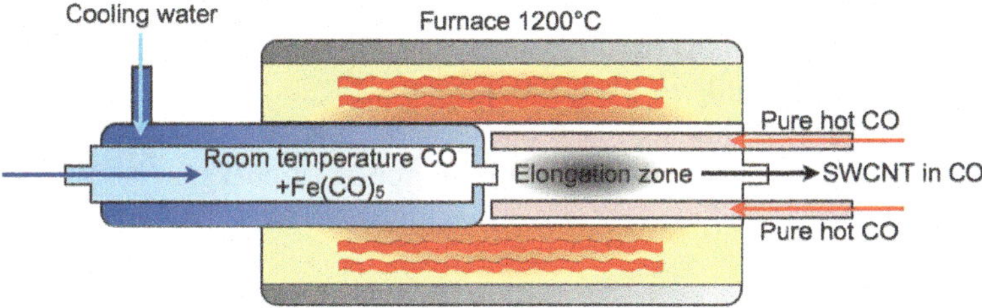

Fig. 6 Schematic representation of a HiPco furnace

2.4 *Solar Energy*

The solar energy method emerged originally for the production of fullerenes, until 1998, when Laplaze et al. brought some variations to the process in order to launch carbon nanotubes synthesis [25].

The peculiarity of this technique is to employ solar rays to mimic the laser ablation process, obtaining gram scale production of nanotubes relatively inexpensively and in a more controllable manner.

The principle of the synthesis is based on the sublimation of a graphite composite rod induced by sunlight, which is collected and reflected by a parabolic mirror onto the target surface.

The high temperature (2,600–2,700°C) causes both the carbon and the catalyst to vaporize and afterwards to condense onto the walls of a water-cooled screen.

MWCNTs and SWCNTs grow preferentially in relation to both catalyst composition and pressure: using Ni/Co mixtures, at low pressure, the sample collected mainly contains MWCNTs, whereas at higher pressures only bundles of SWCNTs are obtained. Luxembourg et al. reported the mass production of SWCNTs using a 50-kW solar reactor, able to synthesize 0.1–100 g/h of highly pure material [26]. It has also been observed that the best quality product resulted from using 15 cm long graphitic rods with 2% of Ni/Co catalyst, under He atmosphere.

2.5 *Electrolysis*

Another novel and low-consumption method for nanotube production is electrolysis, developed by Hsu et al. in 1995 [27]. The novelty of this approach consists in the generation of nanotubes by passing a current through an ionic salt medium between two electrodes. Alkali metals such as Li, Na, or K dissociate from their chloride salts on a graphite cathode, which is consumed during the electrolysis with subsequent nucleation of nanotubes. The material so formed, containing a mixture of CNTs and nanoparticles, is separated from the ionic salt by dissolution in distilled water followed by filtration.

Generally this method favors the growth of MWCNTs; however, Bai et al. described specific conditions (810°C using Ar as inert gas) to promote the electrolytic conversion of graphite to SWCNTs [28].

Clear advantages of this approach are the use of mild reaction conditions, the cheap materials employed, and the low consumption of energy.

3 Properties

3.1 Electronic Properties

As previously explained, carbon nanotubes can be thought of as a graphene sheet which has been rolled up to form a cylinder. Their structure is mathematically described by the chiral vector C_h, which expresses their circumferential periodicity, together with the translation vector T, which is along the nanotube axis. In this way, C_h and T defines a one-dimensional rectangle of unrolled tube, represented by the shaded areas in Fig. 7.

Each possible wrapping of graphene into the tubular form could be predicted knowing the values of the n,m integers, which are in turn connected to the a_1 and a_2 basic graphite vectors, $C_h = na_1 + ma_2$.

For instance, in Fig. 7a, a (3,4) nanotube is under construction. The corresponding final wrapping identifies a "chiral" tube, since $m \neq n$. In turn, the resulting tube will be "armchair" when $m = n$ and "zigzag" when $m = 0$, as described in Fig. 7b, c.

The same differentiation into three classes is achievable knowing the chiral angle $\theta = \tan^{-1}[\sqrt{3}(n/(2m + n))]$ (= 30° for armchair, = 0° for zigzag, $0° < \theta < 30°$ for chiral nanotubes).

The theoretical categorization of nanotubes on the basis of the geometric structure is of fundamental importance for the prediction of their electronic properties [6]. Although graphene is a zero-gap semiconductor (Fig. 8), carbon nanotubes can be metals or semiconductors, depending on the helicity and diameter.

To understand why, we should consider that in an isolated sheet of graphite two energetic bands with linear dispersion (π and π*) meet at the Fermi level, generating a K point in the Brillouin zone. Hereby, the whole Brillouin region is delimited by a determined set of K, Γ, and M points, while the Fermi surface of an ideal graphite layer consists of the six corner points K (Fig. 9).

When forming a tube, only a certain set of states of the planar graphite sheet is allowed, that may not include the K points. Whenever the K point is comprised into the Brillouin area, the nanotubes are metallic (semimetals with zero band gap); otherwise they are semimetals with tiny band gaps when $(n-m)$ is a multiple of 3, or semiconductors with large gaps in all the other cases.

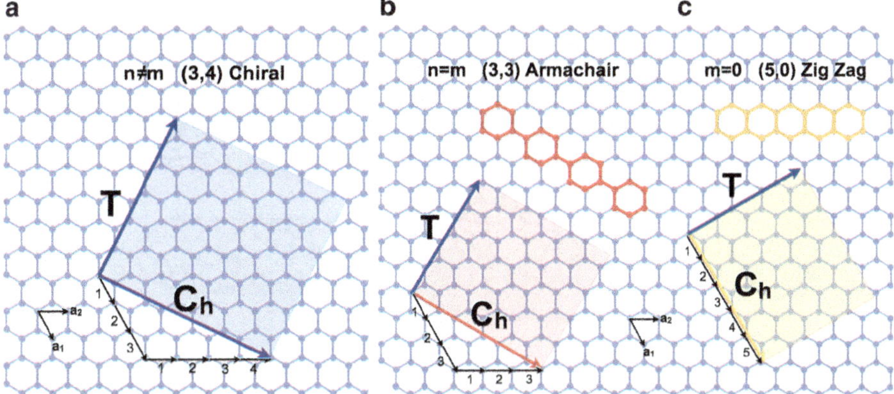

Fig. 7 Schematic diagram showing the possible wrapping of the 2D graphene sheet into tubular forms: (**a**) (3,4) chiral tube; (**b**) (3,3) armchair tube; (**c**) (5,0) zig zag tube

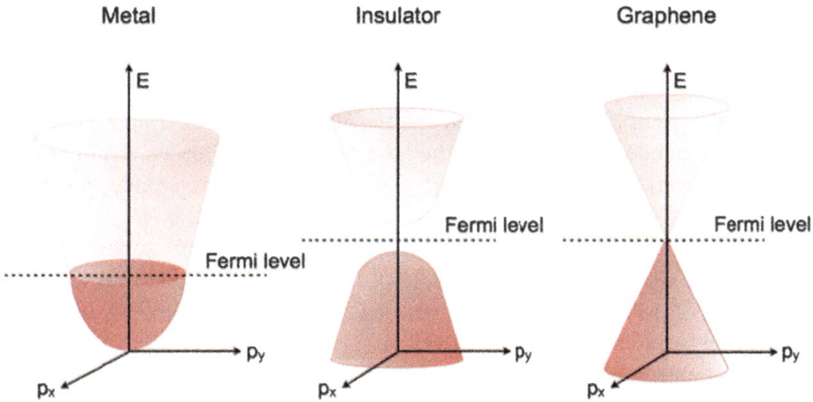

Fig. 8 Metals, band insulators and graphene electron densities at the Fermi level

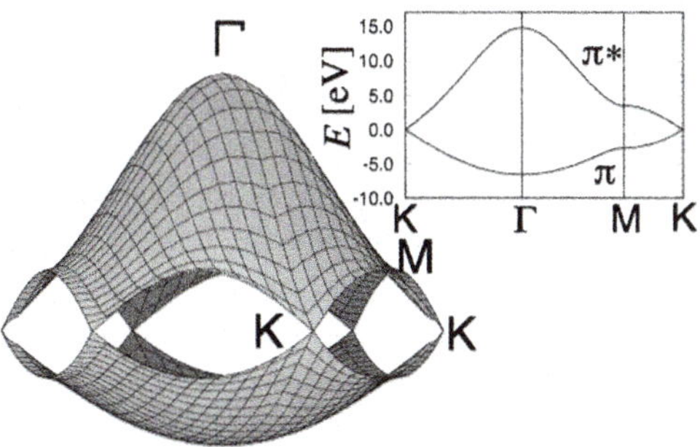

Fig. 9 Representation of the electronic energy states of graphene, showing the 3D view of the π/π^* bands. The *insert* shows the energy dispersion along the high-symmetry lines between the Γ, M, and K points. Reprinted with kind permission of Springer Science + Business Media from [29]

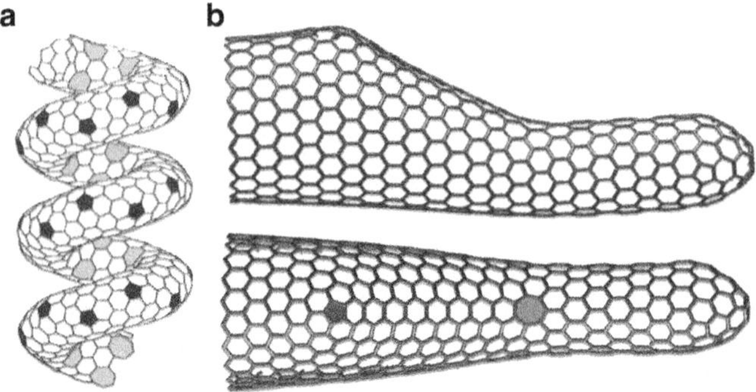

Fig. 10 (**a**) Molecular model of a spiral SWCNT, obtained by insertion of pentagons and heptagons within the hexagonal network. (**b**) +30° pentagon declination and −30° heptagon inclination results in different local curvatures. Reprinted with permission from [37]

In summary, (n,n) armchair tubes are always metallic, (n,m) chiral and $(n,0)$ zig-zag tubes in which $(n-m)$ is a multiple of 3 are semimetals displaying metallic behavior, while chiral and zig-zag tubes where $(n-m)$ is not a multiple of 3 are always semiconductors.

Besides the helicity, the diameter also affects the electronic behavior of CNTs [30]. It has been shown that as the tube radius (R) increases, semiconductors and semimetal band gaps decrease with a $1/R$ and $1/R^2$ dependence, respectively [31, 32].

Ultra-small-diameter SWCNTs have been found to exhibit superconductivity properties after confinement inside inert porous $AlPO_4^{5-}$ zeolite crystals (with inner diameter of 0.73 nm), as a demonstration of the relationship between diameter and electronic behavior of the tubes [33].

Another remarkable research investigation performed by Odom et al. in 1998, permitted resolution by scanning tunneling microscopy (STM) of the hexagonal-ring structure of the walls, revealing for the first time the explicit relationship between the structure and the electronic properties. The local density obtained from conductance measurements and the theoretical predictions was indeed found to be in very good agreement [34].

Several studies have also confirmed that CNTs are not as perfect as they were supposed to be. Different geometries like squares, pentagons, heptagons, and many more spangle the honeycomb network, modifying drastically the electronic properties of the whole structure. As a consequence, the explicit introduction of defects (not necessarily non-hexagonal rings, but also vacancies or doping agents) into the walls could be an interesting way to tailor the intrinsic properties of CNTs.

It has been shown that the pentagon–heptagon pair (Fig. 10) is the smallest topological defect inducing minimal local curvature and preventing at the same time the tube from collapsing or closing. Its incorporation within the hexagonal network of a single carbon nanotube changes its helicity, and therefore the energy gap between the electronic bands, providing the basis for the development of new diodes for nano-electronics [35, 36].

3.2 Mechanical Properties

Carbon nanotubes are expected to have high axial strength as a result of the sp^2 carbon–carbon chemical bond of the rolled up graphene layer, which is one of the strongest bonds known in nature for an extended system.

A great number of theoretical and experimental studies has been dedicated to the investigation of mechanical properties of carbon nanotubes with the aim to develop novel reinforced and stiff materials.

The mechanical properties of a material are described by the definition of its Young's modulus (or tensile modulus) along a given direction, which is expressed in Pascal (Pa) and defined as

$$Y = \frac{1}{V_{eq}} \left(\frac{\partial^2 E}{\partial \varepsilon^2} \right)_{\varepsilon=0} \tag{1}$$

where E is the total energy of the system, V_{eq} is the volume at the equilibrium, and ε is the strain. The Young's modulus defined for carbon nanotubes will be

$$\widetilde{Y} = \frac{1}{S_{eq}} \left(\frac{\partial^2 E}{\partial \varepsilon^2} \right)_{\varepsilon=0} \tag{2}$$

where S_{eq} is the nanotube surface area at zero strain that is determined unambiguously. A frequently applied convention is that $V_{eq} = S_{eq}h$, with $h = 0.34$ nm (the interlayer spacing in graphite).

In other words, the tensile modulus is a measure of the stiffness of a material, expressed by dividing the tensile stress exerted over an object under tension by the tensile strain, which in turn is defined as the amount by which the dimension of the object changes because of the stress. In the light of this, materials such as rubber or polystyrene present pretty low Young's modulus (0.1 and 3 GPa, respectively), while graphene's and diamond's tensile modulus lie in the region of 1,000 GPa.

The elasticity of a body is directly related to the chemical bonding of its constituent atoms, the reason why we expected CNTs' Young's modulus to equalize that of graphene when the diameter is not too small to distort the C–C bond significantly.

Much effort has been made towards the exact determination of the tensile modulus of nanotubes. The first attempt to achieve a reliable value for MWCNTs was reported in 1996 by Treacy et al., who measured the oscillation amplitude of the intrinsic thermal vibrations of an MWCNTs' series by means of a TEM. They found very high Young's moduli, with a mean value of 1.8 TPa [38].

Two years later the same team reported a subsequent study based on the same experimental approach, but employing a larger sample of nanotubes. This time a mean value of 1.25 TPa was achieved, which is in better agreement with the C_{11} graphite basal plane tensile modulus [39].

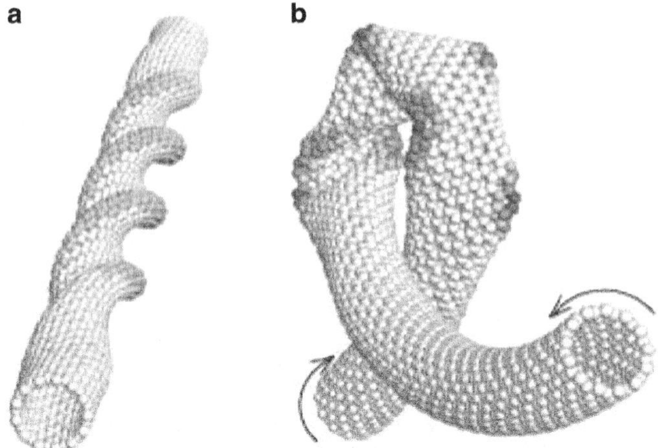

Fig. 11 Simulation of torsion (**a**) and collapse (**b**) as functions of the torsion angle. Reprinted with kind permission of Springer Science + Business Media from [29]

The similar value of 1.28 ± 0.5 TPa was also found by Wong et al., who took advantage of an AFM to pin MWCNTs at one end of the tip and measured the bending force vs. displacement along the unpinned ends [40].

Other approaches have given Young's modulus in the range of 10–50 GPa, reflecting the dependence upon the degree of defects within the tubes network [41].

Concerning the resilience, both theory-simulations and experimental evidence suggested the capacity of nanotubes to change their shape and absorb energy without creating a permanent distortion.

Although resilience is unexpected for a graphite-like material, carbon nanotubes under compression or bending form ripples and under torsion flatten into a helix, without irreversible atomic rearrangements. This property has been attributed to the small dimension of the tubes and to the competition between van der Waals and elastic energies [31].

Iijima et al. reported high-resolution electron microscopy observations of the occurrence of kinks on the sidewalls of SWCNTs under high bending angles, which were quantified by means of atomistic simulations [42]. The remarkable flexibility of the hexagonal grid, up to very high strain values, could be connected to the ability of the sp^2 network to rehybridize into sp^3 proportionally to the local curvature out of the plane (Fig. 11).

Concerning tensile strength, which is the maximum amount of tensile stress that a material can support without fracture, CNTs have been estimated to come close to values of 300 GPa, gaining the title of materials with the highest tensile strength.

By virtue of their outstanding mechanical properties combined with low density, CNTs are promising materials for the strengthening of fibers, replacing carbon and glass fibers, and for the development of high-performing composites with polymers, ceramics, or epoxides [43].

3.3 Defects

The term defect refers to any interruption of the perfect crystallographic structure periodicity in a regularly-patterned material. Defects in CNTs occur in several forms, affecting unambiguously their chemical and physical properties. Frequently the presence of defects is deliberately induced to tune chemical reactivity, charge injection, metallic behavior, etc.

There are many ways to break the regular hexagonal pattern of the nanotubes, which are mainly classified into four groups: *topological*, when non-hexagonal rings are introduced, *rehybridization*, if the carbon atom rehybridizes from sp^2 to sp^3 configuration, *incomplete bonding defects*, when vacancies or dislocation are found, and finally *doping*, when other elements are introduced.

For instance, the presence of the pentagon–heptagon pair changes dramatically the electron densities of states, generating helicity and yielding a metallic tube from a semiconducting one [44].

Carbon nanotubes have also been found to reconstruct spontaneously as a reflex to uniform atom loss on the surface [45]. When controlled electron irradiation is applied to SWCNTs, atom removal occurs at a slow rate, leading to the formation of holes that are mended mainly by dangling bond saturation, forming a highly defective layer cylinder of smaller diameter. Experiment has shown the formation of rings such as squares, pentagons, heptagons, and unstable high-membered rings that have been observed to disappear leaving five- and seven-membered rings (Fig. 12).

The insertion of the heptagon–pentagon pair into the sp^2 network has been exploited to design a new class of nanotubes with intrinsic metallic behavior, independent of chirality and tube diameter. The controlled combination of five-, six-, and seven-membered rings leads to defective structures called Haeckelite CNTs (Fig. 13), in virtue of the similarity to the radiolaria protozoa drawings by Haeckel, that exhibit unusual high-intensity peaks at the Fermi level indicating possible superconductivity properties.

Other nanostructures have been developed covalently connecting crossed single-wall carbon nanotubes by means of electron beam welding at high temperatures [46]. Stable junctions between nanotubes have been created in situ with “Y”, “X”, and “T” geometries, finding that heptagons played a key role in the topology (Fig. 14).

Also in this case the formation of junctions involving seven- or eight-membered rings occurs as a consequence of the cross-linking of dangling bonds after vacancies promotion.

Most of these observations have been supported by molecular dynamics studies, which suggest a strong metallic character in the crossing points and localized donor states caused by the presence of the heptagons.

The possibility of handling molecular connections between nanotubes opens the door to the design of new prototypes for nanoelectronics, where different tubular

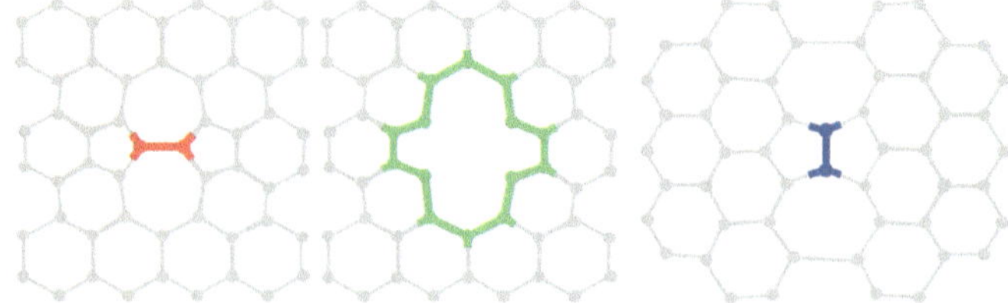

Fig. 12 Typical defects observable on nanotubes' surface

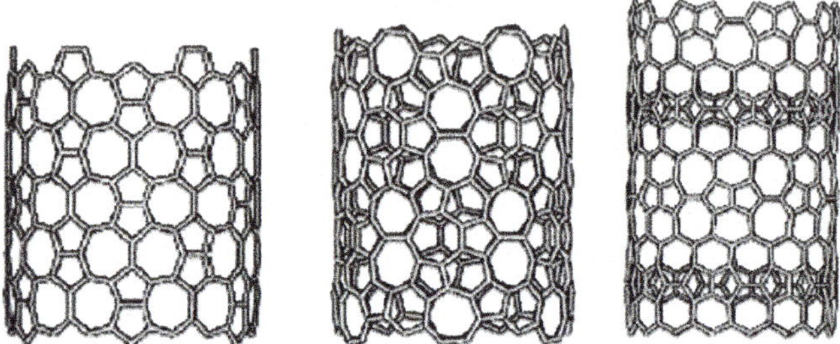

Fig. 13 Non-chiral Haeckelite nanotubes of different diameter containing different combinations of pentagons, hexagons, and heptagons. Reprinted with permission from [44]. Copyright 2002 American Chemical Society

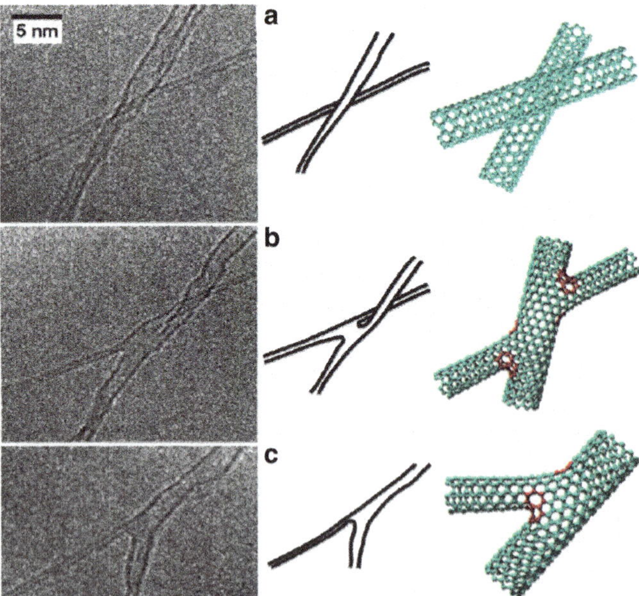

Fig. 14 Transmission electronic microscope images of "X" and "Y" junctions. Reprinted with permission from [46]. Copyright 2002 American Physical Society

junctions could act as multiterminal electronic station or building blocks for three-dimensional nanotube networks.

4 Chemistry of Carbon Nanotubes

The outstanding properties exhibited by CNTs make them exceptional candidates for materials and biomedical applications. However, there is a big difference between the promising potential of a new material and its practical application on a widespread basis. As-produced CNTs are insoluble in most organic or aqueous solvents and tend to aggregate as a result of strong van der Waals interactions between individuals tubes, rendering difficult their manipulation and limiting their processability. Thus, for example, in biomedical application a valid solution is the modification of the nanotube surface in order to prepare aqueous solutions of nontoxic and biocompatible systems.

Another key problem is that, according to the procedure used to synthesize CNTs, mixtures of metallic and semiconducting CNTs coexist with their correspondingly different electronic structures. The presence of these mixtures limits the applications of CNTs as versatile building blocks for molecular electronics. Impurities also interfere with most of the desired properties as well as with biocompatibility. For these reasons, purification and separation of the tubes has been a matter of intensive study.

Hence, given all these challenges, the use of chemistry has emerged as one of the most effective means for manipulating CNTs and even separating metallic from semiconducting species. A large number of chemical treatments that include both covalent and noncovalent approaches have been described, giving scientists the ability to combine these structures with other types of materials.

In the next two sections we will explore the advances in the formation of nanotube derivatives and the combination of techniques used for their characterization. Because of the huge amount of work in this area, this will not be a comprehensive review but will at least be representative of the progress in this field.

4.1 The Covalent Approach

The hexagonal network of sp^2-hybridized carbons of CNTs can be covalently functionalized by taking advantage of the curvature of the tubular walls. Similar to fullerenes, there is indeed a local strain relief enhancing carbon nanotubes reactivity, which is induced by the change of pyramidalization and π-orbitals misalignment of conjugated carbon atoms [47, 48]. Although fullerenes are much more reactive than CNTs as a consequence of the spherical geometry and the smaller diameter, a big set of reactions is allowed for both allotropes [49–51].

More precisely, in each single nanotube it is possible to distinguish two regions: the tips and the sidewalls. The tips resemble a hemisphere, the reason why they are expected to undergo typical fullerene reactions with comparable reactivity. On the other hand, the side walls are characterized by carbon atoms presenting lower pyramidalization angles and therefore exhibiting different chemical behavior.

The way the tube is wrapped also influences the chemical reactivity, rendering metallic CNTs more reactive than the semiconductor analogues as a consequence of slightly less aromaticity of the former [52].

The diameter is another important element affecting reactivity: smaller tubes present more enhanced pyramidalization angles and orbital misalignment, being therefore more susceptible to chemical modification.

Finally, the presence of structural defects in the hexagonal pattern plays a fundamental role in the covalent attachment of molecules. Indeed, the insertion of the pentagon–heptagon pair significantly increases the local curvature of the lattice, creating preferential points for addition reactions.

4.1.1 Derivatization of CNTs Following Oxidation

Oxidation of CNTs has been typically used to remove the traces of metallic catalysts employed in the fabrication process. The purification process occurs in strong acidic and oxidative conditions, such as mixtures of concentrated acids combined with heating or sonication, and affords shortened open tubes decorated with oxygenated functions, predominant on the tips [53]. A completely different way to produce short and open tubes has been described by our groups using a planetary mill in solvent-free conditions. The process avoids the use of strongly acidic conditions and produces highly dispersible SWCNTs that can be efficiently functionalized in a variety of synthetic ways [54].

Among the groups introduced on the tubular grid (carbonyl, carboxyl, hydroxyl, etc.), carboxylic functions are particularly exploited as anchor points for amidation and esterification reactions that are easily carried out by means of acyl chloride intermediates or carbodiimide derivatives, commonly 1-ethyl-3-(3-dimethylaminopropyl)carbodiimide (EDC) and hydroxybenzotriazole (HOBt) (Fig. 15).

Several examples of amidation reactions on shortened tubes are reported in the literature, leading to highly soluble and dispersible derivatives suitable for a great number of applications. Haddon and co-workers were pioneers of this approach, since they first exploited amidation of the carboxylic functions to conjugate nanotubes with several solubilizing agents [55]. Among these, the conjugation of SWCNTs with poly(aminobenzene sulfonic acid) (PABS) and polyethylene glycol (PEG) to form water-soluble graft copolymer was particularly successful, yielding highly-loaded SWCNTs (30% of PABS and 71% of PEG, estimated by thermogravimetric analysis) which resulted in a final water solubility of 5 mg/mL [56] (Fig. 16).

The general synthetic pathway introduced by Hamon et al., involving acylation of the carboxylic groups via an acyl chloride intermediate, was followed by several

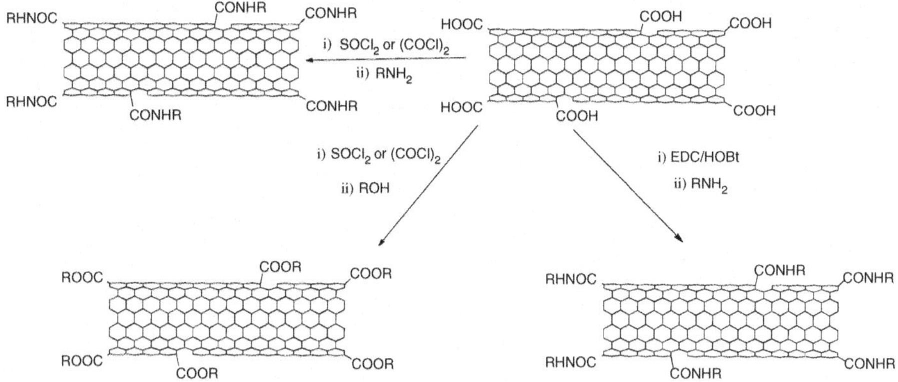

Fig. 15 Amidation and esterification reaction on CNTs after oxidative treatment

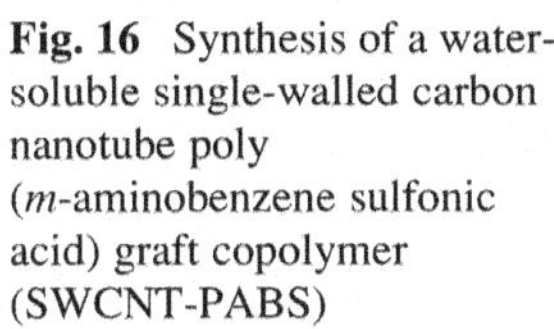

Fig. 16 Synthesis of a water-soluble single-walled carbon nanotube poly (*m*-aminobenzene sulfonic acid) graft copolymer (SWCNT-PABS)

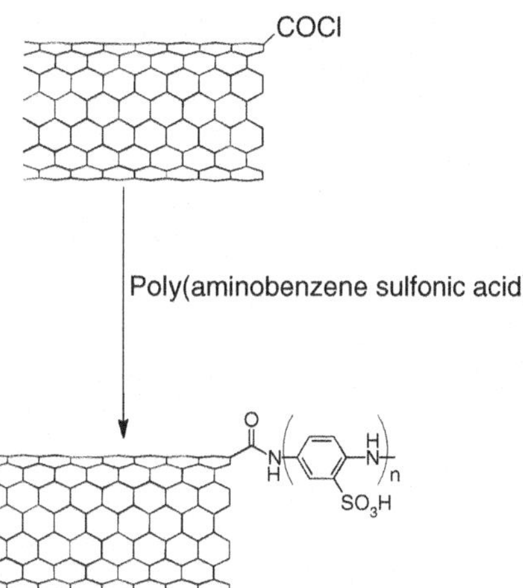

groups in order to enhance CNTs properties by combination with other interesting materials. In this way, Delgado et al. first reported the synthesis of the SWCNT-fullerene chimera and subsequent works demonstrated the ground state electronic communication occurring from SWCNTs and C_{60} moieties [57, 58].

Much effort has also been made towards the preparation of novel CNTs-based bio-conjugates for biological and bio-sensing purposes. CNTs have been covalently linked to DNA, proteins, enzymes (Fig. 17), and antibodies [59–63].

Simplicity and reliability of the amidation strategy have inspired the covalent attachment of an extended amount of molecules on CNTs (dendrimers, porphyrins, fluorescent probes, crown ethers, gold nanoparticles, etc.), making this approach one of the most representative of CNT covalent chemistry [64–69].

An alternative to the amidation reaction is represented by the esterification of the carboxyl functions. Also in this case carboxylic acids are rendered suitable for

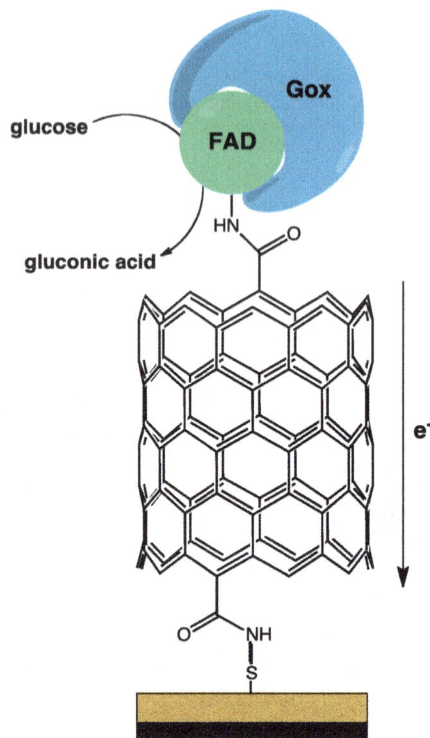

Fig. 17 Assembly of the SWCNT electrically-contacted glucose oxidase electrode. SWCNT was firstly immobilized via carbodiimide chemistry on the electrode surface, and then the amino derivative of the flavin adenine dinucleotide (FAD) cofactor was covalently attached to the carboxyl groups at the SWCNT tips [62]

the alcohol nucleophilic substitution by previous reaction with an acyl chloride. The literature offers a wide range of examples along this line, describing CNTs variously conjugated with porphyrins, lipophilic and hydrophilic dendrons, polyols, etc. [70–73].

The formation of carbon nanotube-inorganic hybrids using this approach have been widely explored and gives rise to a new strategy for building metallo-nanomaterials, which could find potential applications in nanoscale electronic devices [74, 75] (Fig. 18).

Au dendrimer encapsulated nanoparticles (DENs) have also been immobilized onto the surface of MWCNTs functionalized with carboxylic acid groups. This acidic functionalization leads to debundling of MWCNTs and provides surface sites for interactions with amino groups present on the surface of the dendrimer [76] (Fig. 19).

4.1.2 Covalent Sidewall Chemistry

In contrast to the esterification or amidation reaction of the acid groups in oxidized CNTs, the functionalization via direct reactions is more difficult to accomplish

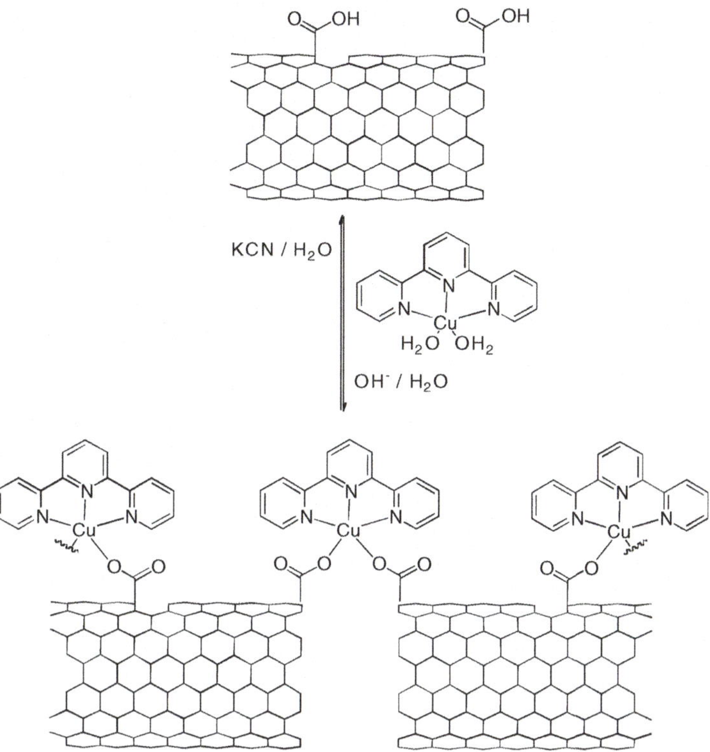

Fig. 18 Formation and disassembly of terpyridine Cu(II)-mediated nanocomposites of SWCNTs

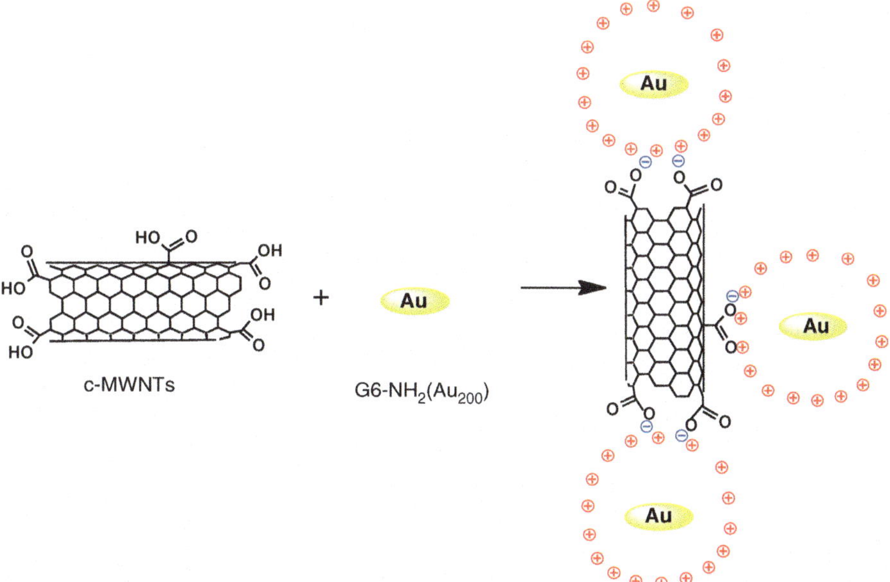

Fig. 19 Gold DENs attachment on the MWCNT surface under mild conditions

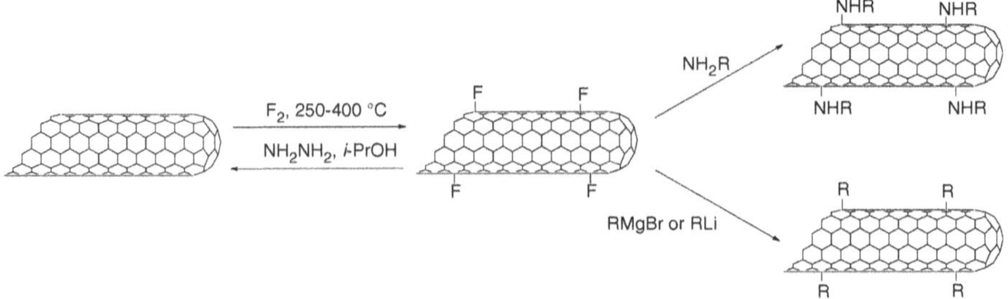

Fig. 20 Scheme illustrating fluorination on the nanotubes allowing further functionalization

as it requires the use of highly reactive species. In general, addition reactions to carbon–carbon double bonds cause a transformation of sp^2-hybridized into sp^3-hybridized carbon atoms. Such changes are associated with the modification of the predominantly trigonal-planar local bonding geometry into a tetrahedral geometry. As previously explained, this process is energetically more favorable in the cap region where the curvature is more emphasized; it is also probable that reactions preferably occur in the proximity of structural defects sites.

Addition reactions on CNTs can be performed at different steps of functionalization and do not necessarily require previous oxidation, making CNT derivative properties easier to tune.

Fluorination

The addition of fluorine to CNTs can be achieved by applying high temperatures (between 250 and 400°C) and using hydrofluoric acid as catalyst. The method was firstly reported by Mickelson et al. with best yields allowing the insertion of one fluorine atom every two carbon atoms [77].

Fluorination is of great interest since it facilitates further functionalization on the tubes by nucleophilic substitutions [78]. Taking advantage of the weakness of the C–F bond, organolithium and Grignard reagents are typically employed to afford alkyl-modified nanotubes [79].

Fluorinated nanotubes were reported to have moderate solubility in alcoholic solvents, which is by the way significantly improved in the alkylated tubes. One advantage of this method is that pristine nanotubes could be effectively recovered by using anhydrous hydrazine in isopropanol [80] (Fig. 20).

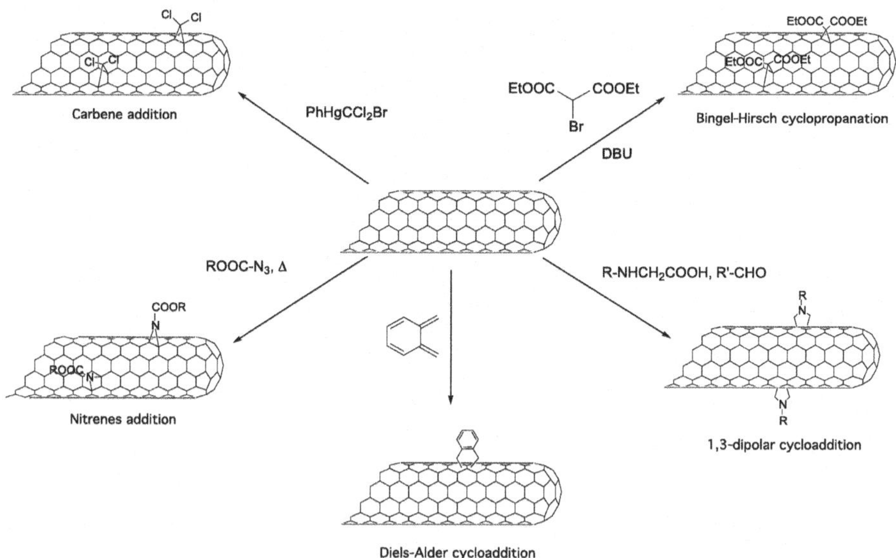

Fig. 21 Summarizing scheme of cycloaddition reactions on carbon nanotubes

Cycloadditions

The group of cycloadditions comprises a large number of reactions that are largely employed for CNTs sidewall functionalization. Among the first developed with this aim we find carbene addition, which is shown to yield the derivative depicted in Fig. 21 after the addition of phenyl (bromodichloromethyl) mercury to previously oxidized tubes [81]. The resulting saturation of the carbon atoms replaced with a cyclopropane ring functionality provoked dramatic changes in the optical spectra and at the Fermi level transitions. In this case, thermal treatment of the derivative above 300°C determined the rupture of the Cl–Cl bonds, but the electronic structure of the initial tubes was not restored.

In a similar fashion, the Bingel–Hirsch [2 + 1] cyclopropanation reaction describes the use of diethylbromomalonate as a carbene precursor in order to obtain a cyclopropane ring bearing two ester appendages, both available for subsequent modifications on the tubes (Fig. 21). The degree of functionalization was estimated to be about 2% [82–84].

Other reactive species suitable for sidewall modification are nitrenes, as reported by Hirsch and co-workers who reacted SWCNTs with different alkyl azidoformates at 160°C leading to N_2 extrusion and subsequent formation of alkoxycarbonylaziridino-modified tubes [85].

Impressive work based on nitrene cycloaddition was reported by Yinghuai et al., who functionalized SWCNTs with C_2B_{10} carborane cages and found specific tumor-targeting activity after administration of the derivative to mice [86].

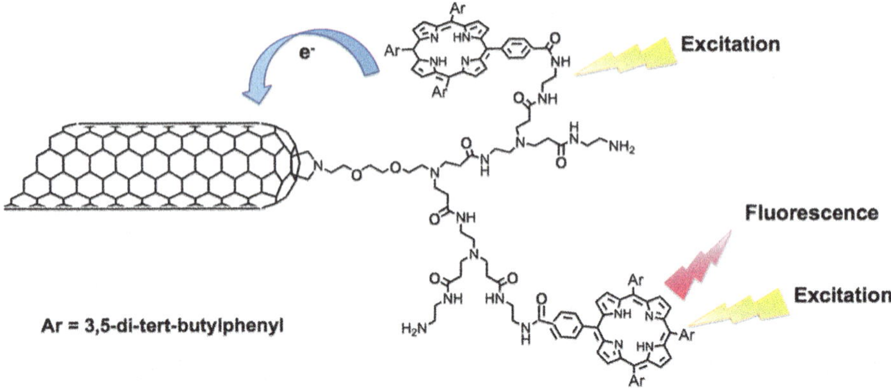

Fig. 22 Dendrimer-functionalized SWCNT. The dendrimers are linked directly to the SWCNT surface using a divergent methodology

Highly functionalized CNTs can be obtained via 1,3-dipolar cycloaddition of azomethine ylides (Fig. 21), generated by thermal condensation of alfa-aminoacids and aldehydes [87]. This strategy leads to the attachment of a wide variety of substituents on the sidewalls, according to the nature of the aldehyde and the aminoacid chosen for the generation of the ylide. In this way, covalent linkage of several molecular species has been described, such as triethylene glycol chains, ferrocene moieties, phenol groups, or polyamidoamine dendrimers directly synthesized onto the surfaces of pyrrolidino-functionalized tubes (Fig. 22) [49, 88–91].

Due to the lower toxicity of functionalized CNTs as compared to pristine CNTs, these systems have shown great potential in nanomedicine. Thus, using azomethine ylide chemistry, CNTs have been modified with aminoacids and peptides, antibodies, and other biological entities, suggesting that functionalized tubes have potential as drug delivery vehicles [92, 93].

Organic functionalization via azomethine ylides has also been exploited for different purposes, such as the purification of the raw material from the byproduct impurities, which is performed by annealing the functionalized tubes at high temperatures (300°C) yielding a final material free of amorphous carbon and almost free of metal content [94].

Despite the high versatility of sidewall functionalization methods, a severe limitations are the great amount of solvents needed to disperse CNTs and the long reaction times necessary to perform the reactions. An alternative approach is the use of microwave irradiation to activate CNT reactivity. Strong absorptions observed after CNTs exposure to microwave radiations have been explored by different groups in order to assist CNT functionalization [95].

Langa and co-workers described the 1,3-dipolar cycloaddition of nitrile imines and nitrile oxide to the sidewalls of modified SWCNTs using microwave activation [96]. Our group has shown that a solvent-free technique, combined with microwave irradiation, produces functionalized nanotubes by using 1,3-dipolar cycloaddition of aziridines in just 1 h of reaction, an extraordinary time reduction when compared

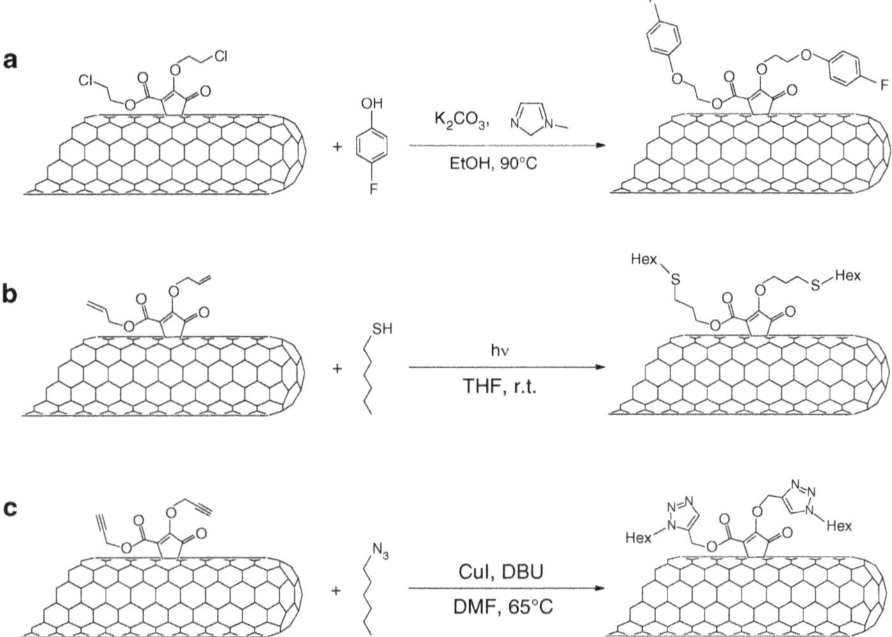

Fig. 23 Functionalization of CNTs from well-defined precursors: (**a**) via S_N2 reaction; (**b**) via thiol addition; (**c**) via click chemistry

with conventional heating. Moreover, the solvent-free conditions pave the way to green protocols and large-scale functionalization [97].

Microwave irradiation has also been successfully applied for the functionalization of CNTs by using the Bingel reaction and it was found that, by changing the output power, the amount of covalently linked substituents could be controlled [66].

In general, [4 + 2] or Diels–Alder cycloadditions to CNTs have been less used because of the low stability of the final products. The first example in this direction was reported by Langa and co-workers, who performed a MW-assisted Diels–Alder cycloaddition by reaction of ester-functionalized SWCNTs with *o*-quinodimethane, generated in situ from 4,5-benzo-1,2-oxathin-2-oxide [98].

Finally, a novel cycloaddition approach using zwitterions has been reported. The reaction involves the initial formation of a positively charged five-membered ring resulting from attack of highly nucleophilic species in the presence of an electrophile. It was first described by the addition of 4-dimethylaminopyridine (DMAP) to dimethyl acetylenedicarboxylate (DMAD). In the final step the addition of a second nucleophile replaced DMAP with an alkoxy group to yield the functionalized CNTs [99].

Following this zwitterionic approach, cobalt porphyrin-functionalized CNTs have been prepared, showing, when mixed with Nafion, excellent catalytic performances for oxygen reduction in acidic media at room temperature. Further derivatives obtainable through this synthetic protocol are reported in Fig. 23.

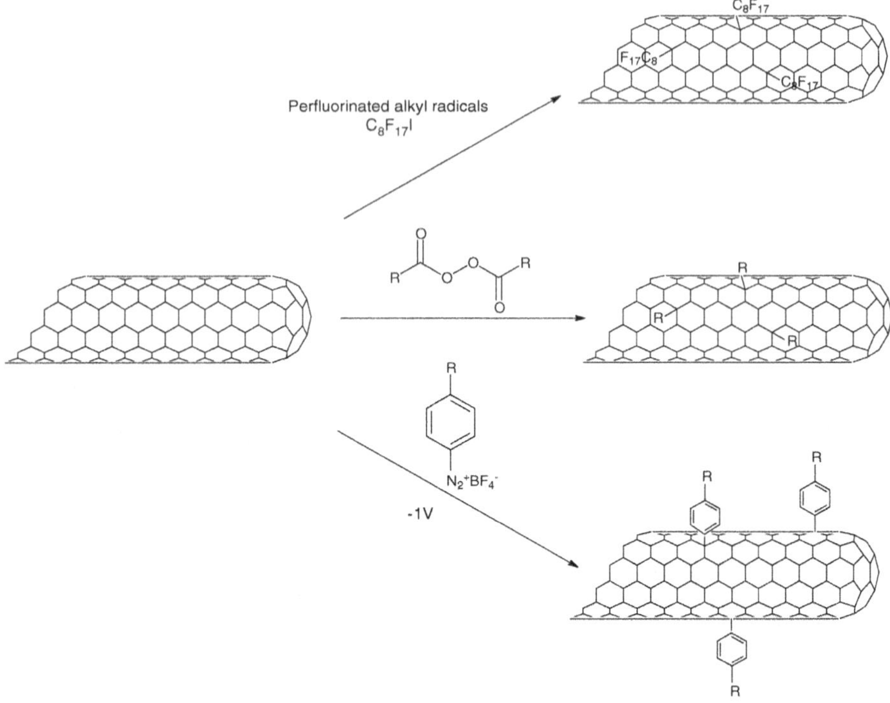

Fig. 24 Radical additions on CNTs

Radical Additions

This strategy takes advantage of the high reactivity of radical species produced either photochemically or thermally. Photolysis of heptadecafluoro-n-octyl iodide generating the correspondent alkyl radical was exploited by Holzinger et al. (Fig. 24) for covalent functionalization of CNTs with perfluorinated alkyl chains [100].

The same type of derivatization is achievable by reaction of the nanotubes with organic peroxides under thermal conditions or with aromatic ketones photoreduced by alcohols [101, 102].

A further example of versatile and widely diffused sidewalls functionalization strategy is the so-called Tour reaction (Fig. 24), in which aryl groups can be appended on nanotubes by means of electrochemical reduction of the correspondent diazonium salts [103]. The degree of functionalization achievable with this approach corresponds to 1 functional group out of 20 carbon atoms of the carbon framework. Several examples have been reported by the same group with the aim of improving this methodology and rendering the procedure rapid and useful. In subsequent works the authors described how to obtain the same derivatives by direct treatment of the nanotubes with aryl diazonium tetrafluoroborate salts,

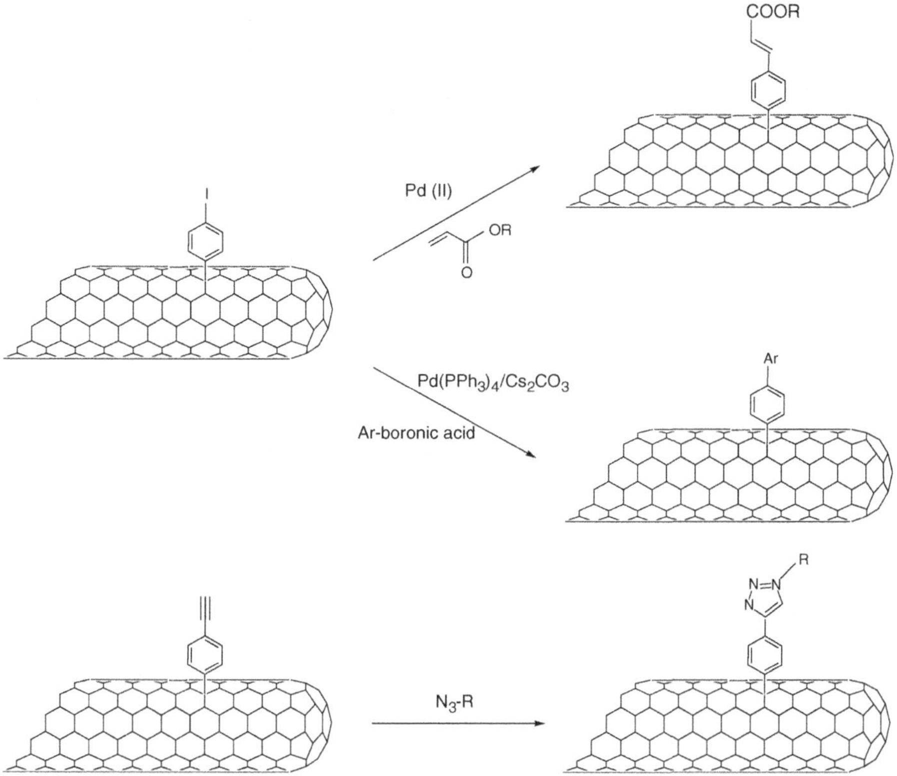

Fig. 25 Modification of functionalized CNTs by diazonium chemistry via Heck, Suzuki and click chemistry

avoiding electrochemically induced reactions. The same reaction was successfully performed "on water" in the presence of a substituted aniline and an oxidizing agent [104].

A further variation was introduced to this approach, involving the use of ionic liquids (imidazolium based ionic liquids) and K_2CO_3 as reaction media, which were ground together with SWCNTs and aryl diazonium salts in a pestle and mortar at room temperature [103].

Functionalized CNT production by diazonium chemistry has been subjected to further modification using different reactions. Campidelli and co-workers introduced via "click chemistry" porphyrin dendrons, while iodophenyl-functionalized CNTs have been used as precursors of Suzuki and Heck coupling reactions [105–107]. The final derivatives are potentially useful in materials and science applications (Fig. 25).

Free radicals have also been generated from organic hydrazines, either by thermal air oxidation or under microwave conditions [108, 109]. The microwave-assisted approach

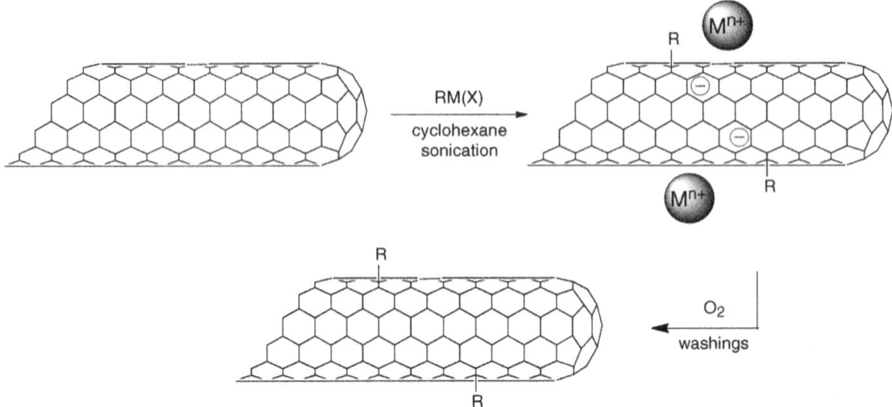

Fig. 26 Nucleophilic alkylation-reoxidation functionalization of CNTs

affords CNT-derivatives with a similar yield of functionalization but with an optimal reaction time of 5 min, whereas thermal treatments take days.

Reductive alkylation of CNTs also yields sidewall functionalized tubes. In these reactions, CNTs are treated with lithium, sodium, or potassium in liquid ammonia, followed by the addition of alkyl/aryl iodides or alkyl/aryl sulfides [110, 111]. The debundling observed in liquid ammonia is associated with the fast electron transfer process from the metal to the nanotubes: charged carbon nanotube intermediates are exfoliated due to electrostatic repulsion. Addition of the alkyl halide leads to the formation of a radical anion that dissociates to yield halide and the alkyl radical, which can then react with the CNT sidewalls. Some authors have shown that reductive alkylation is dependent on the tube diameter and shows some selectivity of metallic toward semiconducting tubes [112]. Similar arguments have been made for diazonium reactions and the reaction could be used as a potential way for the separation of metallic vs semiconducting nanotubes.

Nucleophilic and Electrophilic Additions

The nucleophilic addition of carbenes to CNTs was reported by Hirsch and co-workers [85]. The same group has also described nucleophilic additions using organolithium and organomagnesium compounds, producing highly functionalized alkyl nanotube derivatives, or with amine-based nucleophiles leading to amino-functionalized CNTs which exhibit enhanced solubility in many solvents [113–115] (Fig. 26).

Non-conventional techniques such as microwave irradiation or ball milling have been used for the electrophilic addition of alkyl halides [116]. Other electrophilic reactions have also been reported upon mechanochemical reactions in the presence of a Lewis acid and $CHCl_3$ [117].

4.2 Noncovalent Methods

Noncovalent functionalization of CNTs is based on their indirect modification with different chemical species, mainly by means of π–π and hydrophobic interactions.

The exploration of alternative approaches to the covalent modification arises from the necessity to exfoliate CNTs bundles. In the search for a non-invasive and efficient way to exfoliate bundles, a large number of molecules has been considered and classified into four different groups: (1) surfactants, (2) small aromatic molecules, (3) biomolecules, and (4) polymers.

(1) Nanotubes can be suspended in aqueous media in the presence of surfactants, such as sodium dodecylsulfate (SDS) or sodium dodecylbenzene sulfonate (SDBS), whose hydrophobic tail is believed to interact with the tubular graphene surface. The interaction with the amphiphile is even stronger when the hydrophobic portion is aromatic [118, 119]. As a matter of fact, (2) pyrene and porphyrin derivatives do interact with the sidewalls of CNTs by means of π–π interactions. In addition, it was demonstrated how the noncovalent association opens the door to a novel family of electron donor–acceptor nanohybrids [120].

(3) Biomolecules can be associated supramolecularly with CNTs in both aqueous and organic solutions. A remarkable increase of CNT water-solubility is produced due to the formation of wrapped complexes in which nanotubes are embedded.

The presence of hydrophobic regions on the proteins plays a fundamental role in their adsorption on nanotubes, as shown for the immobilization of metallothionein and streptavidin [121, 122].

An important proof of concept was given by Erlanger et al. who reported the spontaneous recognition between a monoclonal antibody against C_{60} and SWCNTs, efficiently characterized by the analysis of the antibody binding sites, made by a cluster of hydrophobic aminoacids [123].

(4) Polymers have also been employed to wrap CNTs. The choice of using a polymer is motivated for several reasons. First, the formation of supramolecular complexes with CNTs occurs almost spontaneously in organic solvents such as $CHCl_3$, avoiding time-consuming procedures. Second, atomic force microscopy (AFM) has shown not only that polymers homogeneously coat the tubes all over their surface but also that they mostly isolate each nanotube from the others, breaking up the bundles. Last, water-stable solutions of SWCNT-polymer complexes are simply achievable by virtue of the ability to append polar side-chains on the polymer (such as polyvinylpyrrolidone -PVP- or polystyrenesulfonate -PSS-) [124].

Among the polymers, one of the most widely used for wrapping CNTs is poly (*m*-phenylvinylene) (PmPV), since this polymer contains aromatic repetition units that strengthen the interaction with the graphitic layer [125].

Noncovalent functionalized SWCNTs with lipopolymers covalently attached to siRNA with a cleavable disulfide bond (Fig. 27) were used to transfect human cells in order to silence the expression of HIV specific cell surface receptors [126].

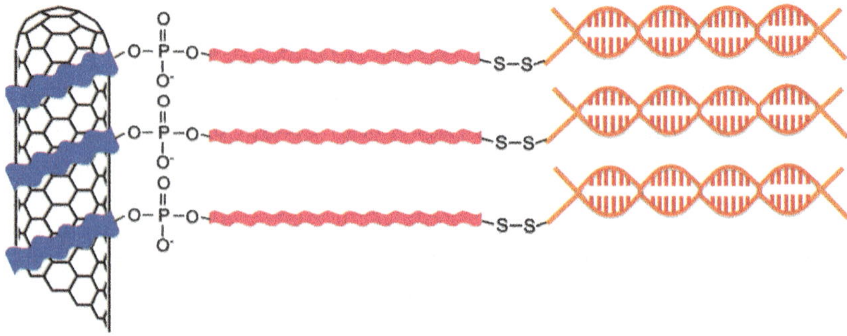

Fig. 27 Functionalization of SWCNTs with a PEG phospholipid for the conjugation of thiol–siRNA through disulfide linkages

4.2.1 Endohedral Filling

Another interesting possibility that CNTs offer is that they can encapsulate molecules or ions in their inner cavity and confine them. This fact could be useful, for example, to protect the encapsulated materials from reactions in contact with the atmosphere or, in a different approach, CNTs could behave as nanoreactors where confined reactions can occur.

CNTs were first filled by fullerenes, azafullerenes, and endohedral metallofullerenes producing peapod structures. Besides fullerenes, other organic molecules have been confined such as conjugated dyes, ionic liquids, and polymeric species. Moreover, inorganic substances such as iodide salts have been encapsulated. An interesting example is the confinement of Gd^{3+} ion clusters within ultra-short SWCNTs; these new derivatives have been explored as "nanocapsules" for Magnetic Resonance Imaging (MRI), exhibiting efficiencies 40–90 times larger than any Gd^{3+}-based contrast agent (CA) in current clinical use [127].

4.3 Multifunctionalized Carbon Nanotubes

While scientists have shown that CNTs can be chemically modified, the integration of these systems into functional materials is still a challenging task. The advantage of a noncovalent attachment is that the perfect structure of the CNT is not damaged and their properties remain intact. On the other hand, the covalent functionalization offers the possibility to have functional groups strongly attached on the nanotube surface, and therefore avoids the risk of desorption. Still, for some applications, CNTs must be modified with more than one group and in these cases simultaneous optimization of several functionalities is a demanding issue.

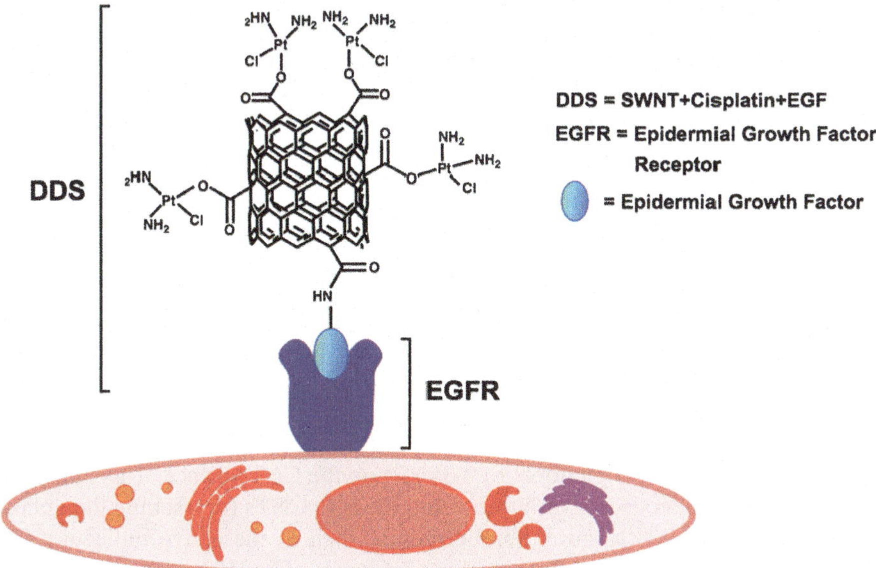

Fig. 28 Multifunctionalized CNTs conjugated with EGF target the cell-surface receptor (EGFR)

Multifunctional CNTs are very attractive for drug delivery applications, sensors, or in the preparation of composites [12, 13, 92].

Both covalent and noncovalent strategies have been used in order to conjugate drugs to CNTs and at the same time to increase biocompatibility and decrease toxicity. Targeted drug delivery with CNTs involves the conjugation of drugs but also of targeting and/or imaging agents to the same tube and different strategies have been carefully designed for this purpose.

The approach of coating CNTs with PEGylated phospholipids (PL-PEG) has been developed by the Dai group. Using this strategy the targeted delivery of aromatic drugs such as doxorubicin is possible. The aromatic drugs are loaded directly onto the nanotube surface via π–π staking and the functional groups on the SWCNT coating molecules (PL-PEG) can be conjugated with targeting molecules such as peptides [128, 129].

In another approach, folic acid as targeting ligand has been linked to a Pt(IV) prodrug compound and then conjugated to PEGlylated SWCNTs [130].

On the other hand, covalently modified CNTs possess the advantage of having functional groups attached on the nanotubes and therefore problems like molecule desorption are avoided. Thus, Bhirde et al. have used the chemical modification of carboxylic groups at the nanotube tips and sidewalls, introduced after strong acid treatment, to attach anticancer agents such as cisplatin and epidermal growth factor (EGF) to target specifically squamous cancer [131] (Fig. 28).

An alternative strategy for introducing two different and orthogonal functionalizations to CNTs is to perform 1,3-dipolar cycloaddition on the oxidized tubes.

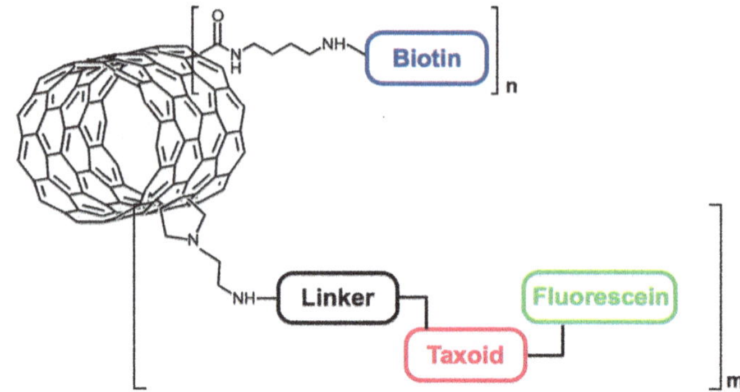

Fig. 29 SWCNT covalently attached to an anticancer drug and biotin as tumor-targeting system

In this way, two distinct functionalities coexist on the CNT surface. This approach enables the simultaneous link of fluorescent probes to CNTs for tracking the uptake of the material as well as an antibiotic moiety such as the active molecule. For this purpose, our group chose fluorescein and amphotericin B (AmB), respectively. AmB is considered to be a most effective antibiotic in the treatment of chronic fungal infections [132]. The results were very promising as it was observed that appropriate conjugation can increase the effectiveness of AmB, while decreasing its toxicity to mammalian cells. Using a similar approach a biotin-SWCNT-linker taxoid drug delivery system has been prepared. SWCNTs were covalently attached to an anticancer drug and biotin as the tumor-targeting module (Fig. 29). The anticancer drug, taxoid, is readily released because of the linker that possesses a disulfide bond that can be cleaved by endogenous thiols [133].

The covalent approach permits other interesting possibilities. Our group has prepared multifunctionalized CNTs, using a combination of two different addition reactions: the 1,3-dipolar cycloaddition of azomethine ylides and the addition of diazonium salts both via a simple, fast, and environmentally friendly method. The two reactions can be performed in series [134]. Results show that the radical arene addition saturates more reactive sites than cycloaddition. Therefore, to attach two different functional groups to SWCNTs, the order of reaction should be first the cycloaddition and then the arene addition. This gives rise to comparable functionalization degrees for the two additions. If only a few functional groups are required along with an excess of other organic/inorganic groups, then the functionalization order can be reversed. This latter could be the case of some biomedical applications, where, for example, many more drug molecules are needed with respect to a contrast agent (Fig. 30).

Finally, it was shown that nucleophilic additions to the sidewalls of carbon nanotubes are reversible and thus retrofunctionalization of these CNT derivatives can be achieved under mild conditions. This equilibrium can be used in transfunctionalization reactions, where one functional entity is partly exchanged for another, giving access to CNTs with different functionality profiles [135, 136] (Fig. 31).

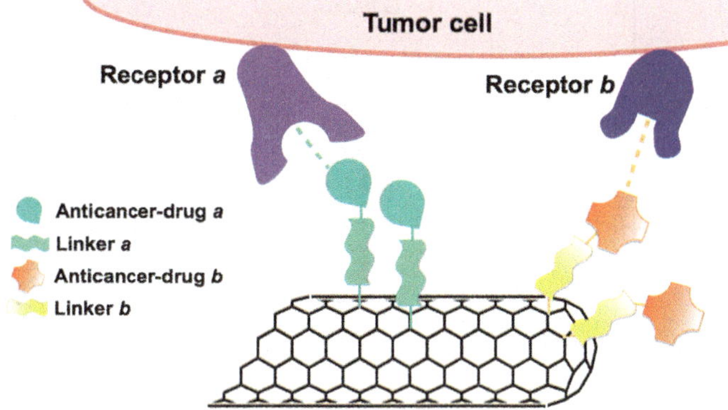

Fig. 30 Representation of multifunctionalized CNTs bearing chemical appendages covalently linked through varied chemistry for eventual multireceptor recognition

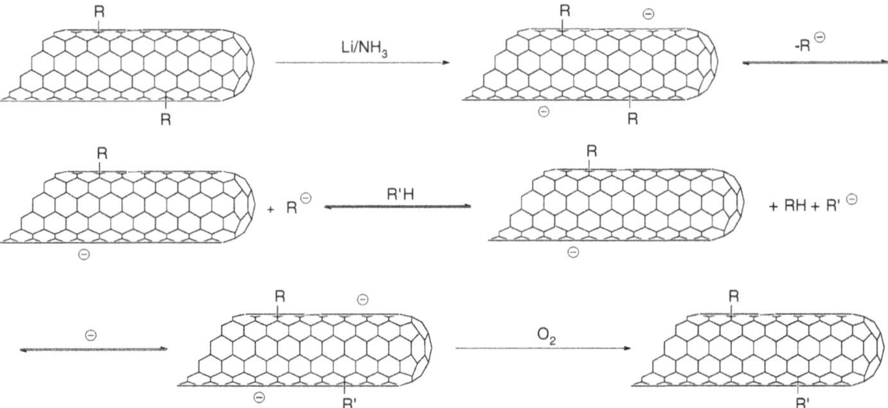

Fig. 31 Exchange of functional groups R, R′ (R = *n*Bu; R′ = *n*PrNH) based on the deprotonation of $nPrNH_2$ by *n*Bu, which is generated by reduction of *n*Bu-functionalized SWCNT

5 Characterization of CNTs

Many of the most common characterization techniques employed in classic organic chemistry, such as ^{1}H-NMR, ^{13}C-NMR, mass spectrometry, and FT-IR, give incomplete information when applied to CNTs, since no details about structure, morphology, purity, and chemical modification can be extracted.

Furthermore, CNTs cannot be conclusively described as a consequence of the production methods bringing about different diameters, length distribution, helicity, and catalyst residues. Also, purification processes and chemical modifications of CNTs induce additional variations to the material, removing the tips, reducing the lengths, and introducing defects on the sidewalls.

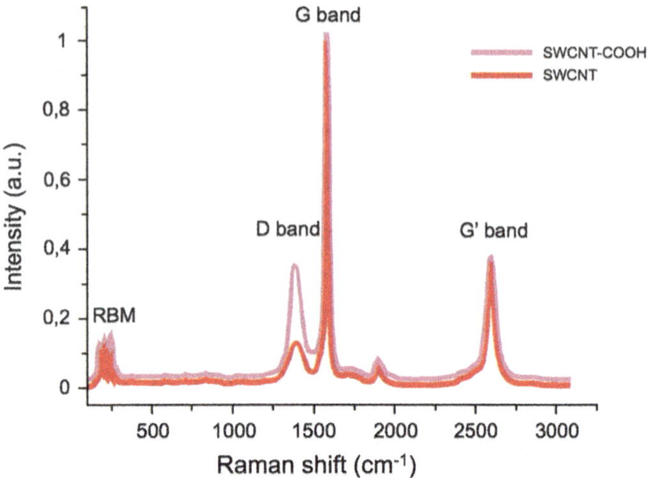

Fig. 32 Raman spectra of pristine and oxidized SWCNTs

Nevertheless, although unambiguous characterization of CNTs is not achievable with just a single technique, the synergy of different methodologies can supply relevant information on their structure, purity, and defects, before and after chemical manipulation.

Hence, we will report here the most common techniques helping in the characterization of CNTs.

5.1 *Spectroscopies*

The most common spectroscopic techniques used in the CNT field are UV–vis-NIR, IR, and Raman.

These techniques are very useful as they help to elucidate the helicity, the diameter, and the presence of structural defects.

5.1.1 Raman Spectroscopy

Raman spectroscopy has emerged as a very useful diagnostic tool to probe purity, structure, and degree of functionalization of carbon nanotubes [137].

Carbon nanotubes present a rich Raman spectrum (Fig. 32). Major features are the G band (G-graphite), the D band (D-diamond or disorder), and the RBM (radial breathing modes).

The G band, localized in the spectrum between 1,500 and 1,600 cm^{-1}, represents the high frequency tangential mode (Fig. 33) and results from a multi-peak combination. The two most relevant G peaks observed are the G^+ and G^-, referring

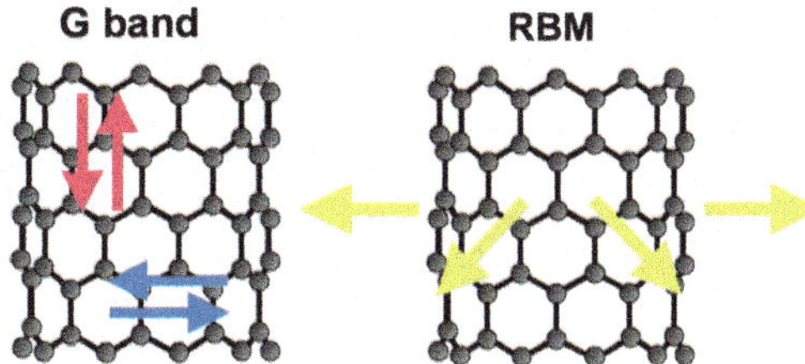

Fig. 33 Different atomic displacements on nanotubes producing the corresponding Raman signals

respectively to the atomic displacement along the tube axis and along the circumferential direction. When broadened, the G^{-} peak identifies metallic tubes, while it presents typical Lorentzian-like shape in the case of semi-conducting ones [138].

The D band, located in the 1,200–1,400 cm^{-1} region, is indicative of the amount of defects in a sample, and therefore it may represent the abundance of sp^3-hybridized atoms present in the sample, on both nanotubes and other carbonaceous residues. Being associated with the sp^3 carbon, introduced after chemical modification of the nanotubes, the D band is an important tool for estimating the degree of functionalization of a sample. According to the increased level of disorder, the signal is very pronounced in MWCNTs, so that its intensity cannot be taken into account for the evaluation of the organic functionalization. The G' band, which can be observed between 2,500 and 2,900 cm^{-1}, is an overtone of the D-band. This band is not actually connected to any increment of disorder, indeed it is still present in Raman spectra of defect-free CNTs. Whereas the G' peak is of fundamental importance for graphene analysis, since it is indicative of the level of exfoliation of graphite into individual flakes, its contribution to CNTs characterization is weak.

In contrast, the RBM is a peculiarity of CNTs describing their isotropic radial expansion (or contraction) through prominent and sharp peaks localized from 160 and 300 cm^{-1}. Its frequency depends on the tube diameter, to which is inversely proportional [139].

The RBM peaks also give important information concerning the environment of the tubes since shape and position are modified according to the presence of bundles or layers of graphite (MWCNTs).

5.1.2 UV–Vis-NIR Absorption Spectroscopy

Carbon nanotubes can be considered as mono-dimensional objects since the ratio between their length and diameter is high (aspect ratio). In carbon nanotubes the density of states (DOS) are non-continuous functions characterized by the presence of spikes, called van Hove singularities, that are peculiar for each (n,m) carbon nanotube. In more detail, CNTs are characterized by different optical absorptions, according to the transitions between different sets of van Hove singularities, expressed in Fig. 34 as S_{11} or S_{22} for semiconducting tubes and as M_{11} for metallic

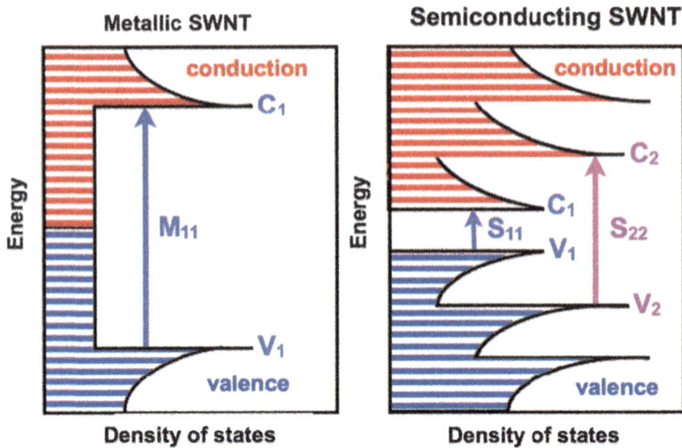

Fig. 34 Densities of states diagrams of metallic SWCNT (*left*) and semiconducting SWCNT (*right*)

ones. Therefore the chirality of the tubes can be studied by their absorption spectra (Fig. 35), where transitions between the valence and conductive bands are well distinguished [140].

Beside the position of the peaks, the shape also affords significant information. When bundled, the vis-NIR spectrum of CNTs is characterized by broad and low-defined van Hove singularities, which become sharp and well-separated when nanotubes are individualized by means of surfactants [141]. The absorption features even disappear with a high degree of functionalization, as a consequence of the π-conjugation that is progressively destroyed, changing dramatically the electronic arrangement of the tubes.

Although this technique provides exhaustive indications about CNTs chirality and debundling degree, it also presents some drawbacks, which limit its effective routine employment as a characterization tool. Among these, the most difficult to achieve is to obtain a stable dispersion of nanotubes, from which impurities have been totally removed [142]. Indeed, at present, a 100% pure reference sample of nanotubes is not yet available.

5.1.3 Infrared Spectroscopy

CNTs possess weak dipole moments, resulting in Infrared (IR) vibrational modes that are difficult to detect. For this reason IR spectroscopy is much less used than Raman spectroscopy for nanotube characterization.

Nevertheless, this technique can produce relevant information on CNTs purity, electronic structure, and degree of functionalization with good reliability. Earlier reflectance observations of SWCNTs as nonpurified powder were able to reveal

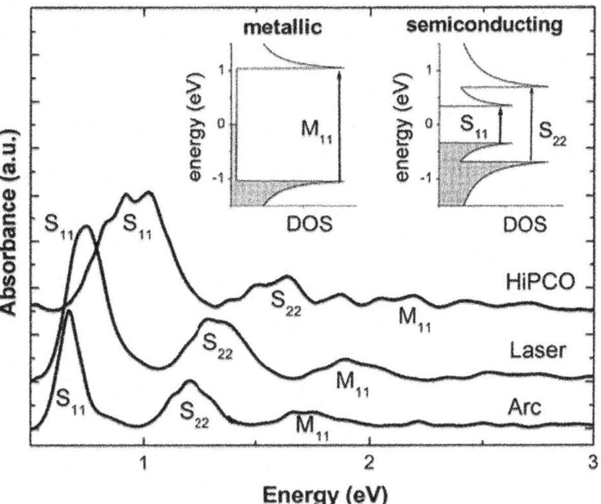

Fig. 35 Absorption spectra of SWCNTs produced by HiPco, laser, and arc methods. Reprinted with permission from [140]. Copyright 2001 American Chemical Society

two main features at 868 cm^{-1} and 1,590 cm^{-1}, very close to the A_{2u} and E_{1u} modes of graphite, respectively [143]. In the last decade, having available theoretical calculations on freestanding CNTs, it has been possible to assign IR active phonon modes to the correspondent chiral indices, highlighting defined CNT structures. For instance, semiconducting chiral nanotubes present one typical phonon mode at 5,386 cm^{-1} (0.67 eV), while metallic tubes are characterized by two signals at 9,791 cm^{-1} (1.21 eV) and 9,172 cm^{-1} (1.14 eV) [144].

Concerning the analysis of functionalized CNT samples, IR is particularly useful for the characterization of oxidized tubes, since carboxylic acid or carboxylate C=O stretching vibrations are characterized by strong absorptions at 1,719 cm^{-1} and 1,620 cm^{-1} respectively. Others chemical fragments easily identified by IR spectroscopy on CNTs are the amide group (between 1,660 and 1,640 cm^{-1}), the C–H bond with a vibration band around 2,900 cm^{-1}, the C–Cl stretching mode peak at 798 cm^{-1}, and the peak between 3,200 and 3,500 cm^{-1} correspondent to the O–H stretching [145].

Apart from the introduction of peaks typical of the organic appendage, other relevant modifications can appear in the IR spectrum after CNTs functionalization. For instance, covalent chemistry on the nanotubes perturbs the periodicity of the lattice, opening gaps at the Fermi level and changing a metal CNT to a semiconductor. As a consequence, the simultaneous weakening in the strength of the interband transitions is itself a signal of covalent functionalization, both in semiconducting and in metallic CNTs. Alternatively, ionic doping of CNTs with electron acceptors leads on one hand to the complete removal of the peak at 5,386 cm^{-1} and on the other to enhanced metal-like intraband transitions visible in the far-infrared (FIR), by creating partially filled metallic bands [145].

5.2 *Microscopy Techniques*

Microscopy techniques permit one to acquire important qualitative information about the average length, morphology, thickness, and purity of CNTs. These techniques are extensively used because of the large amount of data simultaneously obtainable from a single image.

5.2.1 Atomic Force Microscopy

The major advantage of this technique is to provide accurate three-dimensional topographic information.

AFM is suitable for the analysis of CNTs when they are homogeneously dispersed and spin-coated in a convenient solvent at very low concentrations (1 μg/mL) in order to allow proper individualization and reliable measurement of the size distribution (diameter and length).

AFM studies have allowed differentiation between SWCNTs and DWCNTs by measuring the compressed height and the mechanical forces of the tubes squeezed between an AFM tip and a silicon surface [146]. Moreover, several species attached to CNTs, covalently or not, have been successfully identified using AFM: gold colloids and silver clusters, magnetic nanoparticles encapsulated in carbon nanotubes, biomolecules such as enzymes, BSA, DNA, etc. [60, 62, 147–149].

Among the many applications of this technique, one particularly interesting relates to the use of CNTs as longlasting and mechanically non-invasive AFM tips, opening the door to non-conventional microfabrication rules for better lateral resolution [150].

5.2.2 Transmission Electron Microscopy

When employed for the characterization of nanotubes, transmission electron microscopy (TEM) reveals important details about dispersibility, purity grade, size dimension, and a crude assessment of the functionalization in some cases [60]. With the development of high-resolution TEM (HRTEM) techniques, progress has been achieved towards the accurate assignment of chiral indices of SWCNTs (Fig. 36), which can be analyzed on the basis of the intensity distribution on the main layer lines [151].

Impressive images provided by HRTEM allowed the identification of one-dimensional arrays of C_{60} inside SWCNTs, the so-called peapod structures, and more recently, conformational changes of small hydrocarbon molecules confined in carbon nanotubes were observed [152, 153].

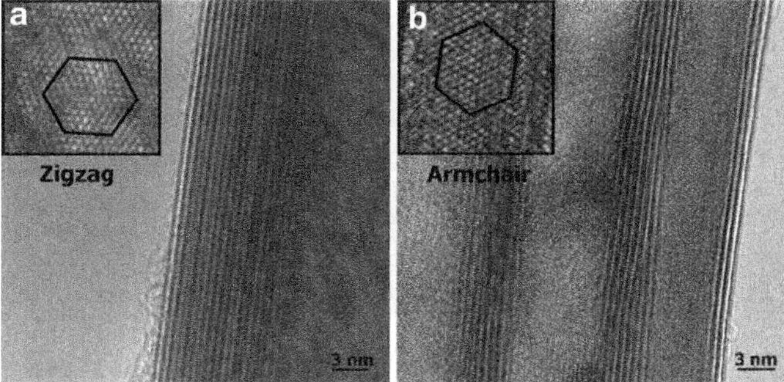

Fig. 36 HRTEM images of carbon nanotubes. The schematic hexagons identify zigzag (**a**) and armchair (**b**) morphologies. Reprinted with permission from [154]. Copyright 2008 National Academy of Sciences, USA

5.2.3 Scanning Electron Microscopy

Scanning electron microscopy (SEM) provides information about the external morphology of the sample, its composition, orientation, and electrical conductivity. Concerning the applicability on CNTs, it is a powerful tool when an assessment of the purity and three-dimensional morphological data are needed. In this direction, SEM is the instrument for the visualization of patterned surfaces where different CNTs topologies are grown [155]. On the other hand, the relatively low resolution cannot provide information about functionalization on processed tubes or accurate details about their structure. This drawback has been overcome by the development of high resolution SEM (HRSEM), through which a resolution of 1.5 nm can be achieved. In addition, the system is optimized for operations at low kV, allowing uncoated and insulating materials to be examined with minimum charging (Fig. 37).

5.2.4 Scanning Tunneling Microscopy

STM is a powerful technique that allows the observation of the very intimate hexagonal skeleton of carbon nanotubes as well as the measurement of their chiral angles and diameter.

In this context, major contributions have been given by the group of Cees Dekker, who confirmed the correlation between the electronic properties of CNTs and their (n,m) indices. The same group examined the bandgaps of both metallic and semiconducting tubes and found them in good agreement with theoretical predictions [30, 157].

More recently our group has also demonstrated the utility of this technique for the characterization of chemically modified nanotubes [158]. STM images have indeed revealed the appendages of aliphatic chains attached to SWCNTs via amide

Fig. 37 SEM images of aligned carbon nanotubes. *Left*: CNT-column array; *right*: higher magnification on the CNT-columns. Reprinted from [156], Copyright 2012, with permission from Elsevier

reaction and localized them both at the tips and at the sidewalls. In a similar approach, *tert*-butyl groups could be clearly visualized by STM [113].

A mapping study of functionalized carbon nanotubes, showing the possibility of combining STM topography to tunneling conductance mapping was reported, in order to highlight the dependence of functionalization on crystallographic orientation [159].

5.2.5 Thermogravimetric Analysis

The employment of thermogravimetric analysis (TGA) for CNTs characterization is so widespread that it deserves a separate section for its description.

This technique is fundamental for the evaluation of the functionalization degree of a sample according to the weight loss that occurs after heating up to 1,000°C [90].

Owing to the different thermal stability between the nanotube carbonaceous skeleton and the volatile chemical species appended on the sidewalls, it is possible to evaluate the amount of functionalities introduced on the tubes, which are usually expressed in μmol/g.

Generally the instrument is composed of a high-precision balance with a pan, commonly made of platinum, on which the samples are placed. The pan is loaded into an electrically heated oven equipped with a thermocouple, which accurately measures the variation of temperature inside the chamber. The measurement can be carried out in an inert gas, such as N_2, He, or Ar, or in oxygen atmosphere.

Thermogravimetric analysis could be exploited to characterize nanotubes at the raw stage, after purification and when chemically processed. When the pristine material is analyzed in the presence of oxygen, it is possible to obtain information about its residual mass and the oxidation temperature T_0, whereas under an inert atmosphere the carbon structure is maintained and the weight loss indicates unequivocally the amount of organic material.

6 Conclusions

Carbon nanotubes are being extensively investigated, owing to their fascinating electronic and mechanical properties and great perspective for practical applications. Though CNTs are very difficult materials to manipulate and characterize, major recent efforts, including new and versatile functionalization methodologies, have made it possible to implement CNTs in many functional devices. New interesting uses can be envisaged, as our knowledge on these novel materials progresses, so that we may expect to see a bright future and a wide range of exciting applications.

References

1. Kroto HW, Heath JR, Obrien SC, Curl RF, Smalley RE (1985) Nature 318:162–163
2. Radushkevich LV, Lukyanovich VM (1952) Zurn Fisic Chim 111:24
3. Iijima S (1991) Nature 354:56–58
4. Iijima S, Ichihashi T (1993) Nature 363:603–605
5. Bethune DS, Kiang CH, De Vries MS, Gorman G, Savoy R, Vazquez J, Beyers R (1993) Nature 363:605–607
6. Hamada N, Sawada S, Oshiyama A (1992) Phys Rev Lett 68:1579–1582
7. Saito R, Fujita M, Dresselhaus G, Dresselhaus MS (1992) Phys Rev B 46:1804–1811
8. Saito R, Fujita M, Dresselhaus G, Dresselhaus MS (1992) Appl Phys Lett 60:2204–2206
9. Jorio A, Dresselhaus MS, Dresselhaus G (2008) Carbon nanotubes: advanced topics in the synthesis, structure, properties and applications. Springer, Berlin
10. Cao Q, Rogers JA (2009) Adv Mater 21:29–53
11. Endo M, Strano MS, Ajayan PM (2008) Top Appl Phys 111:13–61
12. Kostarelos K, Bianco A, Prato M (2009) Nat Nanotech 4:627–633
13. Fabbro C, Ali-Boucetta H, Da Ros T, Kostarelos K, Bianco A, Prato M (2012) Chem Commun 48:3911–3926
14. Hirsch A (2002) Angew Chem Int Ed 41:1853–1859
15. Ebbesen TW, Ajayan PM (1992) Nature 358:220–222
16. Szabò A, Perri C, Csatò A, Giordano G, Vuono D, Nagy JB (2010) Materials 3:3468–3517
17. Farhat S, de La Chapelle ML, Loiseau A, Scott CD, Lefrant S, Journet C, Bernier P (2001) J Chem Phys 115:6752–6759
18. Journet C, Maser WK, Bernier P, Loiseau A, de la Chapelle ML, Lefrant S, Deniard P, Lee R, Fischer JE (1997) Nature 388:756–758
19. Guo T, Nikolaev P, Rinzler AG, Tomanek D, Colbert DT, Smalley RE (1995) J Phys Chem 99:10694–10697
20. Yacamàn MJ, Yoshida MM, Rendon L, Santiesteban JG (1993) Appl Phys Lett 62:202–204
21. Stadermann M, Sherlock SP, In J-B, Fornasiero F, Gyu Park H, Artyukhin AB, Wang Y, De Yoreo JJ, Grigoropoulos CP, Bakajin O, Chernov AA, Noy A (2009) Nano Lett 9: 738–744
22. Choi HC, Kim W, Wang DW, Dai HJ (2002) J Phys Chem B 106:12361–12365
23. Nikolaev P, Bronikowski MJ, Bradley RK, Rohmund F, Colbert DT, Smith KA, Smalley RE (1999) Chem Phys Lett 313:91–97
24. Isaacs JA, Tanwani A, Healy ML, Dahlben LJ (2010) J Nanopart Res 12:551–562
25. Laplaze D, Bernier P, Maser WK, Flamant G, Guillard T, Loiseau A (1998) Carbon 36: 685–688
26. Luxembourg D, Flamant G, Laplaze D (2005) Carbon 43:2302–2310

27. Hsu WK, Hare JP, Terrones M, Kroto HW, Walton DRM, Harris PJH (1995) Nature 377:687
28. Bai JB, Hamon AL, Marraud A, Jouffrey B, Zymla V (2002) Chem Phys Lett 365:184–188
29. Dresselhaus MS, Dresselhaus G, Avouris P (2011) Top Appl Phys 80:273–286
30. Wildoer JWG, Venema LC, Rinzler AG, Smalley RE, Dekker C (1998) Nature 391:59–62
31. Dresselhaus MS, Dresselhaus G, Charlier JC, Hernández E (2004) Phil Trans R Soc Lond A 362:2065–2098
32. Rao AM, Richter E, Bandow S, Chase B, Eklund PC, Williams KA, Fang S, Subbaswamy KR, Menon M, Thess A, Smalley RE, Dresselhaus G, Dresselhaus MS (1997) Science 275:187–191
33. Wang N, Tang ZK, Li GD, Chen JS (2000) Nature 408:50–51
34. Odom TW, Huang JL, Kim P, Lieber CM (1998) Nature 391:62–64
35. Yao Z, Postma HWC, Balents L, Dekker C (1999) Nature 402:273–276
36. Zhang M, Li J (2009) Mater Today 12:12–18
37. Terrones M (2003) Annu Rev Mater Res 33:419–501
38. Treacy MMJ, Ebbesen TW, Gibson JM (1996) Nature 381:678–680
39. Krishnan E, Dujardin TW, Ebbesen PN, Yianilos, Treacy MMJ (1998) Phys Rev B 58:14013–14019
40. Wong EW, Sheehan PE, Lieber CM (1997) Science 277:1971–1975
41. Salvetat JP, Briggs GAD, Bonard JM, Bacsa RR, Kulik AJ, Stockli T, Burnham NA, Forro L (1999) Phys Rev Lett 82:944–947
42. Iijima S, Brabec C, Maiti A, Bernholc J (1996) J Chem Phys 104:2089–2092
43. Ajayan PM, Tour JM (2007) Nature 447:1066–1068
44. Charlier J-C (2002) Acc Chem Res 35:1063–1069
45. Ajayan PM, Ravikumar V, Charlier J-C (1998) Phys Rev Lett 81:1437–1440
46. Terrones M, Banhart F, Grobert N, Charlier J-C, Terrones H, Ajayan PM (2002) Phys Rev Lett 89:75505–75509
47. Niyogi S, Hamon MA, Hu H, Zhao B, Bhowmik P, Sen R, Itkis ME, Haddon RC (2002) Acc Chem Res 12:1105–1113
48. Lu X, Chen Z, Schleyer PR (2005) J Am Chem Soc 127:4313–4315
49. Tasis D, Tagmatarchis N, Bianco A, Prato M (2006) Chem Rev 106:1105–1136
50. Karousis N, Tagmatarchis N, Tasis D (2010) Chem Rev 110:5366–5397
51. Singh P, Campidelli S, Giordani S, Bonifazi D, Bianco A, Prato M (2009) Chem Soc Rev 38: 2214–2230
52. Joselevich E (2004) Chem Phys Chem 5:619–624
53. Liu J, Rinzler AG, Dai H, Hafner JH, Bradley RK, Boul PJ, Lu A, Iverson T, Shelimov K, Huffman CB, Rodriguez-Macias F, Colbert DT, Smalley RE (1998) Science 280:1253–1256
54. Rubio N, Fabbro C, Herrero MA, de la Hoz A, Meneghetti M, Fierro JLG (2011) Small 7: 665–674
55. Hamon MA, Chen J, Hu H, Chen Y, Itkis ME, Rao AM, Eklund PC, Haddon RC (1999) Adv Mater 11:834–840
56. Zhao B, Hu H, Yu A, Perea D, Haddon RC (2005) J Am Chem Soc 127:8197–8203
57. Delgado JL, de la Cruz P, Urbina A, López Navarrete JT, Casado J, Langa F (2007) Carbon 45:2250–2252
58. Giordani S, Colomer JF, Cattaruzza F, Alfonsi J, Meneghetti M, Prato M, Bonifazi D (2009) Carbon 47:578–588
59. Baker S, Cai W, Lasseter T, Hammers RJ (2002) Nano Lett 2:1413–1417
60. Jiang K, Schadler LS, Siegel RW, Zhang X, Zhang H, Terrones M (2004) J Mater Chem 14: 37–39
61. Wohlstadter JN, Wilbur JL, Sigal GB, Biebuyck HA, Billadeau MA, Dong L (2003) Adv Mater 15:1184–1187
62. Patolsky F, Weizmann Y, Willner I (2004) Angew Chem Int Ed 43:2113–2117
63. Sardesai N, Pan S, Rusling J (2009) Chem Comm 33:4968–4970

64. Sun Y-P, Zhou B, Henbest K, Fu K, Huang W, Lin Y, Taylor S, Carroll DL (2002) Chem Phys Lett 351:349–353
65. Shi X, Wang SH, Shen M, Antwerp ME, Chen X, Li C, Petersen EJ, Huang Q, Weber WJ, Baker JR (2009) Biomacromolecules 10:1744–1750
66. Umeyama T, Fujita M, Tezuka N, Kadota N, Matano Y, Yoshida K, Isoda S, Imahori H (2007) J Phys Chem C 111:11484–11493
67. Zhu WH, Minami N, Kazaoui S, Kim Y (2004) J Mater Chem 14:1924–1926
68. Kahn MGC, Banerjee S, Wong SS (2002) Nano Lett 2:1215–1218
69. Zhang D, Wang T, Li J, Guo Z-X, Dai L, Zhang D, Zhu D (2003) Chem Phys Lett 367: 747–752
70. Baskaran D, Mays JW, Zhang XP, Bratcher MS (2005) J Am Chem Soc 127:6916–6917
71. Zheng X, Jiang DD, Wang D, Wilkie CA (2007) Chem Commun 46:4949–4951
72. Sun Y-P, Huang W, Lin WY, Fu K, Kitaygorodskiy A, Riddle LA, Yu YJ, Carroll DL (2001) Chem Mater 13:2864–2869
73. Fernandez d'Arlas B, Goyanes S, Rubiolo GH, Mondragon I, Corcuera MA, Eceiza A (2009) J Nanosci Nanotechnol 9:6064–6071
74. Eder D (2010) Chem Rev 110:1348–1385
75. Wang P, Moorefield CN, Li S, Hwang SH, Shreiner CD, Newkome GR (2006) Chem Commun 10:1091–1093
76. Herrero MA, Guerra J, Myers VS, Gómez MV, Crooks RM, Prato M (2010) ACS Nano 4:905–912
77. Mickelson ET, Huffman CB, Rinzler AG, Smalley RE, Hauge RH, Margrave JL (1998) Chem Phys Lett 296:188–194
78. Khabashesku VN, Billups EW, Margrave JL (2002) Acc Chem Res 35:1087–1095
79. Saini RK, Chiang IW, Peng H, Smalley RE, Billups WE, Hauge RH, Margrave JT (2003) J Am Chem Soc 125:3617–3621
80. Mickelson ET, Chiang IW, Zimmerman JL, Boul P, Lozano J, Liu J, Smalley RE, Hauge RH, Margrave JL (1999) J Phys Chem B 103:4318–4322
81. Hu H, Zhao B, Hamon MA, Kamaras K, Itkis ME, Haddon RC (2003) J Am Chem Soc 125: 14893–14900
82. Bingel C (1993) Chem Ber 126:1957–1959
83. Camps X, Hirsch A (1997) J Chem Soc 1:1595–1596
84. Coleman KS, Bailey SR, Fogden S, Green MLH (2003) J Am Chem Soc 125:8722–8723
85. Holzinger M, Vostrowsky O, Hirsch A, Hennrich F, Kappes M, Weiss R, Jellen F (2001) Angew Chem Int Ed 113:4132–4136
86. Yinghuai Z, Peng AT, Carpenter K, Maguire JA, Hosmane NS, Takagaki M (2005) J Am Chem Soc 127:9875–9880
87. Georgakilas V, Kordatos K, Prato M, Guldi DM, Holzinger M, Hirsch A (2002) J Am Chem Soc 124:760–761
88. Guldi DM, Marcaccio M, Paolucci D, Paolucci F, Tagmatarchis N, Tasis D, Vazquez E, Prato M (2003) Angew Chem Int Ed 42:4206–4209
89. Georgakilas V, Bourlinos A, Gournis D, Tsoufis T, Trapalis C, Mateo-Alonso A, Prato M (2008) J Am Chem Soc 130:8733–8740
90. Campidelli S, Sooambar C, Lozano Diz E, Ehli C, Guldi DM, Prato M (2006) J Am Chem Soc 128:12544–12552
91. Herrero MA, Toma FM, Al-Jamal KT, Kostarelos K, Bianco A, Da Ros T, Bano F, Casalis L, Scoles G, Prato M (2009) J Am Chem Soc 131:9843–9848
92. Prato M, Kostarelos K, Bianco A (2008) Acc Chem Res 41:60–68
93. Venturelli E, Fabbro C, Chaloin O, Ménard-Moyon C, Smulski CR, Ros TD, Kostarelos K, Prato M, Bianco A (2011) Small 7:2179–2187
94. Georgakilas V, Voulgaris D, Vazquez E, Prato M, Guldi DM, Kukovecz A, Kuzmany H (2002) J Am Chem Soc 124:14318–14319
95. Vazquez E, Prato M (2009) ACS Nano 3:3819–3824

96. Alvaro M, Atienzar P, de la Cruz P, Delgado JL, Troiani V, Garcia H, Langa F, Palkar A, Echegoyen L (2006) J Am Chem Soc 128:6626–6635
97. Brunetti FG, Herrero MA, Muñoz JM, Giordani S, Díaz-Ortiz A, Filippone S, Ruaro G, Meneghetti M, Prato M, Vázquez E (2007) J Am Chem Soc 129:14580–14581
98. Delgado JL, de la Cruz P, Langa F, Urbina A, Casado J, Lopez Navarrete JT (2004) Chem Commun 1734–1735
99. Zhang W, Swager TM (2007) J Am Chem Soc 129:7714–7715
100. Holzinger M, Abraham J, Whelan P, Graupner R, Ley L, Hennrich F, Kappes M, Hirsch A (2003) J Am Chem Soc 125:8566–8580
101. Peng H, Alemany LB, Margrave JL, Khabashesku VN (2003) J Am Chem Soc 125: 15174–15182
102. Wei L, Zhang Y (2007) Nanotechnology 18:495701–495703
103. Dyke CA, Tour JM (2004) J Phys Chem A 108:11151–11159
104. Price BK, Tour JM (2006) J Am Chem Soc 128:12899–12904
105. Campidelli S, Ballesteros B, Filoramo A, Díaz D, de la Torre G, Torres T, Aminurrahman GM, Ehli C, Kiessling D, Werner F, Sgobba V, Guldi DM, Cioffi C, Prato M, Bourgoin JP (2008) J Am Chem Soc 130:11503–11509
106. Yoo BK, Myung S, Lee M, Hong S, Chun K, Paik HJ, Kim J, Lim JK, Joo SW (2006) Mater Lett 60:3224–3226
107. Gómez-Escalonilla MJ, Atienzar P, Fierro JLG, García H, Langa F (2008) J Mater Chem 18: 1592–1600
108. Liu J, Rodriguez M, Zubiri I, Dossot M, Vigolo B, Hauge RH, Fort Y, Ehrhardt J-J, McRae E (2006) Chem Phys Lett 430:93–96
109. Liu J, Zubiri MR, Vigolo B, Dossot M, Fort Y, Ehrhardt JJ, McRae E (2007) Carbon 45: 885–891
110. Liang F, Sadana AK, Peera A, Chattopadhyay J, Gu Z, Hauge RH, Billups WE (2004) Nano Lett 4:1257–1260
111. Chattopadhyay J, Sadana AK, Liang F, Beach JM, Xiao Y, Hauge RH, Billups WE (2005) Org Lett 7:4067–4069
112. Wunderlich D, Hauke F, Hirsch A (2008) J Mater Chem 18:1493–1497
113. Graupner R, Abraham J, Wunderlich D, Vencelova A, Lauffer P, Ræhrl J, Hundhausen M, Ley L, Hirsch A (2006) J Am Chem Soc 128:6683–6689
114. Wunderlich D, Hauke F, Hirsch A (2008) Chem Eur J 14:1607–1614
115. Syrgiannis Z, Hauke F, Röhrl J, Hundhausen M, Graupner R, Elemes Y, Hirsch A (2008) Eur J Org Chem 2008:2544–2550
116. Xu Y, Wang X, Tian R, Li S, Wan L, Li M, You H, Li Q, Wang S (2008) Appl Surf Sci 254: 2431–2435
117. Tagmatarchis N, Georgakilas V, Prato M, Shinohara H (2002) Chem Commun 2010: 2010–2011
118. Priya BR, Byrne HJ (2008) J Phys Chem C 112:332–337
119. Nish A, Hwang J-Y, Doig J, Nicholas RJ (2007) Nat Nanotech 2:640–646
120. Ehli C, Rahman GMA, Jux N, Balbinot D, Guldi DM, Paolucci F, Marcaccio M, Paolucci D, Melle-Franco M, Zerbetto F, Campidelli S, Prato M (2006) J Am Chem Soc 128: 11222–11231
121. Guo Z, Sadler PJ, Tsang SC (1998) Adv Mater 10:701–703
122. Balavoine F, Schultz P, Richard C, Mallouh V, Ebbesen TW, Mioskowski C (1999) Angew Chem Int Ed 38:1912–1915
123. Erlanger BF, Chen B-X, Zhu M, Brus L (2001) Nano Lett 1:465–467
124. O'Connell MJ, Boul P, Ericson LM, Huffman C, Wang Y, Haroz E, Kuper C, Tour J, Ausman KD, Smalley RE (2001) Chem Phys Lett 342:265–271
125. Curran SA, Ajayan PM, Blau WJ, Carroll DL, Coleman JN, Dalton AB, Davey AP, Drury A, McCarthy B, Maier S, Strevens A (1998) Adv Mater 10:1091–1093
126. Liu Z, Winters M, Holodniy M, Dai H (2007) Angew Chem Int Ed 46:2023–2027

127. Sitharaman B, Kissell KR, Hartman KB, Tran LA, Baikalov A, Rusakova I, Sun Y, Khant HA, Ludtke SJ, Chiu W, Laus S, Tòth E, Helm L, Merbach AE, Wilson LJ (2005) Chem Commun 2005:3915–3917
128. Liu Z, Sun X, Nakayama-Ratchford N, Dai H (2007) ACS Nano 1:50–56
129. Ali-Boucetta H, Al-Jamal KT, McCarthy D, Prato M, Bianco A, Kostarelos K (2008) Chem Commun 4:459–461
130. Dhar S, Liu Z, Thomale J, Dai H, Lippard SJ (2008) J Am Chem Soc 130:11467–11476
131. Bhirde AA, Sousa AA, Patel V, Azari AA, Gutkind JS, Leapman RD, Rusling JF (2009) ACS Nano 3:307–316
132. Wu W, Wieckowski S, Pastorin G, Benincasa M, Klumpp C, Briand JP, Gennaro R, Prato M, Bianco A (2005) Angew Chem Int Ed 44:6358–6362
133. Chen J, Chen S, Zhao X, Kuznetsova LV, Wong SS, Ojima I (2008) J Am Chem Soc 130:16778–16785
134. Brunetti FG, Herrero MA, Muñoz JM, Díaz-Ortiz A, Alfonsi J, Meneghetti M, Prato M, Vázquez E (2008) J Am Chem Soc 130:8094–8100
135. Syrgiannis Z, Gebhardt B, Dotzer C, Hauke F, Graupner R, Hirsch A (2010) Angew Chem Int Ed 49:3322–3325
136. Prato M (2010) Nature 465:172–173
137. Graupner R (2007) J Raman Spectrosc 38:673–683
138. Jorio A, Souza Filho AG, Dresselhaus G, Dresselhaus MS, Swan AK, Ünlü MS, Goldberg B, Pimenta MA, Hafner JH, Lieber CM, Saito R (2002) Phys Rev B 65:155412/1–155412/9
139. Dresselhaus MS, Dresselhaus G, Saito R, Jorio A (2005) Phys Rep 409:47–99
140. Hamon MA, Itkis ME, Niyogi S, Alvaraez T, Kuper C, Menon M, Haddon RC (2001) J Am Chem Soc 123:11292–11293
141. O'Connell MJ, Bachilo SM, Huffman CB, Moore VC, Strano MS, Haroz EH, Rialon KL, Boul PJ, Noon WH, Kittrell C, Ma J, Hauge RH, Weisman RB, Smalley RE (2002) Science 297:593–596
142. Itkis ME, Perea DE, Jung R, Niyogi S, Haddon RC (2004) J Am Chem Soc 127:3439–3448
143. Kuhlmann U, Jantoljak H, Pfander N, Bernier P, Journet C, Thomsen C (1998) Chem Phys Lett 294:237–240
144. Kim UJ, Liu XM, Furtado CA, Chen G, Saito R, Jiang J, Dresselhaus MS, Eklund PC (2005) Phys Rev Lett 95:157402/1–157402/4
145. Kamaras K, Itkis ME, Hu H, Zhao B, Haddon RC (2003) Science 301:1501
146. De Borde T, Joiner JC, Leyden MR, Monit ED (2008) Nano Lett 8:3568–3571
147. Davis JJ, Coleman KS, Azamian BR, Bagshaw CB, Green MLH (2003) Chem Eur J 9:3732–3739
148. Cleuziou JP, Wernsdorfer W, Ondarcuhu T, Monthiouz M (2011) ACS Nano 5:2348–2355
149. Xin H, Woolley AT (2003) J Am Chem Soc 125:8710–8711
150. Wilson NR, Macpherson JV (2009) Nat Nanotech 4:483–491
151. Qin LC (2006) Rep Prog Phys 69:2761–2821
152. Smith BW, Monthioux M, Luzzi DE (1998) Nature 396:323–324
153. Koshino M, Tanaka T, Solin N, Suenaga K, Isobe H, Nakamura E (2007) Science 316: 853–853
154. Sadan MB, Houben L, Enyashin AN, Seifert G, Tenne R (2008) PNAS 105:15643–15648
155. Malik H, Stephenson KJ, Bahr DF, Field DP (2010) J Mater Sci 46:3119–3126
156. Elmer JW, Yaglioglu O, Schaeffer RD, Kardos G, Derkach O (2012) Carbon 50:4114–4122
157. Lemay SG, Janssen JW, vd Hout M, Mooij M, Bronikowski MJ, Willis PA, Smalley RE, Kouwenhoven LP, Dekker C (2001) Nature 412:617–620
158. Bonifazi D, Nacci C, Marega R, Campidelli S, Ceballos G, Modesti S, Meneghetti M, Prato M (2006) Nano Lett 6:1408–1414
159. Nemes-Incze P, Kónya Z, Kiricsi I, Pekker A, Horváth ZE, Kamarás K (2011) J Phys Chem C 115:3229–3235

Top Curr Chem (2014) 350: 111–140
DOI: 10.1007/128_2012_358

Published online: 6 September 2012

Three-Dimensional Aromatic Networks

Shinji Toyota and Tetsuo Iwanaga

Abstract Three-dimensional (3D) networks consisting of aromatic units and linkers are reviewed from various aspects. To understand principles for the construction of such compounds, we generalize the roles of building units, the synthetic approaches, and the classification of networks. As fundamental compounds, cyclophanes with large aromatic units and aromatic macrocycles with linear acetylene linkers are highlighted in terms of transannular interactions between aromatic units, conformational preference, and resolution of chiral derivatives. Polycyclic cage compounds are constructed from building units by linkages via covalent bonds, metal-coordination bonds, or hydrogen bonds. Large cage networks often include a wide range of guest species in their cavity to afford novel inclusion compounds. Topological isomers consisting of two or more macrocycles are formed by cyclization of preorganized species. Some complicated topological networks are constructed by self-assembly of simple building units.

Keywords Aromatic unit · Cage compound · Linker · Macrocycle · Topological isomer

Contents

S. Toyota (✉) and T. Iwanaga
Department of Chemistry, Faculty of Science, Okayama University of Science, 1-1 Ridaicho, Kita-ku, Okayama 700-0005, Japan
e-mail: stoyo@chem.ous.ac.jp

Abbreviations

Bu	Butyl
CD	Circular dichroism
Cp	Cyclopentadienyl
dppp	1,3-Bis(diphenylphosphino)propane
ESI	Electrospray ionization
HPLC	High performance liquid chromatography
^{i}Pr	Isopropyl
MALDI-TOF	Matrix-assisted laser desorption time-of-flight
trityl	Triphenylmethyl

1 Introduction

Aromatic units are fascinating building blocks for the construction of three-dimensional (3D) molecular structures because of their rigid panel-like shapes. There are several approaches to the construction of 3D structures from two-dimensional (2D) planar aromatic units. As mentioned in the previous chapters, the deformation of aromatic units into nonplanar structures, such as twisted, curved, and bowl structures, is a common approach to increase the dimensions [1]. Another promising approach is to connect multiple planar aromatic units to each other with linkers to form *networks*, i.e., interconnected systems. This chapter discusses such 3D aromatic networks consisting of several aromatic units and linkers. An advantage of this molecular design is that higher-ordered or complex structures can be constructed from one or a few simple building unit(s), yielding architectures of various sizes, shapes, packings, and flexibilities. Aromatic compounds occasionally have interesting properties arising from delocalized conjugated π-systems, and some compounds have been applied to functional materials such as electronic devices and molecular sensors [2–5]. One merit of network formation is that the properties of each unit can be accumulated in a single molecular system. However, the effects of the accumulation are not always additive because of intramolecular interactions through bonds or space. In some cases, new properties appear in 3D networks even though each unit exhibits little or no such properties. For example, although large 3D networks tend to form inner cavities to accommodate guest molecules, this property should not be expected for each building unit. These phenomena, called *emergence*, are one of the significant outcomes of in-depth studies of 3D aromatic networks.

2 General Remarks

2.1 Aromatic Units

Any aromatic units can be used as a building unit for the construction of aromatic networks [6]. In the case of aromatic hydrocarbons, benzene is the most frequently used, although other polycyclic arenes, including naphthalene, anthracene, and pyrene, are attractive units for the construction of various 3D structures [7]. Heteroaromatic units, such as pyridine and thiophene, are used to generate novel functions by means of their coordination ability and novel electronic properties. The number of aromatic units (two, three, four units, and so on) is a basic factor that characterizes aromatic networks. Some aromatic networks consist of one kind of aromatic unit, whereas others consist of multiple kinds. In an aromatic network, each unit is connected to other units via two or more linkers in most cases. Therefore, the position of the linkers is another important factor that controls the direction and distance of network extension. For example, benzene offers 60°, 120°, and 180° angles at *ortho*, *meta*, and *para* positions, respectively. Large aromatic units should have more variations in connectivity. For example, naphthalene and anthracene have 10 and 15 combinations, respectively, for the introduction of two identical linkers.

2.2 Linkers

Linkers serve to connect one unit to another in aromatic networks via chemical bonds or attractive interactions. Linkers are classified according to such factors as geometry, length, mobility, kind of bond, and number of bonds. These factors control the arrangement and flexibility of arene units and thus greatly influence the structures and dynamic behavior of aromatic networks in 3D space.

Ditopic linkers are the most common, which connect two units across n atoms or $n + 1$ bonds (denoted as n-atom linkers hereafter). Typical examples of covalent one-atom linkers are methylene ($-CH_2-$) and oxygen (–O–), although other

Scheme 1 Typical two-atom linkers in 3D aromatic networks

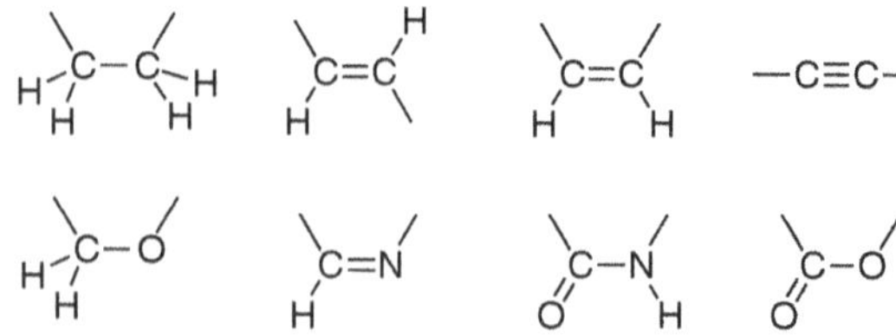

elements can be used. These linkers are not linear and the bond angles at the C or O atom are usually 110–120°. For example, several *m*-phenylene groups are connected to methylene groups as linkers to form 3D cyclic structures in calixarenes [8]. Metal atoms are also used as linkers via coordination bonds with ligands in aromatic units, such as pyridines. For a square-planar metal, *cis* and *trans* junctions offer angles of 90° and 180°, respectively. In order to regulate the geometry, the other two coordination sites are blocked by other ligands, e.g., a bidentate chelate ligand for the *cis* linkage. These one-atom linkers are generally less flexible than saturated longer linkers, as discussed below.

Two-atom linkers that connect two units via three bonds are commonly found in 3D aromatic networks. Typical two-atom linkers are shown in Scheme 1. In ethylene ($-CH_2CH_2-$), a common saturated carbon linker, two bonds connecting aromatic units are separated by ca. 1.54 Å and can extend in various directions depending on the conformation about the central C–C bond. In contrast, unsaturated two-atom carbon linkers have fixed directions. The two bonds extend to the same side or the opposite side of the double bond in *cis*- and *trans*-ethenylene (vinylene, –CH=CH–) linkers. Ethynylene (–C≡C–) is a typical linear linker, in which the distance between terminal arene units is ca. 4.2 Å [9–11]. Ethenylene and ethynylene linkers can extend the π-conjugation between arene units across the double bond and the triple bond, respectively, depending on the conformation. For both linkers the conjugation is maximum in the fully planar conformation and less effective in nonplanar conformations. One carbon atom or two carbon atoms in those two-carbon linkers can be replaced with other elements giving a variety of choices. Some examples are ethers, imines, amides, and esters (Scheme 1). Linkers with more than three atoms or more than four bonds are essentially a combination of the above linkers. Generally, the mobility and flexibility of linker moieties increase with an increase of chain length. However, 1,3-butadiynylene (–C≡C–C≡C–) is a typical rigid linear linker separated by 6.8 Å.

Linkers with two or more atom chains can be distinguished by their orientation. Regarding two-atom linkers, the –X–X– type linkers are symmetrical and have no direction, whereas the –X–Y– type linkers are unsymmetrical and have direction. Simple esters and amides are typical directed linkers. Therefore the use of unsymmetrical linkers considerably increases the number of possible network structures involving constitutional isomers and stereoisomers.

The term tritopic linkers refers to linkers with three positions that enable linkage to arene units. Three tethers extend from a central atom (or an atomic group) in a pyramidal or trigonal shape [12]. Typical pyramidal tritopic linkers involve tetrahedral carbon atoms, pyramidal nitrogen atoms, or other atomic groups as bridgehead moieties. Tetrahedral carbon atoms also function as tetratopic linkers [13]. The structural variation of 3D networks can be dramatically enhanced by using such highly branched linkers.

Not only strong covalent or coordination bonds but also weak interactions, such as hydrogen bonds, can connect linker moieties. Hydrogen bonds are generally too weak to connect two chains firmly, but complementary multi-point hydrogen bonds strengthen the linkage, as found in carboxylic acids or base pairs in nucleic acids [14, 15]. In some molecular networks, weak interactions are utilized as supplemental linkers to support the main framework rather than as main linkers to connect aromatic units. Such interactions often play important roles in controlling the structures and properties of molecular systems.

2.3 Construction of Aromatic Networks

Aromatic networks are constructed from the aromatic units and linkers mentioned above by chemical reactions. In terms of the position of bond formation, there are two approaches to connect two or more aromatic units. One is the formation of bonds between arene units and linker terminals. Cross coupling reactions are commonly used for this purpose because a wide variety of reactions are now available given the recent advances in synthetic methodology [16]. Arenes with halogen substituents (Ar–X) or those with metal substituents (Ar–M) are common synthetic precursors. Conventional aromatic substitution reactions are also practical for the extension of the aromatic network [17]. The second approach, which is valid for only two-atom linkers or longer linkers, is the formation of a bond between atoms in linker chains from substituted arene units. Any bond formation reactions are available depending on the functional groups in the pre-linker substituents. In particular, ring-closing metathesis is a new technique to connect two terminal alkene substituents across arene units [18]. In both connection approaches, each bond formation process should be highly effective to construct large networks in reasonable yields.

Some large aromatic networks are constructed by stepwise reactions and the macrocyclization of linear precursors or the further cyclization of cyclic compounds is often carried out in the last step [19, 20]. In other cases, complicated networks can be formed from simple building units in a single procedure that involves the formation of several bonds. Such composite reactions or self-assembly processes work surprisingly well when the reaction sites are controlled by structural requirement and pre-organization. In particular, the direction of bonds from linkers and arene units is very important to control the shape and size of the network to be formed. This concept, known as "directional bonding approach," is summarized in the review articles by Stang et al. [21, 22].

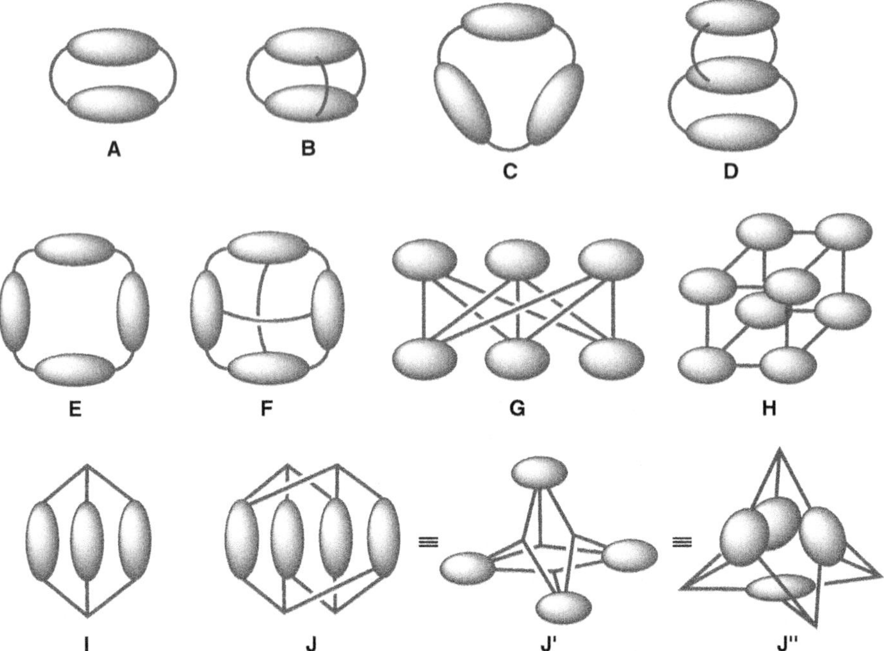

Scheme 2 Various molecular networks consisting of aromatic units (*ellipsoids*) and linkers (*lines or curves*)

2.4 Classification of Aromatic Networks

The modes of network structures can be classified according to theories in related research fields. The graph theory studies structures of graphs consisting of nodes and edges (links), which are equivalent to arene units and linkers, respectively, in aromatic networks [23]. In the field of network topology, topologies are classified into several basic types such as ring, mesh, star, and hybrid. The features of typical networks with at least one ring structure (Scheme 2) are briefly discussed with the aid of those theories.

Aromatic networks are chiefly characterized by the numbers of arene units and linkers and then by the mode of linkage. Network **A** with two arene units and two linkers is the simplest; it represents the structure of cyclophanes, which are fundamental cyclic compounds in aromatic chemistry. Two arene units can be connected by three or more linkers to form multiply bridged cyclophanes, as illustrated in network **B**. Three arene units are connected by three linkers in cyclic form **C**, or by four linkers to form multi-layer architecture **D**. Cyclic structure **E** with four arene units and four linkers is a typical ring-type network and called a simple graph. In contrast, every pair of arene units is connected in network **F** by a linker, and this is called a fully connected mesh or a complete graph.

The number of possible structures increases rapidly with increasing number of arene units. Networks **G** and **H** are examples of structures with six and eight arene units, respectively. In **G**, three units of the first set (top) are connected to all units of the other set (bottom), and this is called a complete bipartite graph. This graph is called utility graph $K_{3,3}$ and features nonplanarity (see Sect. 6.2). Each unit has three linkers in **H** to form a cubic motif, called a 3-regular graph in the graph theory.

A tritopic linker can connect three arene units. Network **I** consists of two tritopic linkers and three arene units. In network **J**, four arene units are connected by four tritopic linkers in all possible combinations. This graph is identical to tetrahedrane motifs **J**′ and **J**″. Arene units and linkers occupy corners and faces of a tetrahedron, respectively, in **J**′ and vice versa in **J**″.

2.5 Other Structural Features

One needs to consider shape, size, flexibility, and chirality in discussing the structural features of aromatic networks, which depend on the number, type, and combination of arene units and linkers. Shape-persistent networks have rigid frameworks and interesting properties, such as self-assembly or molecular recognition. In contrast, flexible networks are conformationally mobile and their observed symmetry depends on the time scale of conformational changes and the methods of observation. Large 3D networks tend to have a cavity within their frameworks. If the cavity is sufficiently large to accommodate foreign molecules, the inclusion phenomenon is expected. Chiral networks are fascinating from the viewpoint of stereochemistry [24]. Such structures can be constructed from achiral building units, where the network itself is inherently chiral. The resolution of chiral 3D network compounds should offer valuable information on the chiroptical properties of novel enantiomers. Another geometrical option for networks with macrocyclic rings is topological isomers [25, 26]. Catenanes and knots have nonplanar topologies, which have afforded new aspects in structural chemistry. These examples will be introduced in Sect. 6.

3 Cyclophanes with Saturated Carbon Linkers

Cyclophanes are a class of cyclic compounds consisting of one or more arene units and saturated or unsaturated linkers. These compounds have been extensively studied in aromatic chemistry because of their unusual structure, conformation, and transannular interactions. The chemistry of cyclophanes has been reviewed in several books and review articles [27–29]. When cyclophanes have multiple arene units, especially arenes larger than benzene, molecules tend to take cage-like structures to form 3D space. In this section we will introduce such cyclophanes with naphthalene or anthracene units and saturated carbon linkers as prototypes of 3D networks.

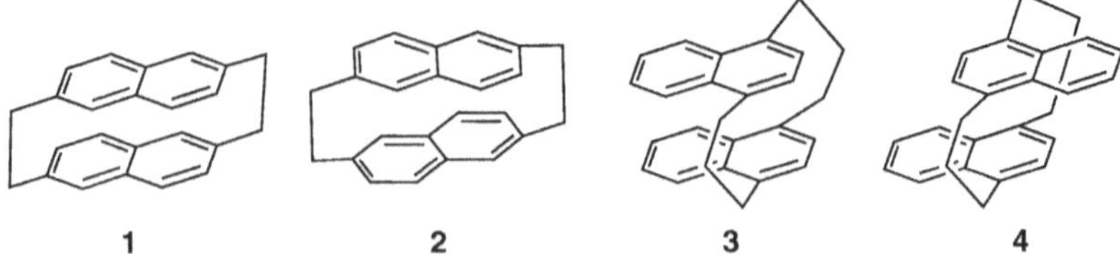

Scheme 3 Typical naphthalenophanes

3.1 Naphthalenophanes

[2.2]Naphthalenophanes (the chain length of each linker in brackets) have a cyclic framework consisting of two naphthalene units and two ethylene linkers. There are several possible structures depending on the bridging positions and the conformations. For example, [2.2](2,6)naphthalenophane (the substituted positions in parentheses) has two isomers **1** and **2** that differ in the direction of two 2,6-naphthylene units (Scheme 3) [30, 31]. Chiral isomer **2** was resolved by the formation of diastereomeric charge transfer complexes. Similarly, chiral and achiral isomers of [2.2](1,5)naphthalenophane were separated [32]. The two aromatic units are so closed that the molecules suffer from severe strains and have little flexibility. Therefore these compounds are good models to examine transannular interactions between naphthalene chromophores at various orientations. Analogous [3.3]naphthalenophanes with two 1,3-propylene linkers are less strained and more flexible than the corresponding [2.2]naphthalenophanes [33, 34]. Two isomers of [3.3](1,4)naphthalenophanes **3** and **4** underwent cycloaddition upon irradiation and these photoproducts reverted to the original compounds upon heating.

3.2 Anthracenophanes

Two anthracene units are linked firmly by two ethylene linkers in [2.2](9,10)anthracenophane **5**, which was readily synthesized from 9,10-bis(chloromethyl)anthracene (Scheme 4) [35]. A photochemical reaction of this compound gave a cyclization product via a biradical intermediate. Two isomers of [2.2](1,4)anthracenophane, **6** and **7**, were synthesized by Hofmann elimination [36, 37]. Upon irradiation at 374 nm, *syn* isomer **7** readily gave [4+4]cycloaddition product **8** at the 9,10-positions and this photoproduct was reverted to **7** by heating or irradiation at 254 nm. The UV spectra of the two isomers indicated that the bathochromic shift due to transannular interactions was more significant in *syn* **7** than *anti* **6** because the two anthracene units were fully overlapped in the former.

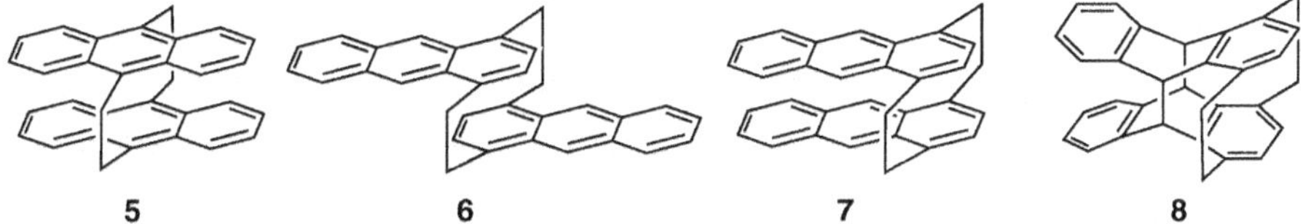

Scheme 4 Typical anthracenophanes and related compound

4 Aromatic Macrocycles with Acetylene Linkers

Macrocyclic compounds with arene units and linear acetylene linkers are novel types of π-conjugated compounds. The shape and conformational freedom of such networks strongly depend on the kind of arene units and the direction of linkers extending from arene units. Not only ethynylene linkers but also 1,3-butadiynylene linkers and longer polyyne linkers can be used for the acetylene linkers. This section shows examples of 3D macrocyclic compounds with various arene units and these linear linkers.

4.1 *Benzene Macrocycles*

For the molecular design of aromatic macrocycles, benzene units (*o*-, *m*-, and *p*-) are most common as aromatic moieties. For example, planar 2D systems can be constructed from six *m*-phenylene units having 120° angular orientation or three *o*-phenylene units having 60° angular orientation [38]. These structures should become nonplanar when the number of units is greater or smaller than the ideal number expected for the planar framework. Selected examples of such nonplanar phenylene macrocycles are introduced in this section.

Four *o*-phenylene units form nonplanar cyclic structures to satisfy the angular requirement. The fundamental compound is tetraphenylene, which has a nonplanar saddle like structure [39]. Several twisted compounds were designed by the insertion of linkers between phenylene units (Scheme 5). Compound **9** was reported in 1968 by Staab et al. [40, 41] and its structure was established by X-ray analysis [140, 141]. Compound **10** with two long linkers and two short linkers also had a helical structure and underwent enantiomerization between chiral helical structures at the barrier of 39 kJ/mol [42]. In the X-ray structure of **11** with longer linkers, two *p*-phenylene units were overlapped with a distance of 3.57 Å [43]. Six *o*-phenylene units form a highly twisted nonplanar structure of **12** [44]. Compounds **13** and **14** are oligomers having both *o*-phenylene and *m*-phenylene units [45]. The macrocyclization of a common precursor with three phenylene units gave two conformational isomers, **13** (C_2) and **14** (C_{2h}). Their macrocyclic frameworks are rigid and stable because no isomerization took place even at 100°C.

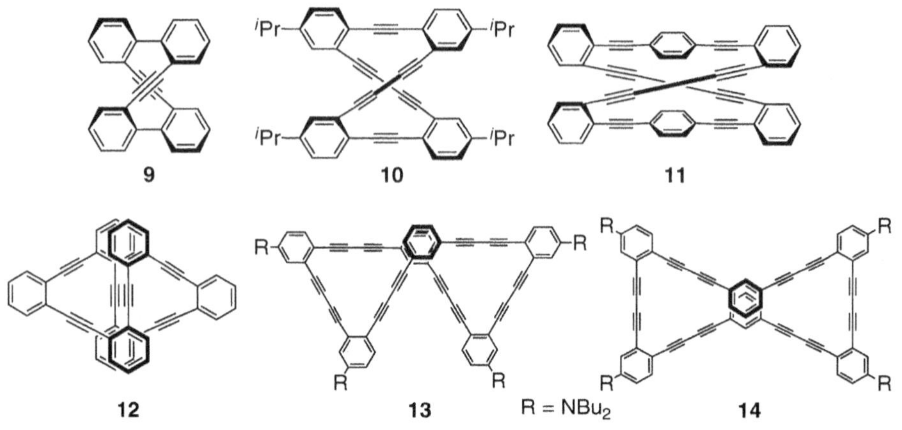

Scheme 5 Macrocycles with benzene units and linear acetylene linkers

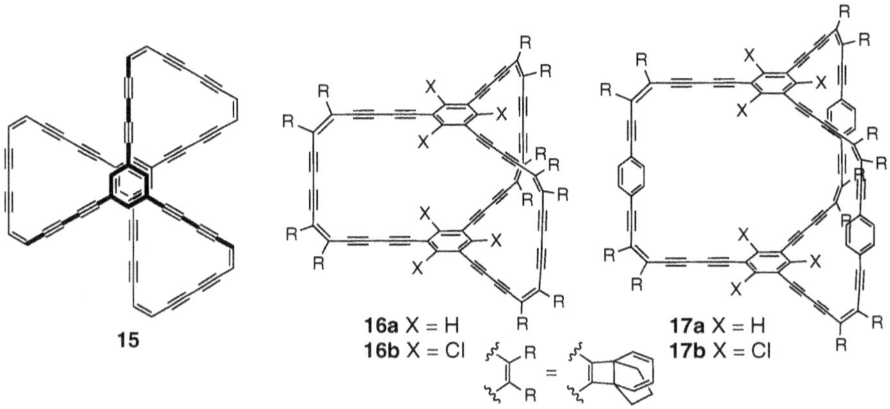

Scheme 6 Macrocycles with two benzene units and three long linkers as fullerene precursors

Compounds **15** and **16**, in which two benzene units are connected by three carbon 16-atom linkers, were designed for precursors of fullerene C_{60} (Scheme 6). These compounds were synthesized by dimerization of terminal alkynes by oxidative coupling. Macrocycle **15** has a twisted D_3 symmetric structure with the two benzene rings separated by 3.3 Å as revealed by X-ray analysis [46, 47]. The calculation suggested that racemization of **15** proceeded rapidly even at low temperature. Compounds **16** gave an ion peak of C_{60}^+ formed via extrusion of indane followed by polyyne cyclization and extrusion of hydrogen or chlorine atoms in their MALDI-TOF mass spectra [48–50]. Compounds **17** have an extra three *p*-phenylene units in the 3D framework [51]. The mass spectrum of chloro derivative **17b** gave an ion peak assignable to C_{78}^-.

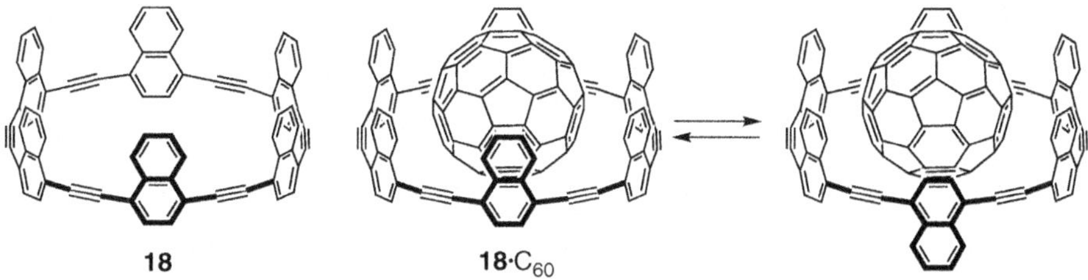

Scheme 7 Naphthalene macrocycle **18** and its C_{60} complex

4.2 *Naphthalene Macrocycles*

Kawase et al. reported various cyclic arylene–ethynylene oligomers as carbon nanorings showing novel structures and properties [52, 53]. Compound **18** is an example of six 1,4-naphthylene units (Scheme 7). This compound forms a complex with C_{60} more tightly than the corresponding phenylene derivative. The NMR spectra of **18**·C_{60} complex revealed the presence of two isomers at −90°C and the major isomer was assignable to the symmetric isomer with all fused benzene groups at the same side. It is noteworthy that the flipping of one naphthylene unit shown in Scheme 7 takes place with maintaining the complexation, as revealed by VT ^{13}C NMR spectra of a complex with ^{13}C-enriched C_{60}.

4.3 *Anthracene Macrocycles*

An anthracene unit possesses a rectangular-like shape and several possibilities of linkage. The mode of substitutions determines the distance and direction of the structural extension. Typical examples are 1,8-subsituted, 1,5-subsituted, and 9,10-subsituted anthracene (hereafter abbreviated as 1,8-A and so on) units with zigzag, crank, and linear connectivities, respectively [54, 55]. In terms of geometry and synthesis, the use of 1,8-A units is convenient for the construction of 3D structures.

Compound **19** with four 1,8-A units and four ethynylene linkers was synthesized by successive cross coupling reactions from 1,8-diiodoanthracene and 1,8-diethynylanthracene (Scheme 8) [56, 57]. X-ray analysis revealed that this tetramer had a 3D diamond-prism structure of D_2 symmetry. Two pairs of anthracene units take a parallel orientation at the interfacial distance of 3.38 Å, which is comparable to the sum of the van der Waals radii of two aromatic carbons (1.7 Å × 2). Tetramer **19** undergoes a dynamic process between the two diamond forms via rotation about the four acetylene linkers. The barrier to this process (38 kJ/mol) was determined by dynamic NMR spectroscopy. A relatively high barrier is attributed to the π···π interactions between the anthracene units stabilizing the original state.

This basic model was modified by two methods, the introduction of substituents on anthracene units or the incorporation of long butadiynylene linkers. Compounds **20** and **21** are such examples having chiral 3D structures [58, 59]. Tetramer **20** has

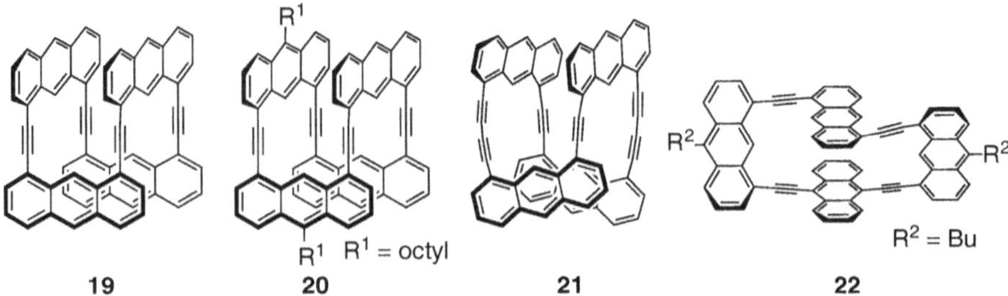

Scheme 8 Macrocycles with four 1,8-anthracene units

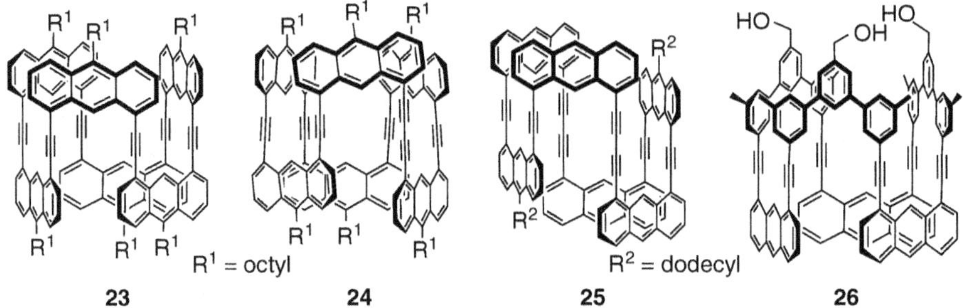

Scheme 9 Macrocycles with six 1,8-anthracene units and analog

two octyl groups at the 10-position of 1,2-alternating 1,8-A units, while tetramer **21** has two long linkers at the opposite corners. The enantiomers of these chiral compounds were resolved by chiral HPLC and showed characteristic CD bands in the UV–vis region. It is notable that spontaneous resolution took place upon crystallization of **21** from chloroform. Namely, each single crystal consisted of either of the enantiomers showing optical activity. Another chiral tetramer **22** consisting of two 1,8-A units, one 1,5-A unit, and one 9,10-A unit was also synthesized and its enantiomers were resolved [60, 61]. The kinetic measurements revealed that the barrier to enantiomerization via rotation about the acetylene axes was 114 kJ/mol at 70°C.

Similarly, cyclic hexamers were constructed from six 1,8-A units. The ^{1}H NMR spectrum of hexamer **23** was very symmetric on the NMR time scale (Scheme 9) [62]. In contrast, the DFT calculation of the alkyl-free derivative of **23** suggested a parallelogram prism structure as energy minimum to maximize intramolecular $\pi\cdots\pi$ stacking. These findings indicate that the molecules undergo facile conformational exchanges between possible parallelogram structures. Hexamer **24** with one long linker and hexamer **25** with two long linkers also prefer to take parallelogram-type structures [62, 63]. Hexamer **24** has a chiral structure, and its enantiomers were resolved by chiral HPLC. Hexameric compound **26**, in which three *m*-terphenyl units were incorporated, was proposed by the group of Sakamoto and Schülter

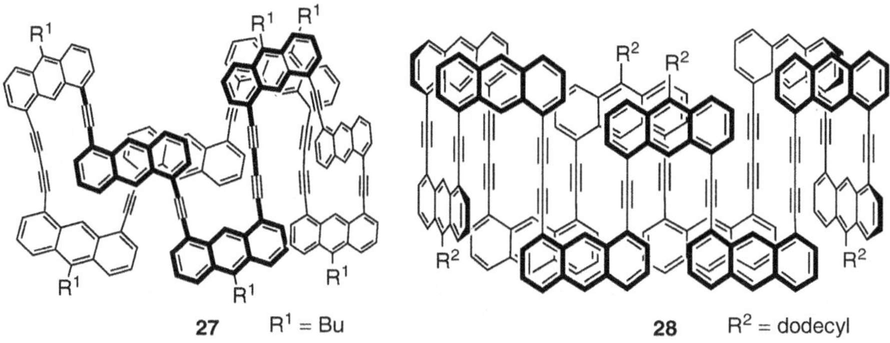

Scheme 10 Macrocycles with many 1,8-anthracene units

[64, 65]. The hydrophilic hydroxyl groups at the terphenyl side assisted the formation of a monolayer at the air–water interface. Recently, 1,3,5-benzenetricarboxylate triester derivative of **26** was utilized for the photochemical formation of a 2D monolayer polymer, where the cyclic structure was fixed into a trigonal prism shape by the extra benzene capping [66].

Larger 1,8-A oligomers were synthesized in a similar manner. Nonamer **27** has six 1,8-A units and three 1,5-units (Scheme 10) [67]. The X-ray analysis revealed a flat and slightly bent structure with two pairs of partially stacked anthracene units. Dodecamer **28** with 12 1,8-A units was obtained with hexamer **25** on cyclization of a hexameric precursor [63]. This macrocycle tends to take highly folded structures according to the DFT calculation. The most stable structure has some internally folded units to yield two pairs of four-layered stackings. The VT ^{1}H NMR spectra showed slow interconversion between folded conformations at low temperature. These compounds offer a new motif of belt-shaped molecules with anthracene units.

Macrocyclic compounds **29–31** have three 9,10-A units in the nearly planar triangle, circle, and hexagon frameworks, respectively (Scheme 11). In **29** and **30**, 9,10-A units are twisted from the macrocyclic plane due to steric hindrance to yield nonplanar structures [68, 69]. It is interesting that compound **29** showed a liquid-crystal phase at high temperature. Because compound **31** has a large macrocyclic framework, three 9,10-A units are coplanar to the macrocyclic plane in the global minimum structure [70]. However, the rotational barrier is so low that this molecule can take nonplanar conformations such as a tube-like shape by rotation of the three units.

5 Cage Type Aromatic Networks

Cage compounds have 3D polycyclic frameworks resembling a cage in structure. When aromatic units are incorporated into the cage structure, these planar units have a function of walls or panels to separate inner and outer space. A large cage

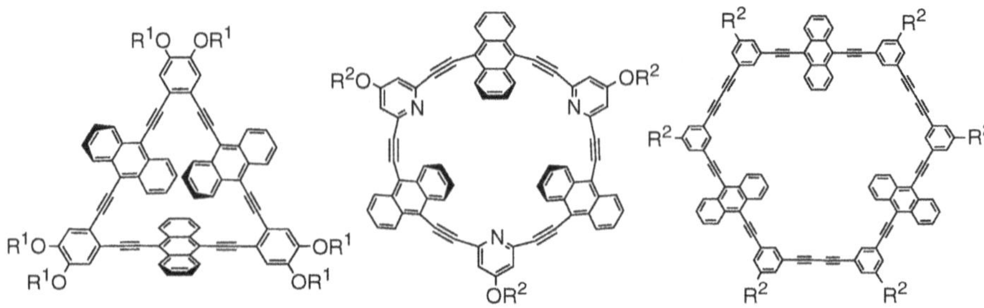

Scheme 11 Macrocycles with 9,10-anthracene units

compound having an enclosed cavity within the network often includes one or more guest species (molecules, ions, etc.). Weak interactions between aromatic walls and guest species as well as host–guest compatibility in size and shape play important roles in forming inclusion complexes.

5.1 Cage Networks with Covalent Linkers

Compound **32**, called spheriphane, was synthesized by macrocyclization of a 1,3,5-tris(2-phenylethyl)benzene precursor by Vögtle et al. (Scheme 12) [71]. This compound has a spherical framework, where four benzene units are clamped by six ethylene linkers. The radius of the intramolecular cavity is 0.28 Å, and this space is sufficient for the inclusion of small metal ions. Several spheriphane analogs have been studies as ion receptors and fullerene related compounds. Compounds **33** and **34** are larger spheriphane derivatives with an extra three benzene units [72, 73]. Compound **34** extracted Ag^+ ion from an aqueous solution much more efficiently than **32** and **33**. This remarkable selectivity of **34** was applied to the incorporation in ion-selective electrodes for Ag^+ and Tl^+ ions. In annelated spheriphane **35**, three ethylene linkers are annelated by benzo groups [74]. The fully annelated derivative with ten benzene rings ($C_{60}H_{36}$) should be a potential precursor to C_{60}. Compound **36** can be regarded as spheriphane with three-atom linkers involving heteroatoms [75]. This compound recognized NH_4^+ and K^+ ions under the conditions of positive ESI mass spectroscopy.

Cage-type compounds based on a 1,3,5-tribenzylbenzene subunit were reported by Kim et al. (Scheme 13) [76]. Compounds **37** and **38** have prismatic frameworks with two cap benzene units and three wall benzene units connected by $–CH_2–$ and $–O–CH_2–$ linkers. ESI-mass measurements revealed that these compounds formed a 1:1 complex with various metal ions and NH_4^+. In the presence of Li^+, Na^+, K^+, and NH_4^+, **37** with *m*-phenylene walls binds Li^+ selectively, while **38** with *p*-phenylene wall units binds both Li^+ and NH_4^+. Prismatic cage compound **39** was constructed from enantiopure cap benzene units and three wall benzene units via

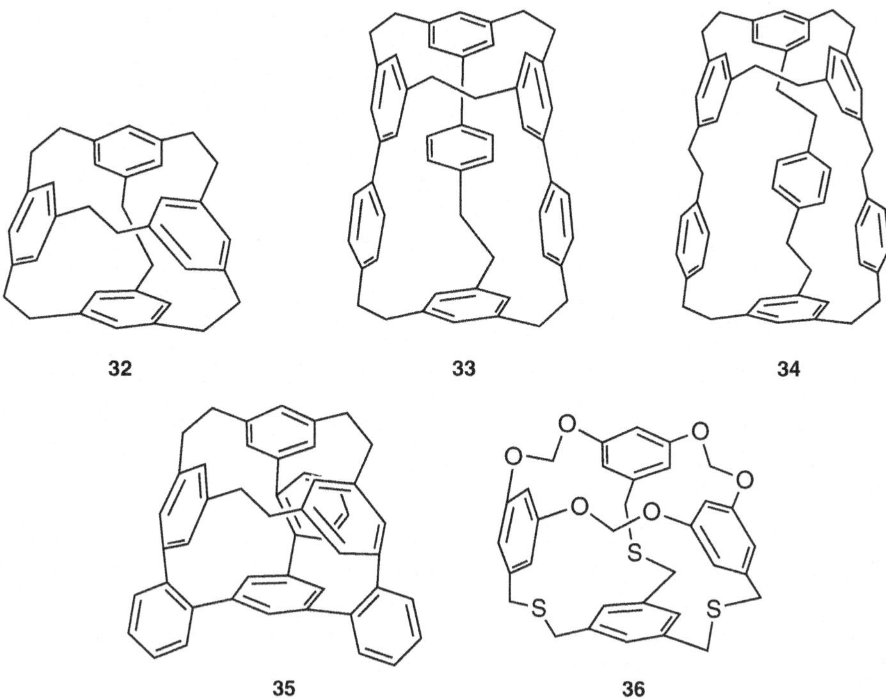

Scheme 12 Spheriphane cage networks with benzene units

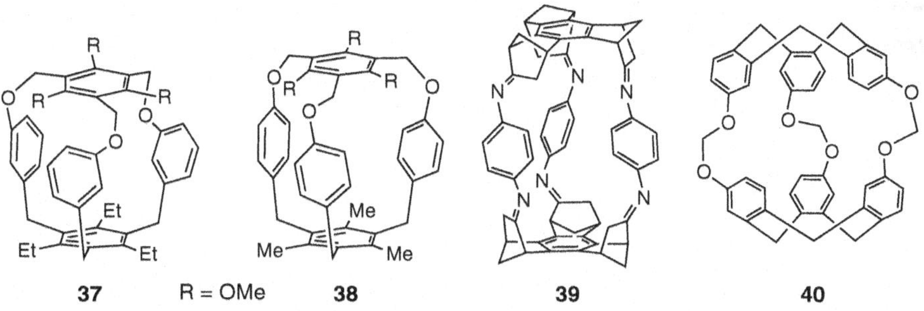

Scheme 13 Prismatic and spherical cage networks with benzene units

imine bonds [77]. In cryptophane-1.1.1 (**40**), two [1.1.1]orthocyclophane units are bridged by three –O–CH_2–O– linkers [78]. Cryptophane derivatives have an inner cavity surrounded by six benzene units and encapsulate various small chemical species. For example, encapsulation of inert guest molecules, Xe, methane, and H_2 in **40** was confirmed by spectroscopic methods [79, 80]. The shape and size of cryptophane 3D networks can be tuned by the length of linkers and the substituents on benzene rings leading to versatile molecular recognition.

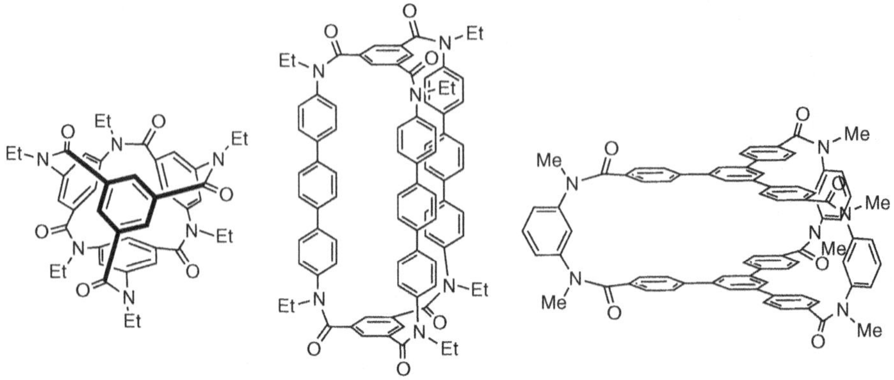

41 **42** **43**

Scheme 14 Cage networks with benzene units and amide linkers

Spherical and prismatic networks can be constructed by amide linkers (Scheme 14). Spherical network **41** consisting of four benzene units and six amide linkers was synthesized by cyclization between a cyclic triamide and 1,3,5-benzenetricaboxylic acid [81]. This molecule is chiral depending on the direction of unsymmetrical amide linkers. The enantiomers were resolved by chiral HPLC. Various prismatic structures can be constructed by formation of amide linkages between a linear diamine and a triangular tricaboxylic acid in a 3:2 ratio. The geometrical shapes are controllable by the length of wall diamine units and the size of cap tricarboxylic acids. Compound **42** with long *p*-terphenyl units takes a helically twisted triangular prism structure due to C–H⋯π interactions [82, 83] between the wall *p*-phenylene units [84]. In contrast, when large 1,3,5-triphenylphenyl units and short *m*-phenylene units are combined, a flat prismatic structure is formed as compound **43** [85].

Garcia-Garibay et al. designed molecular gyroscopes consisting of a 1,4-bis (tritylethynyl)benzene framework where a *p*-phenylene rotor rotated about acetylene axle attaching to two terminal trityl stators [86, 87]. The introduction of three bridges between phenyl groups across the central axle is a rational modification toward facile rotation of the central rotor in the solid state. Triply-bridged derivative **44** with benzophenone-derived chains was synthesized by macrocyclization (Scheme 15) [88]. The central phenylene group is surrounded by the spherical cage with 12 benzene units.

A triptycene has rigid bicyclic structure, where three benzene units are connected by two bridgehead carbons. This structural feature has been utilized to construct 3D molecular cages [89]. Compound **45** was synthesized by Eglinton–Glaser coupling of 2,7,14-tris[(3-ethynylphenyl)ethynyl]triptycene [90]. Its NMR spectra were consistent with the D_{3h} structure, whereas the X-ray structure was slightly twisted from the regular D_{3h} structure. The distance between the inner bridgehead carbons in the two triptycene units reaches 13.0 Å. In the solid state, the

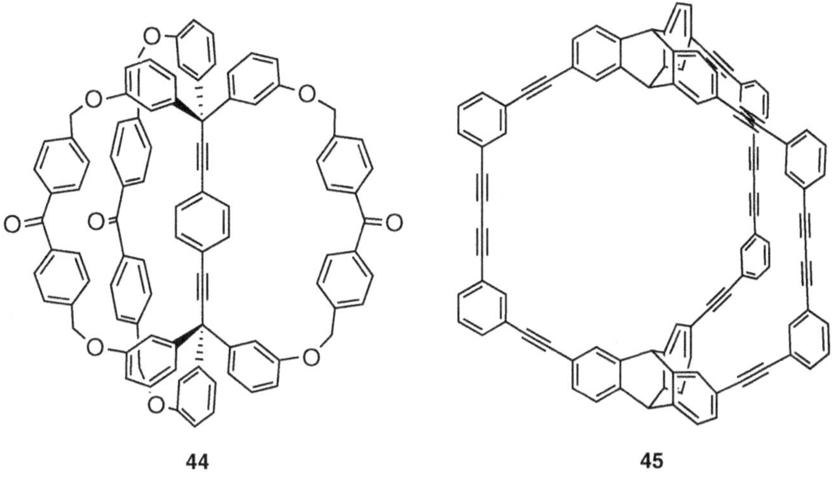

Scheme 15 Macrocyclic cage networks with benzene units and acetylene linkers

cage molecules have a microporous packing. Triptycene units were also incorporated into calixarene macrocycles, and some derivatives could accommodate fullerenes C_{60} and C_{70} in their expanded cavity [91, 92].

5.2 Cage Networks with Coordination Linkers

A variety of 3D cage networks can be constructed by spontaneous self-assembly of rationally designed multitopic ligands with metal ions. Fujita et al. reported the selective formation of octahedron-shaped M_6L_4 complex **46** (M: metal, L: ligand) from tris(4-pyridyl)triazine **47** and *cis*-coordinated Pd(II) complex in water (Scheme 16) [93]. The framework of **46** is ca. 20 Å in diameter and this large cavity accommodates various guest molecules. The cavity size in the octahedral framework was increased by using large ligands with extra linkers between the 4-pyridyl groups and the triazine core. The shape and size of 3D networks can be modified by designing the ligand in terms of orientation, direction, and number of coordination sites [94]. According to these empirical predictions, a giant $M_{24}L_{48}$ polyhedral cage was synthesized from 2,5-bis(4-pyridyl)pyrrole ligand [95]. The cage cavities of these coordination networks are applicable to containers of guest molecules, reaction vessels, and generation of novel molecular species [96–98].

The combination of multitopic ligands and Pt(II)-containing molecules spontaneously generates various 3D cages depending on the geometry of building units [21]. The treatment of anthracene clip **49** with terminal *trans*-coordinated Pt(II) atoms and planar tritopic ligand **50** gave 3D cage network **48** with a trigonal prismatic frameworks (Scheme 17) [99]. The formation of molecular prisms was confirmed by ESI mass spectra. The use of a tripod donor instead of the trigonal

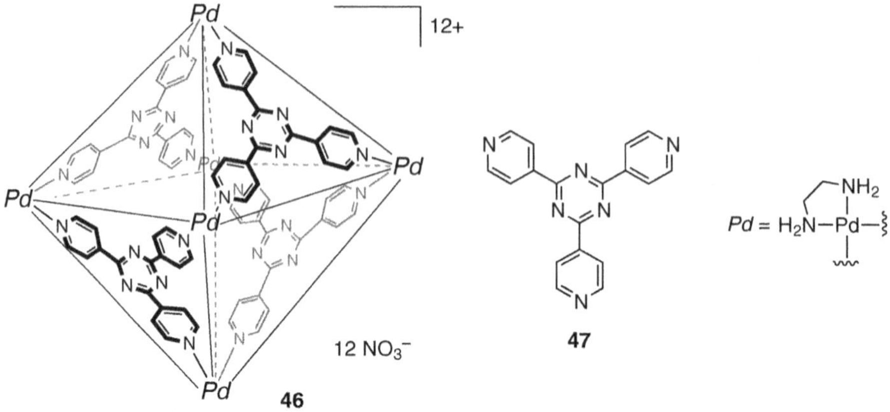

Scheme 16 Cage network with pyridine ligands and Pd metals

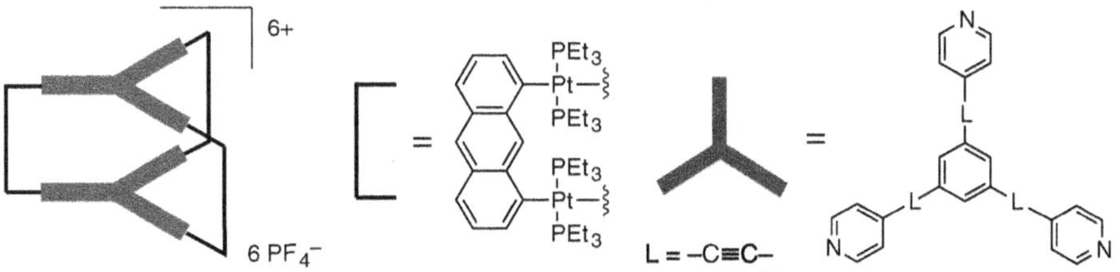

Scheme 17 Prismatic cage networks with anthracene clips and pyridine ligands

donors afforded various polyhedral 3D cages [100]. In most cases the shape of 3D architectures is predictable by the combination of donor and acceptor subunits according to the design principle called "directional bonding approach" [21].

Raymond et al. proposed a molecular design based on bis-catechol ligands and metal ions for the construction of 3D networks (Scheme 18) [101]. Triple helicates were synthesized from three ligands and two octahedral metals. For example, complex **51** consists of Ga(III) metals and ligand **52** to form an M_2L_4 network, which has a helical structure rather than a regular prismatic structure [102]. Tetrahedral M_4L_6 complex **53** was similarly constructed from ligand **54** with a naphthalene core and Ga(III) or Fe(III) metals, where metals and ligands occupied corners and edges of a tetrahedron, respectively [103]. These tetrahedral complexes can encapsulate various chemical species such as tetraalkylammonium ions and cationic Ru complexes. When the tetrahedral Ga_4L_6 network encapsulated CpRu(*p*-isopropyltoluene) cation, the signals due to the two isopropyl-methyl groups in the guest molecule were observed nonequivalently in the ^{1}H NMR spectrum. This diastereotopic environment results from the chiral tetrahedral framework.

Molecular capsules with large anthracene panels were designed by Yoshizawa et al. Bisanthracene ligand **56a** was treated with $Pd(NO_3)_2$ in DMSO in a 2:1 ratio to

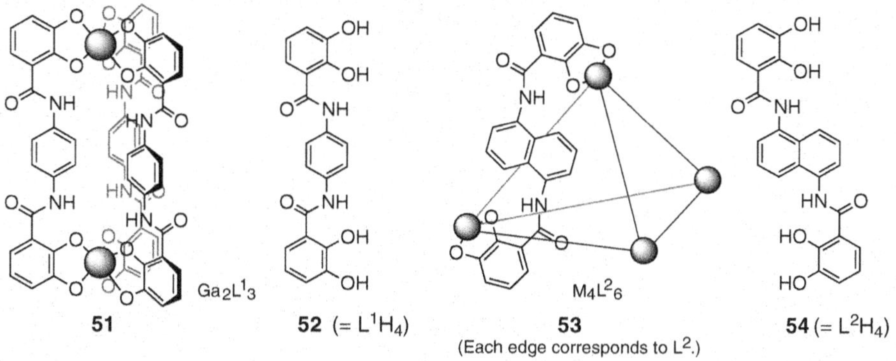

51 $Ga_2L^1_3$ **52** (= L^1H_4) **53** $M_4L^2_6$ (Each edge corresponds to L^2.) **54** (= L^2H_4)

Scheme 18 Cage networks with bis-catechol ligands and metal ions (*gray circles*)

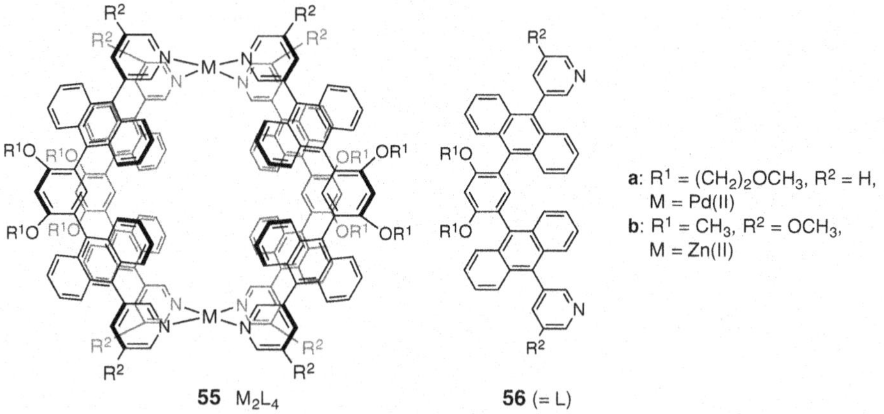

55 M_2L_4 **56** (= L)

a: $R^1 = (CH_2)_2OCH_3$, $R^2 = H$, M = Pd(II)
b: $R^1 = CH_3$, $R^2 = OCH_3$, M = Zn(II)

Scheme 19 Molecular capsules with ligands containing anthracene panels

give self-assembled M_2L_4 molecular capsule **55a** (Scheme 19) [104]. This capsule has a large cavity with a diameter of 10 Å surrounded by large aromatic shells and can encapsulate large molecules such as [2.2]paracyclophane, 1-methylpyrene, and C_{60} as revealed by NMR and mass spectroscopy. The similar M_2L_4 capsule **55b** was also synthesized from ligand **56b** and Zn(II) ions [105]. This complex is highly emissive (Φ_f 0.81) relative to the corresponding Ni(II) and Pd(II) complexes.

Michl et al. proposed prismatic 3D networks consisting of linear rods and star-shaped connectors linked by Pt(II) metals at the corners (Scheme 20) [106]. For example, treatment of biphenyl rods and 1,3,5-tris(4-pyridylethynyl)phenyl caps gave trigonal prism complex **57a** [107]. MM calculation suggested that its framework was slightly twisted from the regular trigonal prism to form a chiral structure. The distance from the prism center to the central CH_2 moiety in the dppp moiety is 15.2 Å. Complex **57b** possesses three 9,10-diethynyltriptycene moieties at the rod

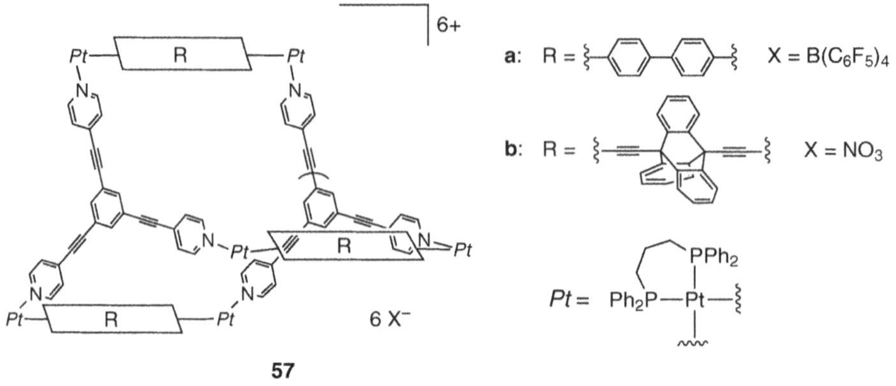

Scheme 20 Cage networks with trigonal prism frameworks

positions [108]. This molecular cage furnished an inner space sufficient for the rotation of the triptycene paddle wheels. Actually, no restricted rotation was observed even at −90°C by NMR spectroscopy.

5.3 Cage Networks with Hydrogen Bonds

Although a large number of aromatic cage networks assembled by hydrogen bonds, such as carcerands and calixarenes, have been explored [26, 109], we introduce resorcinarene capsules as a typical example in this section [110, 111]. Compound **58** is a bowl-shaped cavitand consisting of a calix[4]resorcinarene core and four peripheral imide moieties (Scheme 21). Two cavitand molecules form cylindrical capsule **59** via hydrogen bond networks at the imide moieties. A wide range of cylindrical molecules can be encapsulated within the cavity. For long linear alkanes a single guest molecule enters into the capsule and tends to take a helical conformation rather than an extended conformation to fit into the cavity [112, 113]. The capsule size can be extended by the insertion of glycoluril spaces between the two cavitand molecules, leading to encapsulation of longer alkanes and other long molecules [113–115]. Two resorcinarene cavitands can be held together by ordinary covalent linkers, boronic ester linkers, and metal-ligand coordination linkers [116, 117].

6 Topologically Fascinating Aromatic Networks

Topological isomers refer to isomers with the same connectivities but different topologies. Typical examples are catenanes, knots, and rotaxanes, in which molecules are interlocked by mechanical bonds rather than ordinary chemical

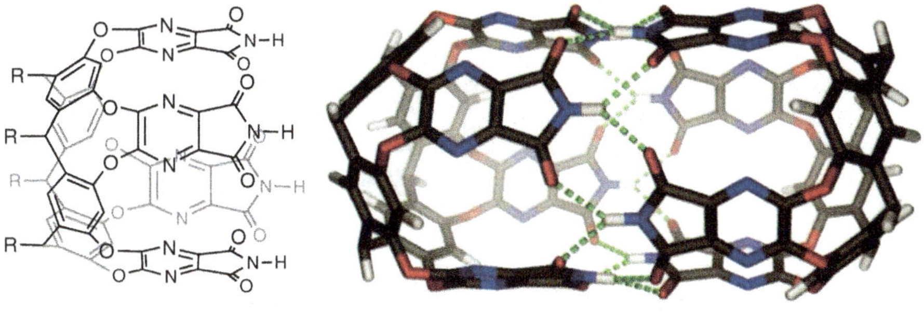

58 R = $(CH_2)_{10}CH_3$ **59** (= 2·**58**)

Scheme 21 Resorcinarene cavitand **58** and dimeric capsule **59** (partly adapted from [113])

bonds. These molecules are fascinating in terms of not only synthetic challenges but also application to functional materials due to their unique geometry [118–123]. Because catenanes and knots have at least one crossing of the ring chain, these molecules have nonplanar topologies. For the selective synthesis of such compounds, aromatic units have often been utilized as rigid building blocks to regulate structural flexibility. In addition, the interactions of aromatic units with other aromatic or nonaromatic units play important roles in preorganizing cyclization steps via π···π, C–H···π, and cation···π interactions, as well as other weak interactions. In this section we will introduce some examples of catenanes and topologically fascinating networks mainly from recent examples.

6.1 Catenanes

Recent advances in chemistry of catenanes, in which two or more macrocycles are mechanically interlocked, are attributed to effective synthesis by intelligent approaches. Template synthesis is the most promising approach, where two or more fragments are organized before cyclization to form catenanes. Such preorganization is occasionally performed by metal coordination [124] or aromatic donor–acceptor interactions [125]. In the former approach, aromatic multidentate ligands such as bipyridines, terpyridines, and phenanthrolines are preferably used to fix the geometry. In the latter approach, for example, interactions between electron deficient and electron rich aromatic units can effectively preorganize building units.

It is interesting that [2]catenane **61** (the number of ring components in brackets) is formed by self-assembly of two isolated rings **60** consisting of two *p*-bis(4-pyridylmethyl)benzene ligands and two Pt(II) ions (Scheme 22) [126]. These methods enabled us to construct not only simple [2]catenanes but also very complicated catenanes with more than two rings or high ordered crossings [127]. Recently, template dynamic synthesis of a [3]catenane was reported by Sanders et al. (Scheme 23) [128]. Two kinds of acyclic units, **63** with two acceptor units

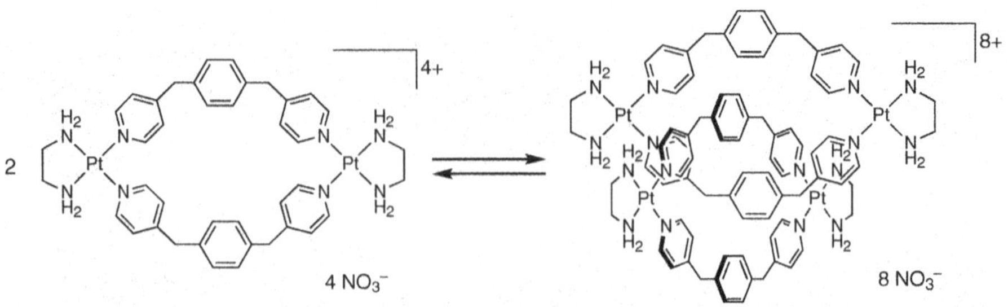

Scheme 22 Equilibrium between monomer ring **60** and catenane **61**

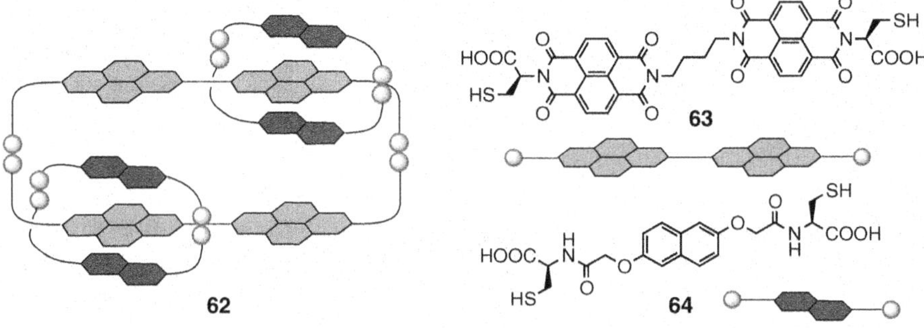

Scheme 23 Template dynamic synthesis of [3]catenane

(naphthalenediimides) and **64** with one donor unit (2,6-dimethoxynaphthalene), were mixed in a high-polarity medium or in the presence of spermine. Then spontaneous self-assembly and cyclization via disulfide bond formation gave [3]catenane **62**. This example provides an alternative route to the efficient synthesis of various catenanes.

Doubly interlocked [2]catenanes, Solomon links, are fascinating synthetic targets because their synthesis is more difficult than that of simple [2]catenanes (Hopf links) [129]. A Solomon link has four crossings, whereas a Hopf link has two crossings. In 1999, Fujita and Sauvage's group jointly reported formation of a Solomon link by combination of metal-template and self-assembly strategies (Scheme 24) [130]. When ligand **66** was treated with Cu(I) and Pt(II) ions, a doubly interlocking [2]catenane **65** formed quantitatively. Covalently linked Solomon link **69** was readily prepared by mixing diaminobipyridine **67** and diformylpyridine **68** with a 1:1 amount of Zn(II)/Cu(II) (Scheme 25) [129]. The linked structure was confirmed by X-ray analysis, and each link **70** contains four benzene units and six pyridine units. A new type of molecular Solomon link was constructed by self-assembly of three units, two bipyridinium units and one macrocycle, with Pd(II) or Pt(II) ions [131]. It is notable that $\pi \cdots \pi$ interactions, hydrogen bonds, and metal coordination cooperatively lead to an unusual cyclization mode.

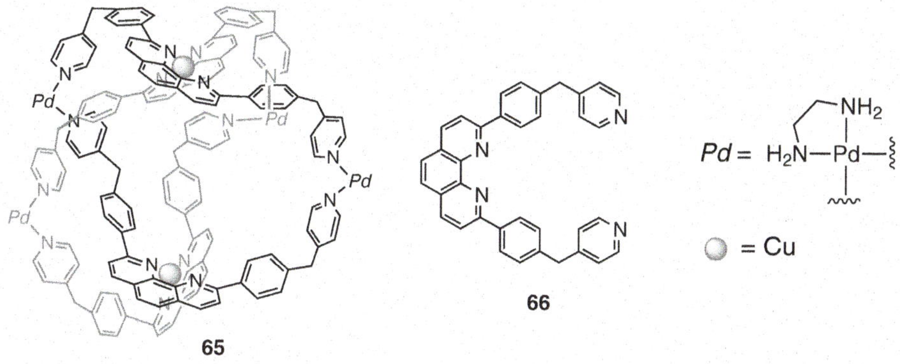

Scheme 24 Metal-template and self-assembly synthesis of Solomon link

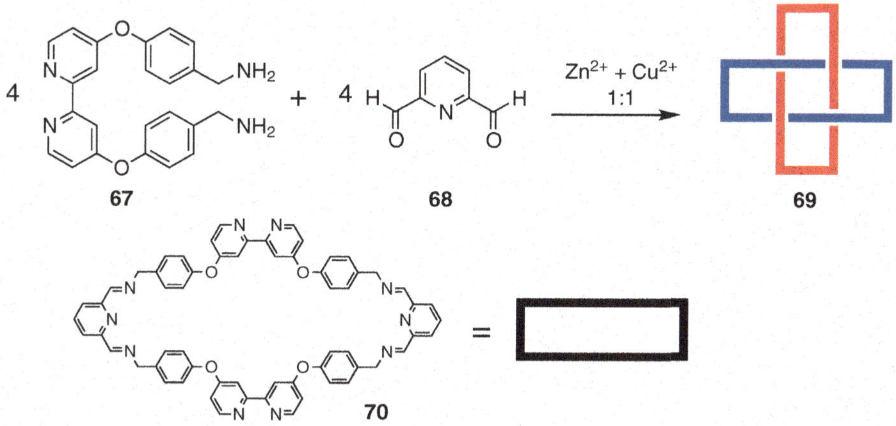

Scheme 25 One-pot synthesis of Solomon link network

Triply interlocked catenanes provide another interesting topological motif, where two 3D cage frameworks are threaded firmly. Catenane **71** was spontaneously formed by the treatment of two kinds of pyridine components **47** and **72** with metal ions (Scheme 26) [132]. A similar interlocked network was constructed from **47**, long pillars, and Pd(II) metals [133]. The interlocked cages establish three compartments, and a triphenylene guest is accommodated in each compartment to form seven-layered aromatic stacks. Triply interlocked catenane **73** was also formed by self-assembly of four tripod-shaped bipyridine ligand **74** and four metal ions (Scheme 27) [134]. Within each catenane cage, all six-chelated Zn(II) centers have the same configuration to have a chiral cage network.

Recently, dimeric interpenetrated cage **75** was obtained by reaction of pyridine ligand **77** with $[Pd(CH_3CN)_4](BF_4)_2$ (Scheme 28) [135]. This system consists of eight ligands and four metal ions and encapsulates one BF_4^- ion between the two inner Pd metals. Upon addition of a halide ion, this cage includes two ions to form a

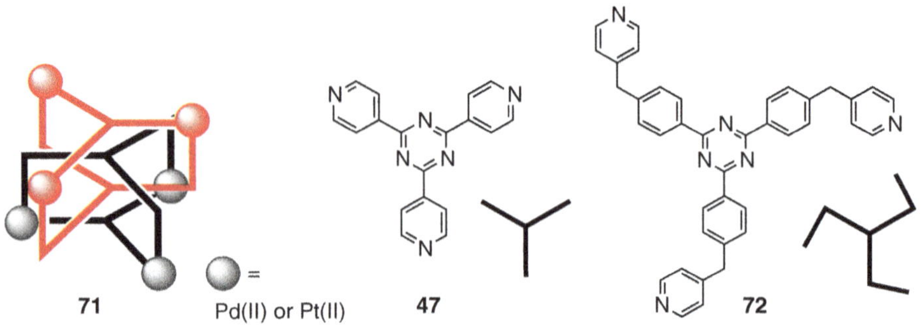

Scheme 26 Triply interlocked catenane network

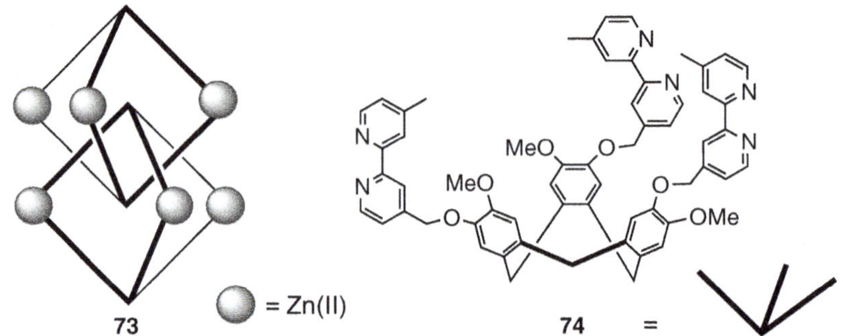

Scheme 27 Triply interlocked chiral catenane network

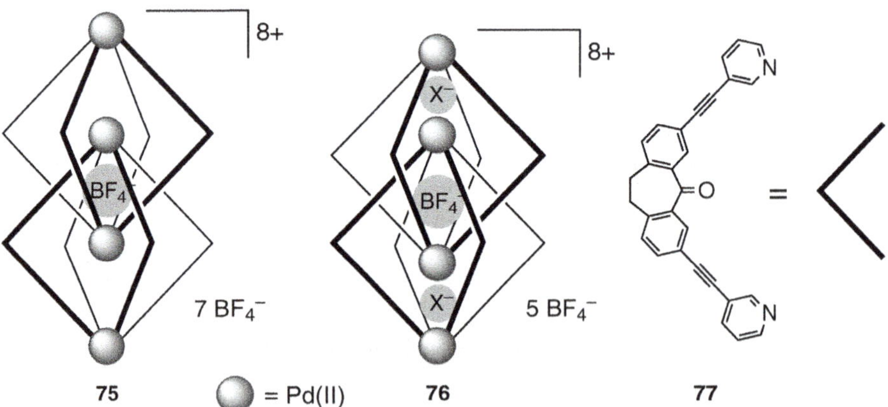

Scheme 28 Dimeric interpenetrated coordination cage

more interpenetrated cage shown as **76**. The high affinity of this cage system toward halide ions is attributed to allosteric binding involving metal–anion interactions and $\pi \cdots \pi$ interactions.

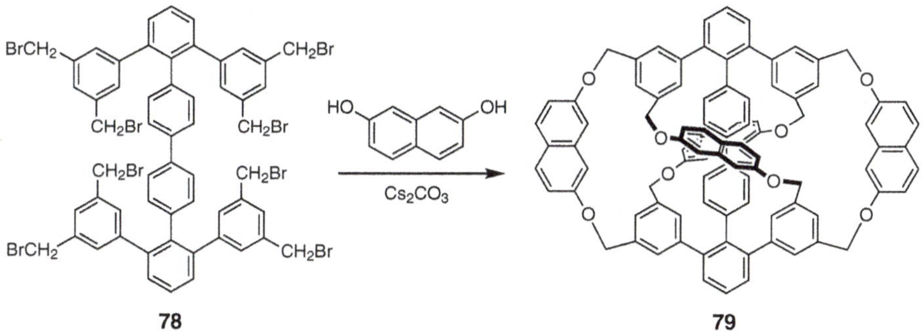

Scheme 29 Formation of Kuratowski-type aromatic network

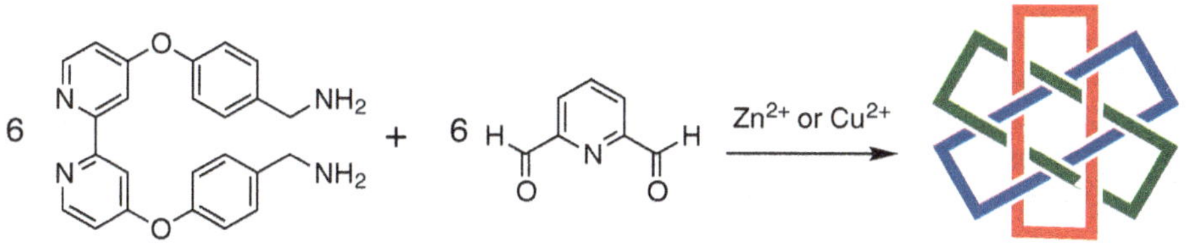

Scheme 30 One-pot synthesis of Borromean link network (see also Scheme 25)

6.2 Other Topological Isomers

A few named topologically interesting molecules are introduced [120]. Multicyclic compounds called Kuratowski cyclophanes have a nonplanar graph, $K_{3,3}$. This graph contains two sets of three points (units) and each in one set is connected to every point in the other set, as illustrated in **G** in Scheme 2. This topology was realized by cyclization of compound **78** and 2,7-dihydroxynaphthalene (Scheme 29) [136]. One of the products was found to be Kuratowski-type cyclophane **79** of D_{2d} symmetry.

A Borromean link consists of three rings, where any of two rings are not catenated to each other but the ring system is held together by the presence of the third ring [137]. The construction of this complicated ring system would require intelligent preorganization. One of the typical strategies is “ring-in-ring strategy,” where a two-ring system with one ring threaded into another ring is used as a template for the third ring formation. For example, Siegel et al. proposed a synthetic approach with polypyridine templates, where diphenylterpyridine units worked as template coordination sites as well as spacers to form a large internal space [138]. Regardless of such high-ordered topology, a molecular Borromean link could be constructed by a one-pot process. Two precursors **67** and **68** used for the Solomon link synthesis were treated with either Zn(II) ions or Cu(II) ions to form Borromean link system **80** in high yield (Scheme 30) [139]. The interlocked structure involving three rings and six metal ions at the coordination sites was unambiguously confirmed by X-ray analysis.

7 Summary and Outlook

The above examples show that aromatic units are attractive building blocks for the construction of various 3D networks. The size, shape, flexibility, and function of 3D aromatic networks can be controlled by the geometry, number, and connection site of aromatic units in addition to the length, direction, and freedom of linker moieties. Small aromatic networks are good models for the evaluation of transannular interactions involving $\pi\cdots\pi$ and C–H$\cdots\pi$ interactions between aromatic units, and occasionally include small metal ions in their cavities to increase cation$\cdots\pi$ interactions as observed for spheriphane derivatives. Typical large aromatic networks are cage compounds having shape-persistent 3D frameworks. Such compounds tend to include various guest molecules in inner cavities. The inner space is also applicable to the threading of other frameworks to construct topologically fascinating interlocked molecular systems. Traditionally, complicated networks have been synthesized by cyclization of acyclic precursors via several steps and the overall yield was not always high. Nowadays there are several examples of self-assembled or spontaneous formation of complicated networks from simple building units. These intelligent approaches to 3D networks depend on principles based on the geometry of aromatic units and linkers. Further optimization of the principle and conditions will enable the construction of very complicated systems including high-ordered topological isomers. Some 3D networks that are structurally related to fullerenes and carbon nanotubes are potential intermediates for the total synthesis of such all-carbon compounds. Because of these structural and electronic features, novel 3D aromatic networks will be attractive research targets in organic chemistry and related areas including supramolecular chemistry and material chemistry.

References

1. Hopf H (2000) Classics in hydrocarbon chemistry. Wiley-VCH, Weinheim (Chap. 12)
2. Klauk H (ed) (2006) Organic electronics: materials, manufacturing, and applications. Wiley-VCH, Weinheim
3. Müllen K, Scherf U (eds) (2006) Organic light emitting devices: synthesis, properties and applications. Wiley-VCH, Weinheim
4. Balzani V, Credi A, Venturi M (eds) (2008) Molecular devices and machines: concepts and perspectives for the nanoworld. Wiley-VCH, Weinheim
5. Müller TJ, Bunz UHF (eds) (2007) Functional organic materials: syntheses, strategies and applications. Wiley-VCH, Weinheim
6. Astruc D (ed) (2002) Modern arene chemistry. Wiley-VCH, Weinheim
7. Harvey RG (1997) Polycyclic aromatic hydrocarbons. Wiley-VCH, New York
8. Gutsche CD (ed) (2008) Calixarenes: an introduction, 2nd edn. RSC, Cambridge
9. Diederich F, Stang PJ, Tykwinski RR (eds) (2005) Acetylene chemistry. Chemistry, biology and material science. Wiley-VCH, Weinheim
10. Hopf H (2000) Classics in hydrocarbon chemistry. Wiley-VCH, Weinheim (Chap. 15.2)
11. Toyota S (2010) Chem Rev 110:5398

12. Ghosh S, Mukherjee PS (2006) J Org Chem 71:8412
13. Sengupta S, Sadhukhan SK (1998) Tetrahedron Lett 39:1237
14. Turner DR, Pastor A, Alajarin M, Steed JW (2004) Struct Bond 108:97
15. Hardie MJ (2004) Struct Bond 111:139
16. Negishi E (ed) (2002) Handbook of organopalladium chemistry for organic synthesis. Wiley, New York
17. Ackermann L (ed) (2009) Modern arylation methods. Wiley-VCH, Weinheim
18. Ma S (ed) (2010) Handbook of cyclization reactions. Wiley-VCH, Weinheim (Chaps. 11 and 12)
19. Frank D, Higson S (2011) Macrocycles: construction, chemistry and nanotechnology applications. Wiley, Chichester
20. Paker D (1996) Macrocycle synthesis: a practical approach. Oxford University Press, Oxford
21. Chakrabarty R, Mukherjee PS, Stang PJ (2011) Chem Rev 111:6810
22. Stang PJ (2009) J Org Chem 74:2
23. Flapan E (2000) When topology meets chemistry. Cambridge University Press, Cambridge (Chap. 5)
24. Mislow K (1999) Top Stereochem 22:1
25. Wolf C (2008) Dynamic stereochemistry of chiral compounds. The Royal Society of Chemistry, Cambridge (Chap. 9)
26. Wisner JA, Blight BA (2008) In: Diederich F, Stang PJ, Tykwinski RR (eds) Modern supramolecular chemistry: strategies for macrocycle synthesis. Wiley-VCH, Weinheim (Chap. 10)
27. Davis F, Higson S (2011) Macrocycles: construction, chemistry, and nanotechnology applications, vol 2. Wiley, Chichester
28. Vögtle F (1993) Cyclophane chemistry synthesis, structures and reactions. Wiley, Chichester
29. Gleiter R, Hopf H (eds) (2004) Modern cyclophane chemistry. Wiley-VCH, Weinheim
30. Haenel M, Staab HA (1973) Chem Ber 106:2203
31. Blank NE, Haenel MW (1983) Chem Ber 116:827
32. Haenel MW (1978) Chem Ber 111:1789
33. Kawabata T, Shinmyozu T, Inazu T, Yoshino T (1979) Chem Lett 8:315
34. Blank NE, Haenel MW (1981) Chem Ber 114:1520
35. Golden JH (1961) J Chem Soc 844
36. Toyoda T, Misumi S (1978) Tetrahedron Lett 19:1479
37. Iwama A, Toyoda T, Yoshida M, Otsubo T, Sakata Y, Misumi S (1978) Bull Chem Soc Jpn 51:2988
38. Haley MM, Pak JJ, Brand SC (1999) Top Curr Chem 201:82
39. Huang H, Stewart T, Gutmann M, Ohhara T, Niimura N, Li Y-X, Wen J-F, Bau R, Wong HNC (2009) J Org Chem 74:359
40. Staab HA, Wehinger E (1968) Angew Chem Int Ed Engl 7:225
41. Staab HA, Wehinger E, Thorwart W (1972) Chem Ber 105:2290
42. Boese R, Matzger AJ, Vollhardt KPC (1997) J Am Chem Soc 119:2052
43. Collins SK, Yap GPA, Fallis AG (2000) Org Lett 2:3189
44. Solooki D, Bradshaw JD, Tessier CA, Youngs WJ, See RF, Churchill M, Ferrara JD (1994) J Organomet Chem 470:231
45. Heuft MA, Collins SK, Fallis AG (2003) Org Lett 5:1911
46. Rubin Y (1997) Chem Eur J 3:1009
47. Rubin Y, Parker TC, Khan SI, Holliman CL, McElvany SW (1996) J Am Chem Soc 118:5308
48. Umeda R, Sonoda M, Wakabayashi T, Tobe Y (2005) Chem Lett 34:1574
49. Tobe Y, Nakagawa N, Naemura K, Wakabayashi T, Shida T, Achiba Y (1998) J Am Chem Soc 120:4544
50. Tobe Y, Nakagawa N, Kishi J, Sonoda M, Naemura K, Wakabayashi T, Shida T, Achiba Y (2001) Tetrahedron 57:3629
51. Tobe Y, Umeda R, Sonoda M, Wakabayashi T (2005) Chem Eur J 11:1603

52. Kawase T (2007) Synlett 2609
53. Kawase T, Oda M (2006) Pure Appl Chem 78:831
54. Toyota S (2012) Pure Appl Chem 84:917
55. Toyota S (2011) Chem Lett 40:12
56. Toyota S, Goichi M, Kotani M (2004) Angew Chem Int Ed 43:2248
57. Toyota S, Goichi M, Kotani M, Takezaki M (2005) Bull Chem Soc Jpn 78:2214
58. Toyota S, Suzuki S, Goichi M (2006) Chem Eur J 12:2482
59. Toyota S, Miyahara H, Goichi M, Yamasaki S, Iwanaga T (2009) Bull Chem Soc Jpn 82:931
60. Ishikawa T, Shimasaki T, Akashi H, Iwanaga T, Toyota S, Yamasaki M (2010) Bull Chem Soc Jpn 83:220
61. Ishikawa T, Shimasaki T, Akashi H, Toyota S (2008) Org Lett 10:417
62. Toyota S, Kawakami T, Shinnishi R, Sugiki R, Suzuki S, Iwanaga T (2010) Org Biomol Chem 8:4997
63. Toyota S, Harada H, Miyahara H, Kawakami T, Wakamatsu K, Iwanaga T (2011) Bull Chem Soc Jpn 84:829
64. Kissel P, van Heijst J, Enning R, Stemmer A, Schlüter AD, Sakamoto J (2010) Org Lett 12:2778
65. Kissel P, Schlüter AD, Sakamoto J (2009) Chem Eur J 15:8955
66. Kissel P, Erni R, Schweizer WB, Rossell MD, King BT, Bauer T, Götzinger S, Schlüter AD, Sakamoto J (2012) Nat Chem 4:287
67. Ishikawa T, Iwanaga T, Toyota S, Yamasaki M (2011) Bull Chem Soc Jpn 84:729
68. Chen S, Yan Q, Li T, Zhao D (2010) Org Lett 12:4784
69. Miki K, Fujita M, Inoue Y, Senda Y, Kowada T, Ohe K (2010) J Org Chem 75:3537
70. Miyamoto K, Iwanaga T, Toyota S (2010) Chem Lett 39:299
71. Vögtle F, Gross J, Seel C, Nieger M (1992) Angew Chem Int Ed Engl 31:1069
72. Gross J, Harder G, Vögtle F, Stephan H, Gloe K (1995) Angew Chem Int Ed Engl 34:481
73. Gross J, Harder G, Siepen A, Harren J, Vögtle F, Stephan H, Gloe K, Ahlers B, Cammamm K, Rissanen K (1996) Chem Eur J 2:1585
74. Nierle J, Kuck D (2006) Synthesis 2914
75. Lélias-Vanderperre A, Chambron J-C, Espinosa E, Terrier P, Leize-Wagner E (2007) Org Lett 9:2961
76. Kim J, Kook Y, Park N, Hahn JH, Ahn KH (2005) J Org Chem 70:7087
77. Shomura R, Higashibayashi S, Sakurai H, Matsushita Y, Sato A, Higuchi M (2012) Tetrahedron Lett 53:783
78. Brotin T, Dutasta J-P (2009) Chem Rev 109:88
79. Fogarty HA, Berthault P, Brotin T, Huber G, Desvaux H, Dutasta J-P (2007) J Am Chem Soc 129:10332
80. Fairchild RM, Joseph AI, Holman KT, Fogarty HA, Brotin T, Dutasta J-P, Boutin C, Huber G, Berthault P (2010) J Am Chem Soc 132:15505
81. Masu H, Katagiri K, Kato T, Kagechika H, Tominaga M, Azumaya I (2008) J Org Chem 73:5143
82. Nishio M, Hirota M, Umezawa Y (1998) The CH/π interaction: evidence nature, and consequences. Wiley-VCH, New York
83. Nishio M (2011) Phys Chem Chem Phys 13:13873
84. Tominaga M, Masu H, Katagiri K, Kato T, Azumaya I (2005) Org Lett 7:3785
85. Tominaga M, Masu H, Katagiri K, Azumaya I (2007) Tetrahedron Lett 48:4369
86. Karlen SD, Garcia-Garibay MA (2005) Top Curr Chem 262:179
87. Khuong T-AV, Nunez JE, Godinez CE, Garcia-Garibay MA (2006) Acc Chem Res 39:413
88. Commins P, Nunez JE, Garcia-Garibay MA (2011) J Org Chem 76:8355
89. Chen C-F (2011) Chem Commun 47:1674
90. Zhang C, Chen C-F (2007) J Org Chem 72:9339
91. Hu S-Z, Chen C-F (2010) Chem Commun 46:4199
92. Tian X-H, Chen C-F (2010) Chem Eur J 16:8072

93. Fujita M, Oguro D, Miyazawa M, Oka H, Yamaguchi K, Ogura K (1995) Nature 378:469
94. Fujita M, Tominaga M, Hori M, Therrien B (2005) Acc Chem Res 38:371
95. Bunzen J, Iwasa J, Bonakdarzadeh P, Numata E, Rissanen K, Sato S, Fujita M (2012) Angew Chem Int Ed 51:3161
96. Yoshizawa N, Klostermanm JK, Fujita M (2009) Angew Chem Int Ed 48:3418
97. Brinker UH, Mieusset J-L (eds) (2010) Molecular encapsulation: organic reactions in constrained systems. Wiley, Chichester
98. Murase T, Nishijima Y, Fujita M (2012) Chem Asian J 7:826
99. Kuehl CJ, Yamamoto T, Seidel SR, Stang PJ (2002) Org Lett 4:913
100. Yang H-B, Ghosh K, Das N, Stang PJ (2006) Org Lett 8:3991
101. Caulder DL, Raymond KN (1999) Acc Chem Res 32:975
102. Caulder DL, Powers RE, Parac TN, Raymond KN (1998) Angew Chem Int Ed Engl 37:1840
103. Fiedler D, Pagliero D, Brumaghim JL, Bergman RG, Raymond KN (2004) Inorg Chem 43:846
104. Kishi N, Li Z, Yoza K, Akita M, Yoshizawa M (2011) J Am Chem Soc 133:11438
105. Li Z, Kishi N, Hasegawa K, Akita M, Yoshizawa M (2011) Chem Commun 47:8605
106. Kottas GS, Clarke LI, Horinek D, Michl J (2005) Chem Rev 105:1281
107. Caskeym DC, Michl J (2005) J Org Chem 70:5442
108. Caskeym DC, Wang B, Zheng X, Michl J (2005) Collec Czech Chem Commun 70:1970
109. Cram DJ, Cram JM (1994) Container molecules and their guests. The Royal Society of Chemistry, Cambridge
110. Conn MM, Rebek J Jr (1997) Chem Rev 97:1647
111. Hof F, Craig SL, Nuckolls C, Rebek J Jr (2002) Angew Chem Int Ed 41:1488
112. Scarso A, Trembleau L, Rebek J Jr (2004) J Am Chem Soc 126:13512
113. Ajami D, Rebek J Jr (2009) Nat Chem 1:87
114. Ajami D, Rebek J Jr (2006) J Am Chem Soc 128:5314
115. Rebek J Jr (2009) Acc Chem Res 42:1660
116. Nishimura N, Yoza K, Kobayashi K (2010) J Am Chem Soc 132:777
117. Kobayashi K, Yamada Y, Yamanaka M, Sei Y, Yamaguchi K (2004) J Am Chem Soc 126:13896
118. Beves JE, Blight BA, Campbell CJ, Leigh DA, McBurney RT (2011) Angew Chem Int Ed 50:9260
119. Stoddart JF (2009) Chem Soc Rev 38:1802
120. Forgan RS, Sauvage J-P, Stoddart JF (2011) Chem Rev 111:5434
121. Sauvage J-P, Amabilino DB (2012) Top Curr Chem 323:107
122. Bruns CJ, Stoddart JF (2012) Top Curr Chem 323:19
123. Sauvage J-P, Amabilino DB (2012) In: Gale P, Steed J (eds) Supramolecular chemistry: from molecules to nanomaterials. Wiley, Chichester
124. Dietrich-Buchecker CO, Sauvage J-P, Kern JM (1984) J Am Chem Soc 106:3043
125. Ashton PR, Goodnow TT, Kaifer AE, Reddington MV, Slawin AMZ, Spencer N, Stoddart JF, Vicent C, Williams DJ (1989) Angew Chem Int Ed Engl 28:1396
126. Fujita M, Ibukuro F, Seki H, Kamo O, Imanari M, Ogura K (1996) J Am Chem Soc 118:899
127. Niu Z, Gibson HW (2009) Chem Rev 109:6024
128. Cougnon FBL, Jenkins NA, Pantos GD, Sanders JKM (2012) Angew Chem Int Ed 51:1443
129. Pentecost CD, Chichak KS, Peters AJ, Cave GWV, Cantrill SJ, Stoddart JF (2007) Angew Chem Int Ed 46:218
130. Ibukuro F, Fujita M, Yamaguchi K, Sauvage J-P (1999) J Am Chem Soc 121:11014
131. Peinador C, Blanco V, Quintela JM (2009) J Am Chem Soc 131:920
132. Fujita M, Fujita N, Ogura K, Yamaguchi K (1999) Nature 400:52
133. Yamauchi Y, Yoshizawa M, Fujita M (2008) J Am Chem Soc 130:5832
134. Westcott A, Fisher J, Harding LP, Rizkallah P, Hardie M (2008) J Am Chem Soc 130:2950
135. Freye S, Hey J, Torras-Galán A, Stalke D, Herbst-Irmer R, John M, Clever GH (2012) Angew Chem Int Ed 51:2191

136. Chen C-T, Gantzel P, Siegel JS, Baldridge KK, English EB, Ho DM (1995) Angew Chem Int Ed Engl 34:2657
137. Cantrill SJ, Chichak KS, Peters AJ, Stoddart JF (2005) Acc Chem Res 38:1
138. Loren JC, Yoshizawa M, Haldimann RF, Linden A, Siegel JS (2003) Angew Chem Int Ed 42:5701
139. Chichak KS, Cantrill SJ, Pease AR, Chiu S-H, Cava GWV, Atwood JL, Stoddart JF (2004) Science 304:1308
140. Irngartinger H (1973) Chem. Ber. 106:761
141. Toyota S, Kawai K, Iwanaga T, Wakamatsu K, Eur J Org Chem, in press. DOI:10.1002/ejoc.201200765

Top Curr Chem (2014) 350: 141–176
DOI: 10.1007/128_2012_387

Published online: 19 July 2012

Antiaromaticity in Nonbenzenoid Oligoarenes and Ladder Polymers

Rebecca R. Parkhurst and Timothy M. Swager

Abstract Polycyclic aromatic hydrocarbons (PAHs) and fully-conjugated ladder polymers are leading candidates for organics electronics, as their inherent conformational rigidity encourages electron delocalization. Many of these systems consist of fused benzenoid or heterocyclic aromatic rings. Less frequently, however, PAHs are reported with character that alternates between the aromaticity of benzene fragments and the antiaromaticity of a nonbenzenoid moiety. This chapter will focus on recent work published on the theory, synthesis, and properties of two such systems: [N]phenylenes containing 4π-electron cyclobutadienoid character, and diaryl[*a,e*]pentalenes containing 8π-electron pentalenoid character.

Keywords [N]Phenylenes · Antiaromaticity · Diaryl[*a,e*]pentalenes · Ladder polymers · Nontraditional graphitic materials

Contents

R.R. Parkhurst (✉) and T.M. Swager
Department of Chemistry, Massachusetts Institute of Technology, Cambridge, MA, US
e-mail: rparkhur@mit.edu; tswager@mit.edu

1 Introduction

The concepts of aromaticity and antiaromaticity have long been employed to explain certain chemical phenomena; however there is still much to be explored regarding the effect of these simultaneously opposing and intertwined theories on the properties of materials. Aromatic molecules are generally characterized by several properties including unexpectedly high thermal stability, nonalternate bond lengths, and diamagnetic anisotropy [1, 2]. Since it was first proposed by Hückel in the 1930s, the theory that these properties belong to cyclic arrays containing $4n + 2$ π-electrons has held true [3]. Alternatively, those species containing 4π-electrons later came to be classified as antiaromatic, and can be high energy and difficult to isolate [4–7].

Organic molecules consisting of multiple fused aromatic rings are commonly known as polycyclic aromatic hydrocarbons (PAHs). Due to the extended conjugation and ring current inherent in these materials, they make a class of important candidates for use in organic electronics [8]. Antiaromatic compounds, on the other hand, are appealing candidates for optoelectronic applications due to small HOMO-LUMO gaps relative to analogous aromatic systems, and theoretically high polarizability, which could improve intermolecular interactions. Although attempts have been made to predict computationally the structure and solid-state properties of polycyclic antiaromatic hydrocarbons (PAAHs), there is still much that is not well understood [9]. In this regard, one area of interest is to study the interplay of these two concepts in polycyclic systems consisting of alternating aromatic and antiaromatic character [10].

2 [N]Phenylenes and Related Materials

The most basic example of antiaromaticity is found in the 4π-electron system of cyclobutadiene (Fig. 1a) [11]. There has been a great deal of theoretical interest in the antiaromaticity of cyclobutadiene; however, experimental analysis has been difficult as a result of the very high reactivity of the parent hydrocarbon.

The [N]phenylenes are a family of PAHs in which benzene rings are fused together by cyclobutadiene rings, resulting in a ladder structure with alternating aromatic and antiaromatic character (Fig. 1b, c) [12]. The simplest member of this family, biphenylene ($n = 2$), was first synthesized in 1941 from 2,2′-dibromobiphenyl after a fairly long history of eluding synthetic chemists [13].

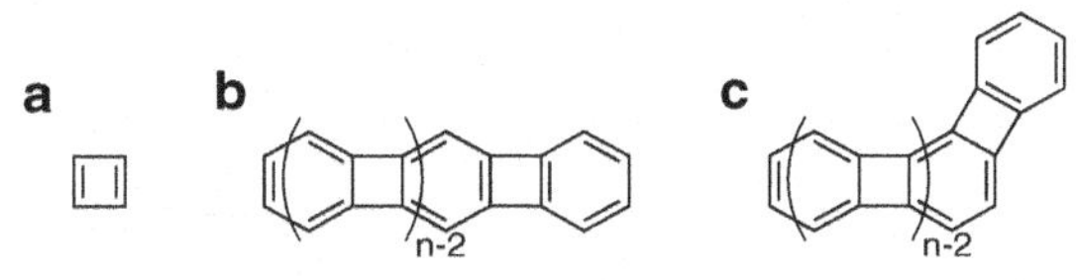

Fig. 1 Structures of (**a**) cyclobutadiene [11]; (**b**) linear and (**c**) angular [N] phenylenes [12]

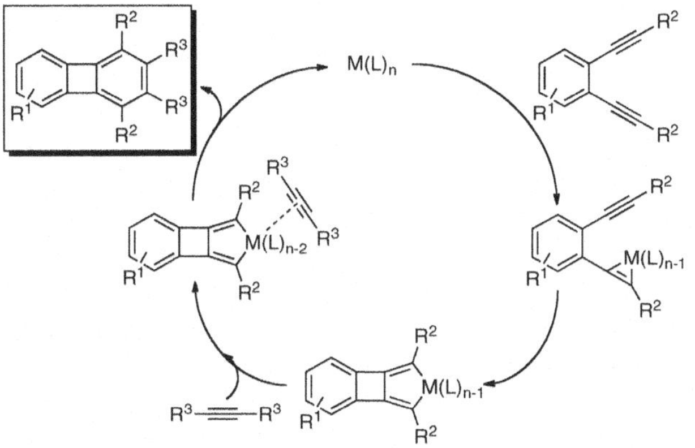

Scheme 1 Catalytic cycle in the metal-catalyzed [2 + 2 + 2] cycloaddition of aryl alkynes to yield [N] phenylenes [12]

Other classic methods to access this scaffold include the dimerization of arynes or contraction of bridged biaryls through extrusion of a small molecule such as N_2, CO_2, or CO [14, 15]. Over the last several decades the synthesis has been dominated by the metal-catalyzed [2 + 2 + 2] cycloaddition methodology (Scheme 1), developed primarily by Vollhardt and coworkers [12]. This approach has improved the synthesis of biphenylene as well as provided access to longer derivatives. It has been demonstrated both experimentally and theoretically that the π-bonds in these systems tend to localize within the six-membered rings, thereby minimizing the 4π antiaromatic character of the cyclobutadiene linkage [16].

In the past the model of conjugated circuits, traditionally employed to determine the resonance energy of PAHs, has failed to explain some of the properties observed in [N]phenylenes [17]. Most significantly, the model does not account for the increased stability of angular [N]phenylenes (Fig. 1c) over linear [N]phenylenes (Fig. 1b) [12, 18–23]. Recently, Randić and coworkers have sought to develop a revised theory to determine the aromaticity of [N]phenylenes that can compensate for this discrepancy [17].

The five possible Kekulé resonance structures of biphenylene are shown in Fig. 2a. According to these structures, the C–C bond connecting the two benzenoid rings should have a bond order of 1½. In reality, this bond has been shown to have even less double bond character than this value suggests. The authors propose excluding the fifth and final structure that does not contain any Clar sextets [24–26] as a contributor to the resonance hybrid. This leads to 1½ bond order for all π bonds except for the connecting CC bond with a bond order of 1. This solution appears to improve the theory for one aspect of the observed bond lengths in biphenylene; however it does not yet account for the decreased bond alternation within the benzenoid rings.

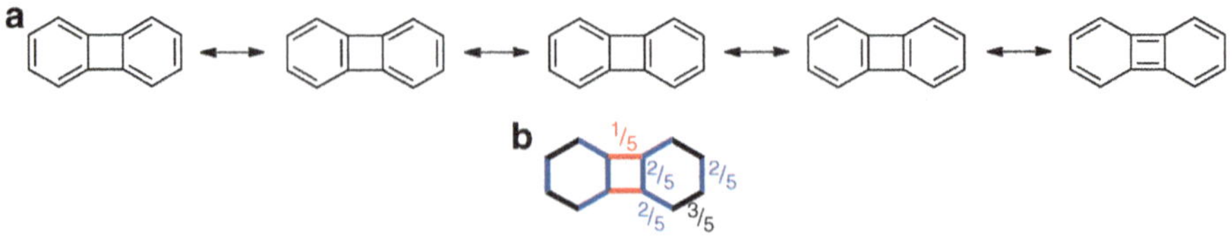

Fig. 2 (**a**) Kekulé structures of biphenylene and (**b**) theoretical π-bond orders based on the above structures [17]

The authors also considered the effect of the proposed reduced set of Kekulé structures on the ring currents of biphenylene, following a structural approach in which aromaticity is a combination of the local ring current of each conjugated circuit in the resonance structures. This model leads to a reduction in the diamagnetic current of benzenoid rings to ½ that of benzene, and a paramagnetic current of ⅓ the magnitude of the same reference in the four-membered rings. Therefore, although fusion of the aromatic and antiaromatic systems results in a reduced contribution by both, the aromatic character of these PAHs is still dominant.

The revised model also provides promising results when applied to longer [N]phenylenes. In the case of linear (Fig. 1b, $n = 3$) and angular [3]phenylene (Fig. 1c, $n = 3$), each have 12 and 13 Kekulé structures respectively. Elimination of those that do not contain at least three Clar sextets leaves only eight possible structures for each. The resonance energy determined from these revised structures (by summing the positive contribution of $4n + 2$ π-electrons and the negative contribution of 4n π-electrons) is indeed higher for angular [3]phenylene, consistent with previous experimental and theoretical findings.

2.1 Recent Synthetic Developments in [N]Phenylenes

As mentioned previously, the most general route to accessing a wide variety of linear, angular, zigzag, and branched [N]phenylenes is via the appropriate phenylethynyl precursor and subsequent [2 + 2 + 2] cycloaddition. This chemistry is relatively well-established through work by Vollhardt and coworkers and has been reviewed previously [12]; therefore this chapter will focus on developments made in the last 10 years.

Several derivatives of bent [N]phenylenes have been reported by Vollhardt and coworkers since 2002. Bent [4]phenylene (**2**), the final isomer of the [4]phenylene family, was synthesized in 2002 (Scheme 2a) [27]. The initial step in the synthesis involved non-selective Pd-catalyzed Sonogashira coupling of 2,3-diiodobiphenylene with *mono*-silated *o*-diethynyl benzene, followed by a second Sonogashira coupling with trimethylsilylacetylene and subsequent deprotection to give the desired triyne precursor **1**. Cobalt-catalyzed [2 + 2 + 2] cyclization of **1** yielded the air-sensitive compound **2a** in 1.7% overall yield over eight steps.

An alternative synthesis, resulting in the relatively more stable bis-(trimethylsilyl)-**2** (**2b**) and involving double [2 + 2 + 2] cycloaddition, was also

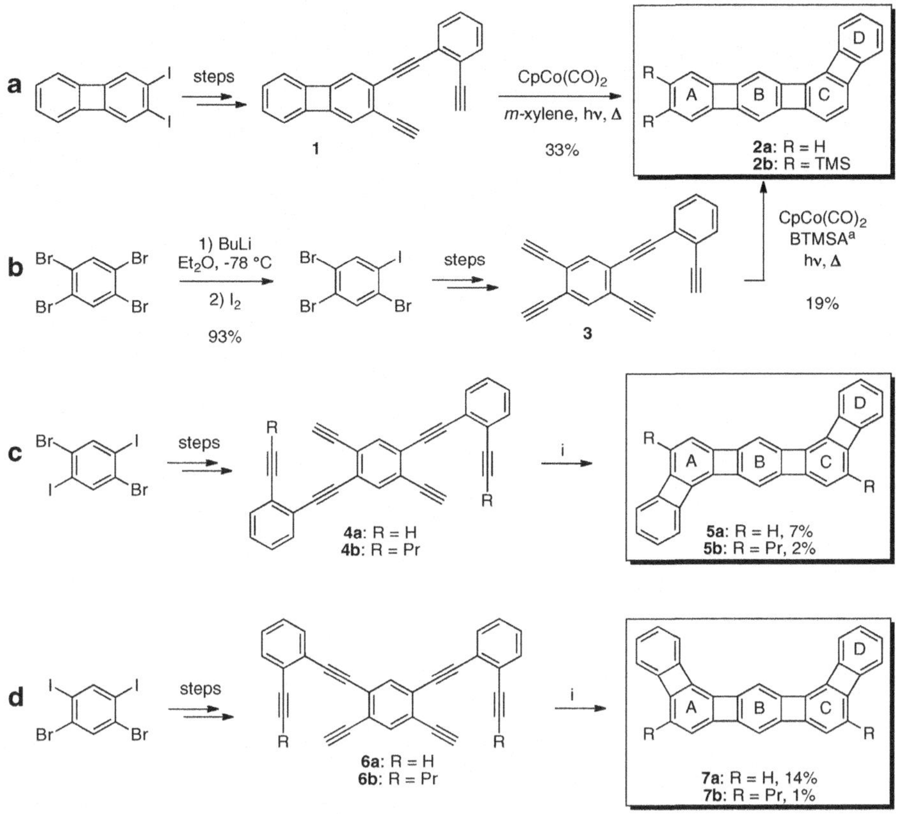

Scheme 2 Synthesis of (**a**) bent [4]phenylene (**2a**) and (**b**) (**2b**) [27]. Synthesis of (**c**) *anti* (**5**) and (**d**) *syn* doublebent [5]phenylene (**6**) [28]. For **5a** and **7a**: (*i*) (1) $CpCo(C_2H_4)_2$, THF, -25 °C (2) 1,3-cyclohexadiene, THF, 110 °C. For **5b** and **7b**: (*i*) $CpCo(CO)_2$, *m*-xylene, hν, Δ. [a]BTMSA = bis(trimethylsilyl)acetylene

reported (Scheme 2b). In this case, 5-iodo-1,2,4-tribromobenzene was generated in high yields by regioselective lithiation of 1,2,4,5-tetrabromobenzene followed by quenching with iodine. Consecutive Sonogashira coupling reactions yielded compound **3**. This pentayne underwent a twofold intra- and intermolecular [2 + 2 + 2] cyclization to yield **2b** in an improved 7% yield over five steps. However, attempts to perform the protodesilation of **2b** to parent hydrocarbon **2a** resulted in decomposition.

In a fashion similar to that discussed above, Vollhardt and coworkers were also able to access *anti* (**5**) and *syn* (**7**) doublebent [5]phenylenes by first constructing the appropriate hexayne precursor (Scheme 2c, d) [28]. Both isomers are accessible through successive Sonogashira couplings of the appropriately substituted 1,2,4,5-tetrahalobenzene, followed by intramolecular [2 + 2 + 2] cyclization.

The simpler singly bent [5]phenylene **10** is also available through a somewhat different route (Scheme 3) [29]. In this case, tetrayne **8** was synthesized via a

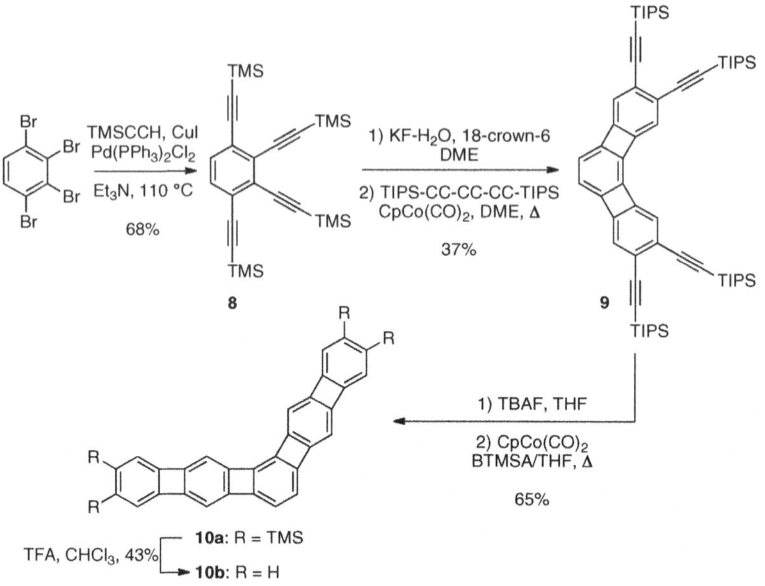

Scheme 3 Synthesis of singly bent [5]phenylene **10** [29]

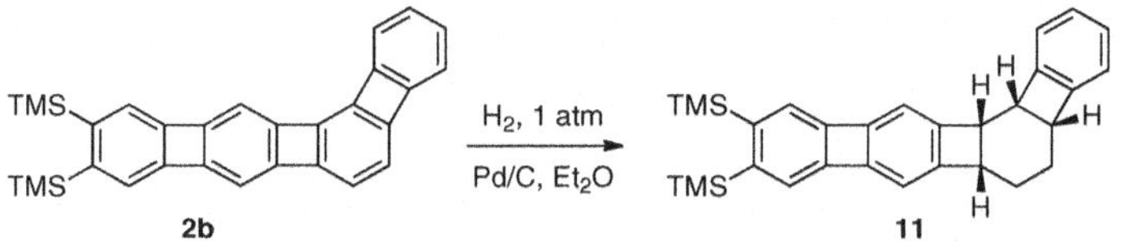

Scheme 4 Hexahydrogenation of ring C of **2b** [27]

fourfold Sonogashira coupling of 1,2,3,4-tetrabromobenzene with trimethylsilylacetylene. Compound **8** was then deprotected in situ followed by a twofold [2 + 2 + 2] cyclization with bis(triisopropylsilyl)-protected 1,3,5-hexatriyne to give angular [3]phenylene-containing tetrayne **9**. A second twofold [2 + 2 + 2] with *bis*(trimethylsilyl)acetylene (BTMSA) and deprotection yielded **10b**. All the bent [5]phenylenes were air-sensitive, and the parent hydrocarbons (**5a**, **7a**, and **10a**) were relatively insoluble.

The aromaticity of the reported [4]- and [5]phenylenes was examined using a number of criteria. For **2**, shielding of the protons bound to ring C relative to those on ring B confirm the increased cyclohexatrienic character of the former and benzenoid character of the latter [27]. As shown in Scheme 4, ring C of compound **2b** can be all *cis*-hexahydrogenated, proving it to be the most activated. This is comparable to the reactivity of the central ring in angular [3]phenylene [12]. In the case of **7a**, deshielding of ring B protons is even more pronounced than in **2a** especially in the bay region; however, chemical shifts of the ring C protons are comparable [28]. The same pattern is seen in comparing the calculated

Table 1 Synthesis of [N]heliphenes ($n = 6$–9). Reaction conditions: $CpCo(CO)_2$, *m*-xylene, hν, Δ [30]

Precursor	[N]Heliphene	Name and yield
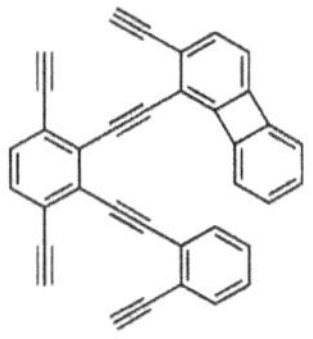	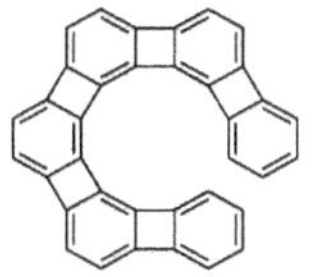	[6]Heliphene (**12**), 12%
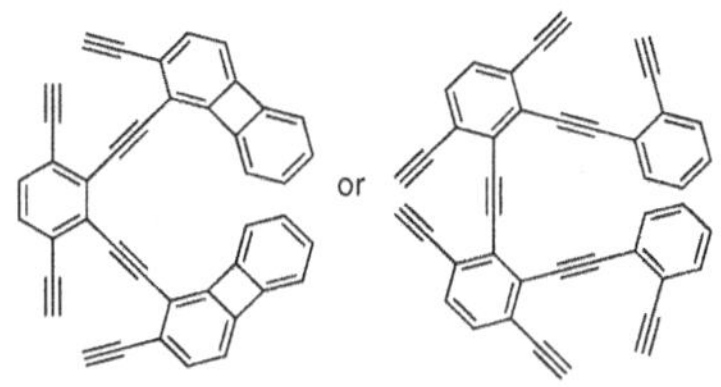 or	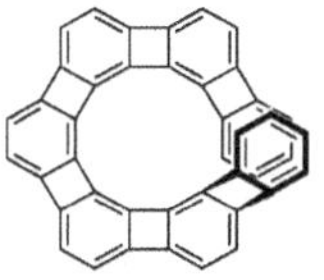	[7]Heliphene (**13**), 8% or 2%
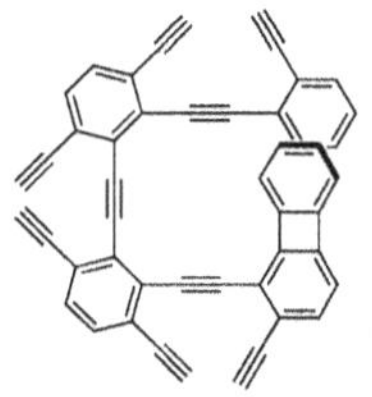	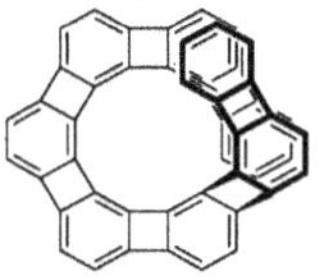	[8]Heliphene (**14**), 2%
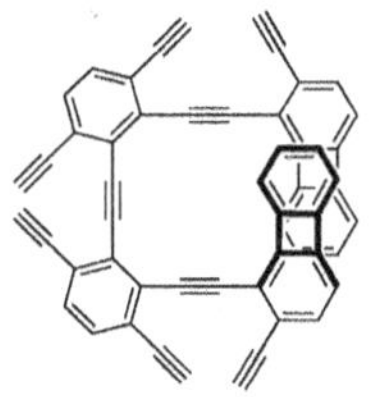	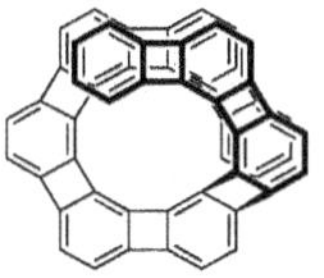	[9]Heliphene (**15**), 2%

NICS (Nucleus Independent Chemical Shifts) values from **2a** and **7a**: the diatropicity of ring B increases from NICS = −6.4 to NICS = −7.5 in going from **2a** to **7a**, whereas for ring C NICS = −2.9 in each case. In **10b** there is even greater shielding of the central ring protons, which is again consistent with decreasing diatropicity of NICS = −2.5 for that ring [29].

The longer angular [N]phenylenes are of both experimental and theoretical interest, becoming helical with $n > 5$ and referred to as [N]heliphenes [23]. Syntheses of angular [6]-[9]phenylenes were reported in 2002 [30, 31]. In each case the [N]heliphene was accessed by a two- or threefold [2 + 2 + 2] cycloaddition of the appropriate ethynyl precursor as outlined in Table 1. Heliphenes **12** and **13** are stable in the solid state while **14** and **15** are only moderately so.

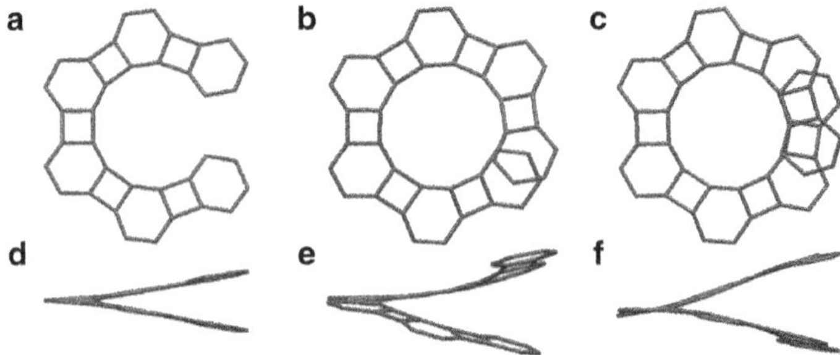

Fig. 3 X-Ray crystal structures: (**a**) Front and (**d**) side views of **12**, (**b**) front and (**e**) side views of **13**, and (**c**) front and (**f**) side views of **14**. Crystal structure data from [30, 31] was gathered from the Cambridge Structural Database (*CSD*)

The ^{1}H NMR spectra of these compounds are consistent with the pattern seen in lower angular phenylenes [19, 32]. The terminal benzenoid rings are the most aromatic, having the most deshielded protons, whereas the penultimate ring has the most shielded protons and therefore the most cyclohexatrienic character. In the case of heliphenes **13–15**, where the termini begin to overlap, only those protons lying over the terminal (aromatic) ring are shielded. The protons that reside over the penultimate ring are actually slightly deshielded. Crystal structures demonstrating the helical turn and overlapping termini of heliphenes **12** through **14** are shown in Fig. 3.

Fewer reports have focused on the development of synthetic methods to access materials containing aromatic segments larger than one benzene ring. Taking advantage of the ability of palladium to catalyze the [2 + 2 + 2] cycloaddition of arynes, the synthesis of tris(benzocyclobutadieno)triphenylene **17** and related phenylene-containing PAHs, **18** and **19**, was reported [33]. Key to this strategy was the synthesis of a 2,3-dehydrobiphenylene precursor **16** from the appropriate 2,3-bis-(trimethylsilyl)biphenylene as demonstrated in Scheme 5. The aryne was then generated by addition of the fluoride ion and followed by either trimerization to give **17** or cyclization with dimethylacetylene dicarboxylate (DMAD) and the appropriate Pd(0) source to give **18** or **19**.

2.2 *Photophysical Properties of [N]Phenylenes*

Examination of photophysical properties of [N]phenylenes of a variety of topologies can also be useful in probing relative levels of conjugation within each chromophore. Consistent with the earlier discussion of relative resonance energies, calculations predict larger HOMO-LUMO gaps for zigzag [N]phenylenes than for the linear isomer due to increased bond alternation [34].

The photophysical behavior of biphenylene (**20**) and both the linear (**21**) and angular (**22**) isomers of [3]phenylene were well-studied prior to the

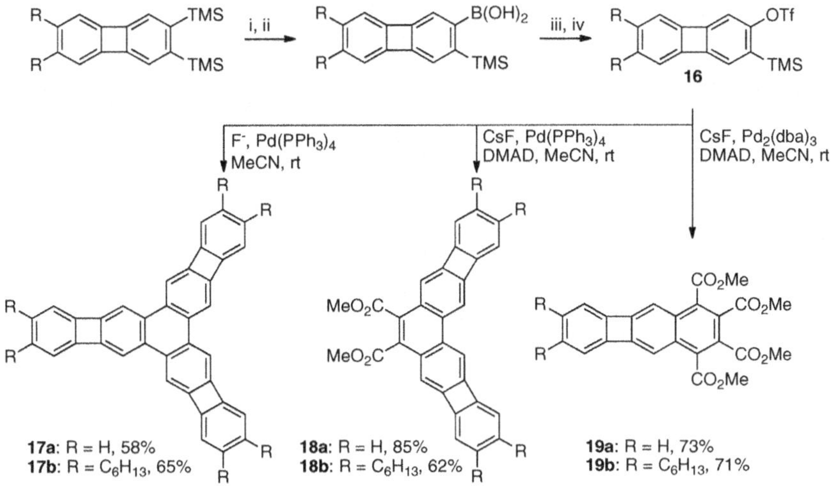

Scheme 5 Pd-catalyzed cyclization of 2,3-dehydrobiphenylene derivatives [33]. Conditions: (*i*) NBS, $AgNO_3$, acetone, rt, (*ii*) (1) BuLi, THF, −78 °C, (2) $B(OMe)_3$, −78 °C, (3) HCl (1 M), (*iii*) oxone, $NaHCO_{3(aq)}$, acetone-H_2O, 0 °C, (*iv*) DIPEA, Tf_2O, CH_2Cl_2, −78 °C. The source of F^- is TBAF for **17a** and CsF for **17b**

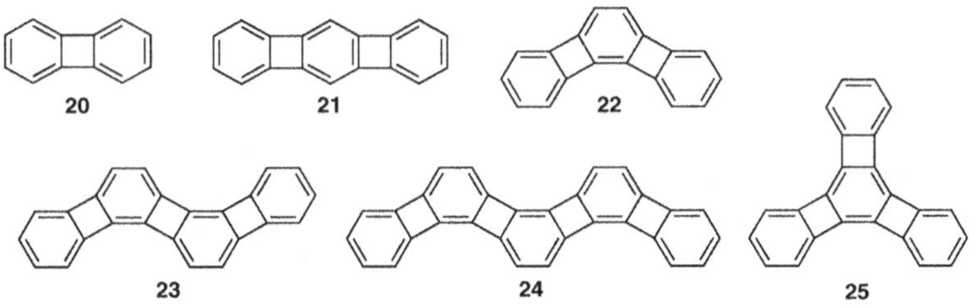

Fig. 4 Structure of [N]phenylenes **20–25** [34]

scope of this review; however, this foundation is important for understanding properties observed in longer isomers. Biphenylene is only weakly fluorescent ($\phi_F = 2 \times 10^{-4}$) [35–37] and only low levels of triplet state population are observed upon direction excitation ($\phi_{ISC} < 10^{-2}$) [38–41].

The values corresponding to the energy of S_1-S_0 transition in compounds **20–22** as well as zigzag [4]- (**23**) and [5]phenylene (**24**), and triangular [4]phenylene (**25**, Fig. 4) are summarized in Table 2 [34]. The bathochromic shift in energy going from **20** to linear **21** is much larger (0.87 eV) than that from **20** to angular **22** (0.13 eV). The rapidly decreasing band gap with linear annulation is consistent with the prediction that linear [N]phenylenes of infinite length could serve as conductive organic wires [20]. Interestingly, angular [3]phenylene is emissive, while linear [3] phenylene is not. The band gap does decrease with angular extension of the [N] phenylene going from **22–24** but at a much slower rate.

Table 2 Photophysical properties of [N]phenylenes in THF [34]

Compound	$E(S_1)$ (eV)	Stokes shift (cm^{-1})	ϕ_F	ϕ_{ISC}
20	2.98	4,750	2.6×10^{-4}	<0.01
21	2.11	–	$<10^{-4}$	<0.01
22	2.85	750	0.07	0.30
23	2.66	70	0.12	0.25
24	2.57	65	0.21	0.45
25	2.81	150	0.15	0.03

Table 3 Lowest energy λ_{max} for some [4]- and [5] phenylenes [27–29, 42, 43]

Compound	Lowest E λ_{max} (nm)	E_g (eV)
2a	486	2.55
26	488	2.54
5a	505	2.46
10b	523	2.37
27	530	2.34

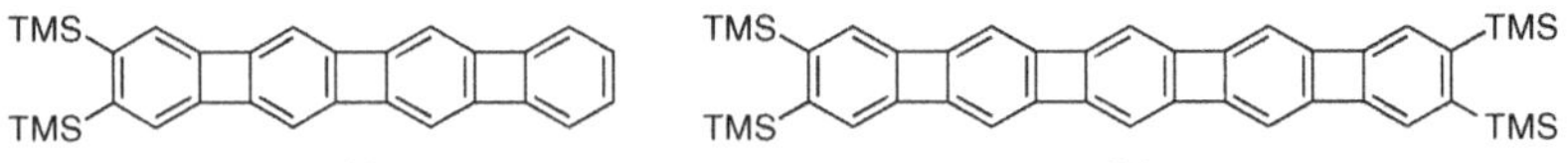

Fig. 5 Structures of [N]phenylenes **26** and **27** [42, 43]

The larger [N]phenylenes **23–25** have very small Stokes shifts (70–150 cm^{-1}) as compared to both **22** (750 cm^{-1}) and **20** (4,750 cm^{-1}), indicating a more rigid structure in which the geometry does not change much in going from S_0 to S_1 [34]. This data together with increased fluorescence quantum yields indicate slower rates of internal conversion (IC) for these species. As indicated in Table 2, the zigzag [N] phenylenes **23** and **24** undergo intersystem crossing (ISC) fairly efficiently with ISC quantum yields ranging from 25% to 30%. Therefore this process is a likely contributor to the deactivation of the S_1 state. However, this is not the case for **20** and **21** for which energy is presumably lost via fast internal conversion (IC).

An interesting pattern also emerges in examination of the band gap (estimated from the lowest energy absorption) of the bent [N]phenylene derivatives discussed in the previous section (Table 3). The lowest energy absorptions of linear 2,3-bis (trimethylsilyl)-[4]phenylene [42] (**26**) and 2,3,9,10-tetrakis(trimethylsilyl)-[5] phenylene [43] (**27**, Fig. 5) are also included in Table 3 for comparison. Bent [4] phenylene **2a**, has almost the same lowest energy λ_{max} as its linear counterpart [27], as does singly bent [5]phenylene **10b** [29]. The λ_{max} of doublebent [5]phenylene **5a**, on the other hand, lies between that of the linear **27** and the higher energy absorption of other known isomers including zigzag (see above) [28]. From examination of this data, it appears that the first angular ring annulation has a stronger effect on the band gap than does a second.

The lowest energy λ_{max} of the heliphenes also increases with length as expected [30, 31]. The bathochromic shift decreases in going from **12** to **13** (12 nm), **13** to **14**

(10 nm) **14** to **15** (9 nm). A plot of λ_{max} vs 1/(N+0.5) provides a straight line that can be extrapolated to 578 nm, possibly indicating the theoretical band gap of polyheliphene.

2.3 *Reactivity of [N]Phenylenes and Related Materials*

Biphenylene (**20**) was long believed to be photoinert; however, evidence of high internal conversion rates indicated that the picture is more complex [44]. Transient absorption measurements of **20** indicate the formation of an initial, vibrationally hot ground state S_0* formed by internal conversion, followed by chemical reaction to form stable products. Correlation between the transient absorbance and laser influence data suggest that hot biphenylene must absorb a second photon to become chemically reactive, further raising its internal energy to compete with collisional deactivation. Several photo-products were identified including phenylacetylene, biphenyl, naphthalene, and acenaphthylene, as well as three unknown. The authors suggest a mechanism involving initial C–C bond cleavage to form the 2,2-biphenyl biradical (Scheme 6).

Work reported in the early 1990s showed that benzo[*a*]pentalene is produced as an intermediate in the Flash Vacuum Pyrolysis (FVP) of biphenylene, further rearrangement leading to acenaphthylene as the major product [45, 46]. The driving force of this skeletal rearrangement is believed to be the relief of strain in the four-membered ring. More recently, however, several groups have revealed another mechanistic pathway involved in rearrangements of larger PAHs containing the phenylene-linkage [47–49]. Phenyl groups are able to migrate by a process in which C–C bond cleavage occurs as described above, followed by several 1,5- or 1,6-hydrogen transfers, and ring closure at a new position and so on (Scheme 7).

Together, the above mechanism and ring contraction/expansion to relieve strain can be used to explain most of the products obtained from FVP of phenylene-containing materials. The major products obtained from FVP of benzo[*b*]biphenylene (**28**), linear [3]phenylene (**21**), and angular [4]phenylene (**29**) are summarized in Scheme 8. Scott and coworkers investigated the FVP of **28** which resulted in two major products: dibenzo[*a,e*]pentalene (**30**) from benzene ring contraction and fluoranthene (**31**) from phenyl migration [47]. Vollhardt and coworkers showed that FVP of **21** resulted in four major PAH products (the same four products as FVP of angular [3]phenylene **22** [50]) as well as 1% of **22** [48]. The conversion of **21** to **22** clearly occurs through the phenyl migration mechanism, whereas the other PAH products appear to result from various benzene ring contraction rearrangements of **22** once it has formed. Finally, **29** is converted to the major product biphenylene dimer **32**, by a mechanism similar to phenyl migration, as well as two additional PAHs [49].

As mentioned previously, it has been shown that angular annulation causes greater bond alternation and decreased diatropicity in the benzene ring at the bend (ring C, Schemes 2 and 3), increasing the reactivity of those double bonds.

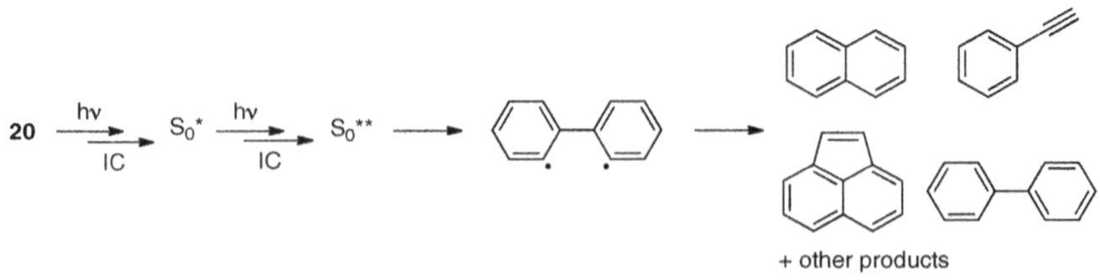

Scheme 6 Proposed mechanism of two-photon photoreactivity of **20** [44]

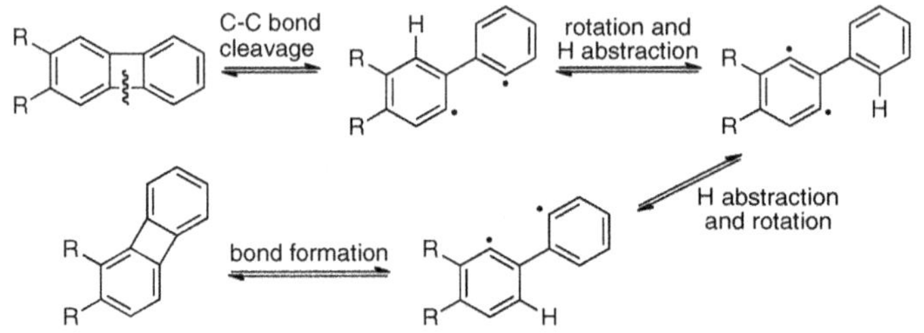

Scheme 7 Mechanism of phenyl group migration in FVP of phenylene-containing PAHs [47–49]

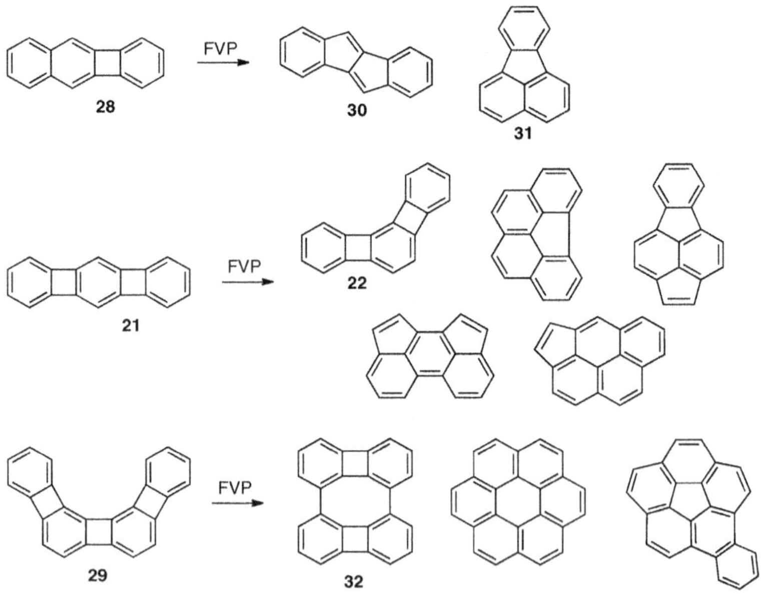

Scheme 8 FVP of **28**, **21**, and **29** [47–50]

In addition to hydrogenation as shown in Scheme 4, it has been demonstrated that the central benzene ring in angular [3]phenylene (**22**) is susceptible to a number of reactions (Scheme 9) [51]. The 2,3,8,9-tetrakis(trimethylsilyl)-substituted **22** can be tris-epoxidized (Scheme 9a) analogously to the reactivity of triangular [4]

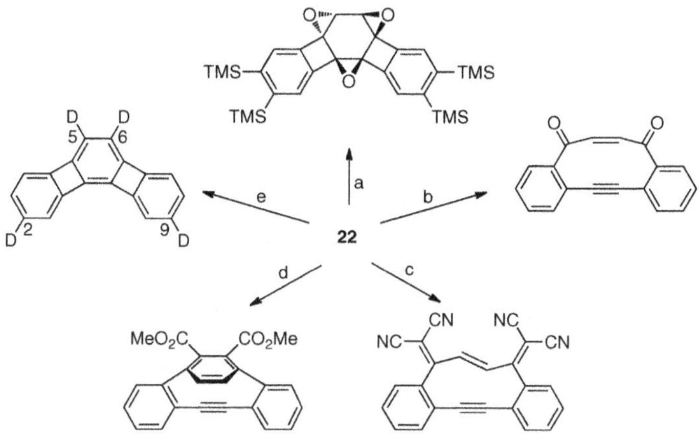

Scheme 9 Reactivity of **22** [51]. Conditions: (*a*) using 2,3,8,9-tetrakis(trimethylsilyl)-**22** as the starting material, dimethyldioxirane (DMDO), acetone, 26%, (*b*) methylene blue (0.9 equiv.), O_2, CH_2Cl_2, hν, 70%, (*c*) TCNE (1 equiv.), CH_3CN, Δ, 78%, (*d*) DMAD (1.6 equiv.), $AlCl_3$ (0.1 equiv.), $PhCH_3$, 74% (*e*) (1) CF_3CO_2D, $CDCl_3$, 60 °C, 80% (2) CF_3CO_2D, $CDCl_3$, 100 °C, 82%

phenylene (**25**) [52]. Epoxidation of the unsubstituted **22**, however, only proceeded twice, leaving the bay region double bond unreacted [51]. The diene composed of the more reactive double bonds underwent [4 + 2] cycloadditions with either singlet oxygen ($^1O_2$), tetracyanoethylene (TCNE), or DMAD, followed by a second electrocyclic reaction to give the product shown (Scheme 9b–d).

Lastly, the deuteration of **22** was attempted in order to examine more generally its reactivity towards electrophiles. Biphenylene [14] and **25** [53] both undergo deuteration exclusively at the β position (C2/9). However, for **22** deuteration occurs preferentially at the α position (C5/6) and then secondarily at the β position under harsher conditions (Scheme 9e) [51]. Calculations (HF/6-31G*) of the protonated intermediates indicated that attack at C5/6 results in the fewest resonance forms with cyclobutadienoid character. Additionally, this reactivity trend corresponds to the relative size of the HOMO coefficients (HF/STO-3G) at those positions.

2.4 *Applications of Phenylene-Containing Materials*

There is a fair amount of interest in the effect of incorporating antiaromatic moieties into systems that normally rely on aromatics to perform a function. In these cases, biphenylene is often turned to as a stable substitute for cyclobutadiene. Previous work has clearly demonstrated that this scaffold is not purely antiaromatic; however, the model systems have had some informative results.

In several cases, π-complexes of [N]phenylenes have been shown to undergo intramolecular inter-ring haptotropic rearrangements (IRHR). Depending on the

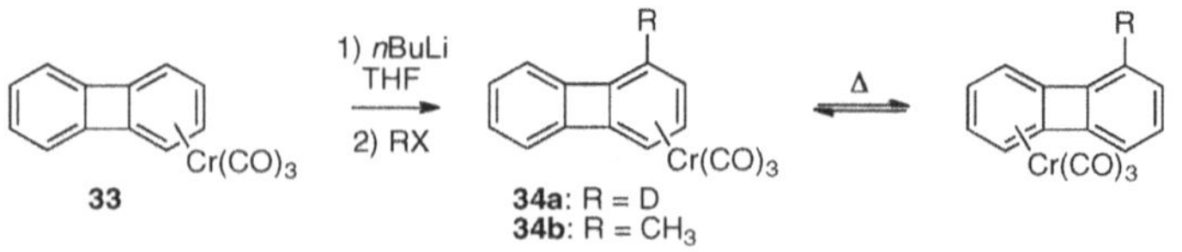

Scheme 10 Synthesis of unsymmetric, substituted η^6-$Cr(CO)_3$-complexes of biphenylene [54]

metal and ligand, the metal atom migrates between either benzene (η^6-η^6) [54] or cyclobutadiene (η^4-η^4) [55, 56] ligands. In both the η^6-η^6 and η^4-η^4 cases of IRHR in [N]phenylenes, DFT calculations show that the migration occurs through mechanism in which the metal atom "walks" the periphery of the PAH.

In the case of the former, the mono-chromium tricarbonyl complex of biphenylene **33** can be selectively lithiated on the complexed benzene ring and subsequently substituted with an electrophile to give two non-degenerate ligand sites on biphenylenes **34a/b** (Scheme 10) [54]. The equilibrium of IRHR of these species was studied by NMR at 130 °C. Whereas **34a** equilibrates to a ratio of 1:1 of the two ligand sites after about 20 h, decomposition of the complex competes with IRHR in **34b** and after 56 h equilibrium had still not been reached. The activation for IRHR in **34a** was determined to be 32.2 kcal mol^{-1}, which expectedly lies between the analogous values in naphthalene (28.5 kcal mol^{-1}) and biphenyl (35.1 kcal mol^{-1}).

More recently it has been shown that cyclobutadiene-metal complexes of the longer linear [N]phenylenes can undergo photo-thermal reversible IRHR [55, 56]. Systems such as these have potential for a variety of applications including solar energy storage and switches. The cyclopentadienyl-cobalt {CpCo} complexes of substituted linear [3]-, [4]-, and [5]phenylenes can be isolated directly from the [2 + 2 + 2] cycloaddition of the appropriate precursor with BTMSA, under modified conditions employing stoichiometric [$CpCo(CO)_2$] and THF as a co-solvent. As indicated by deshielding in the NMR and calculated NICS values, complexation of {CpCo} has an aromatizing effect on the entire PAH; however this is most strong on the metal-bound cyclobutadiene ring.

When irradiated with light [N]phenylene complexes **35–37a** undergo IRHR to give the higher energy haptomers **35–37b** (Scheme 11) reaching a photostationary state as indicated in Table 4. The calculated heats of activation for the thermal reversal of the unsubstituted [N]phenylenes are all very close in value, around 27 kcal mol^{-1}. However, the exothermicity increases going down the series from [3]- to [5]phenylene. This trend can be attributed to the decreasing HOMO-LUMO gap and increasing antiaromaticity along the series, which result in stronger {CpCo} bonds.

The π-stacking interaction of aromatic materials in the solid state is a key variable to consider in achieving high charge mobilities for electronic applications. The [2.2]paracyclophane motif, with an inter-ring distance shorter than the 3.4 Å necessary for effective π-stacking, has been a useful tool for investigating the effects of stacking on optoelectronic properties [57]. In 2005 Leung and coworkers

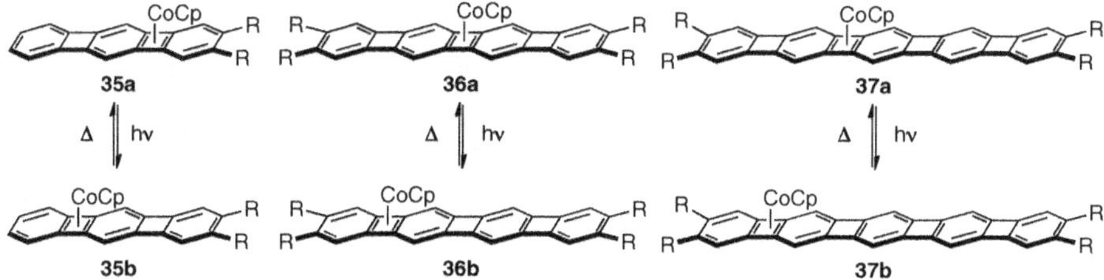

Scheme 11 Photo-thermal IRHR of η^4-CpCo-complexed linear [N]phenylenes, for **35** and **36** R = TMS, for **37** R = H or TMS [55]

Table 4 Photostationary state of IRHR of [N]phenylenes **35–37** [55]

[N]Phenylene-CpCo	Haptomer **a**	Haptomer **b**	$\Delta H^{\neq(c)}$	ΔH^{c}
35[a]	1	1	26.9	0
36[a]	1	2	27.1	−7.6
37[b]	7	3	27.3	−9.7

Irradiated with [a]350 nm lamp, [b]310 + 365 nm lamp
[c]Calculated values for R = H in kcal mol^{-1}

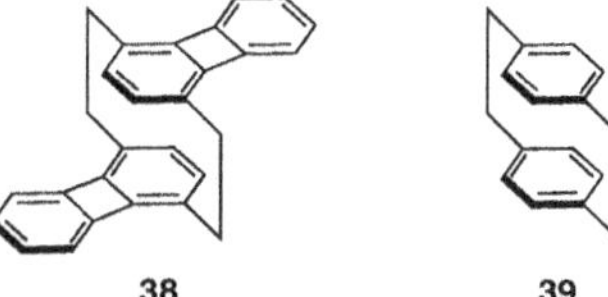

Fig. 6 Structures of *anti*-[2.2]-(1,4)-biphenylenophane **38** and parent [2.2] paracyclophane **39** [57]

Table 5 Photophysical and electrochemical properties of **38**, **39**, and **20** [57]

Compound	Lowest E λ_{max} (nm)	λ_{em} (nm)	τ (ps)	E_{ox} (V vs Ag/Ag^+)
38	378	537	229	1.37, 1.55
39	–	–	–	1.59
20	359	518	194	1.72

synthesized *anti*-[2.2]-(1,4)-biphenylenophane (**38**), employing the biphenylene framework to investigate the phane properties of antiaromatics (Fig. 6). An X-ray crystal structure of **38** confirmed the *anti* relationship of the biphenylene rings. The inter-ring distance of 3.09 Å is identical to that of the parent [2.2]paracyclophane; however, the ethano C–C bridge in **38** is significantly shorter, likely due to the increased flexibility of biphenylene [58]. Additional thermal stability is imparted to **38** by a decrease in the strain of this bond.

The photophysical properties of **38** as compared to those of biphenylene **20** and the electrochemical properties of **38** as compared to **20** and **39** are summarized in Table 5 [57]. A bathochromic shift and spectral broadening in going from **20** to **39** is consistent with ground-state interactions between the biphenylene units in the cyclophane. Since the Stokes shifts in **38** and **20** are

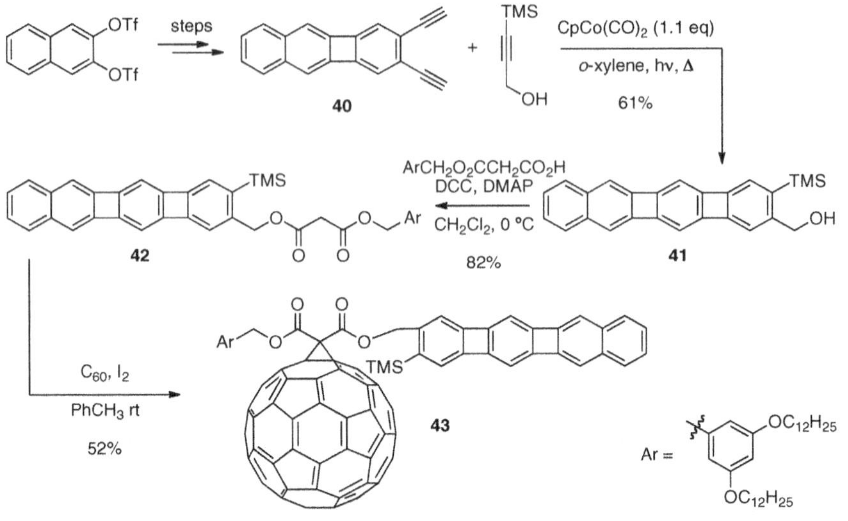

Scheme 12 Synthesis of benzo[3]phenylene-C_{60} dyad **43** [59]

similar, it is difficult to determine whether or not emission in **38** is the result of an excimer-like phane-state. However, the authors suggest that the longer lifetime of **38** implies participation of the phane state. Cyclophane **38** is more easily oxidized than both **39** and **20**.

To investigate the use of [N]phenylenes for photovoltaic applications, the synthesis of linear benzo[3]phenylene-C_{60} dyad **43** was reported in 2005 [59]. The synthesis employed iterative Co-catalyzed [2 + 2 + 2] cyclizations, using 3-trimethylsilylpropargyl alcohol to install a tether in the final cyclization to yield **41**. Finally, diester **42** was reacted with C_{60} using the Bingel conditions [60, 61] to give the desired dyad. The absorption spectrum of dyad **43** closely resembles the sum of its constituents, signifying that there are not significant ground state interactions between the [N]phenylene and fullerene which could otherwise inhibit charge separation (Scheme 12).

As with other linear [3]phenylenes, **41** and **42** are reportedly not fluorescent. This is likely due to high rates of IC, which would make the donor portion of this particular dyad not a good candidate for photovoltaic applications [62]. This led researchers to investigate the photoinduced electron transfer of angular phenylenes, with much slower rates of IC, in dyads. The fluorescence and quantum yield of triangular [4]phenylene dyad **44** displays a solvent polarity dependence characteristic of intramolecular charge transfer; however, levels of the charge-transfer complex are low at equilibrium (Fig. 7).

The small HOMO-LUMO gap of antiaromatics led researchers to investigate the use of biphenylene as a molecular wire in a single molecule device [63]. Research on single molecule devices made from acenes has shown that increasing conductivity can be correlated to decreasing HOMO-LUMO gap and increasing quinoidal stability [64]. Thioether-substituted biphenylene **47** was synthesized according to the procedure outline in Scheme 13 and its conductance compared with that of fluorenes

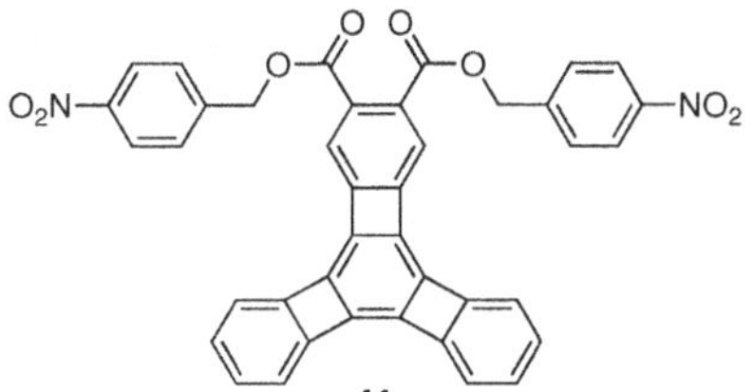

Fig. 7 Triangular [4] phenylene dyad **44** [62]

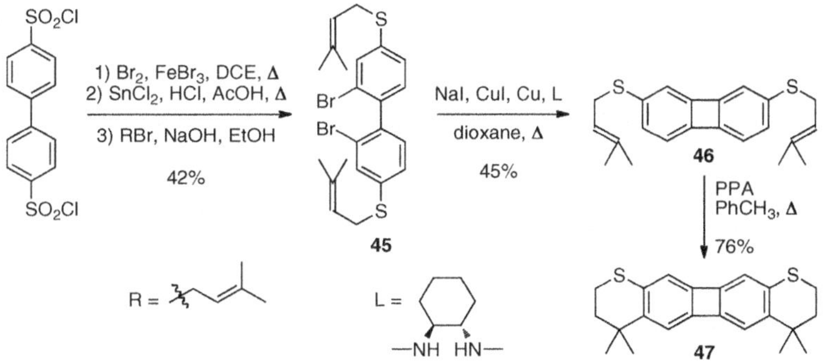

Scheme 13 Synthesis of thioether-substituted biphenylene **47** [63]

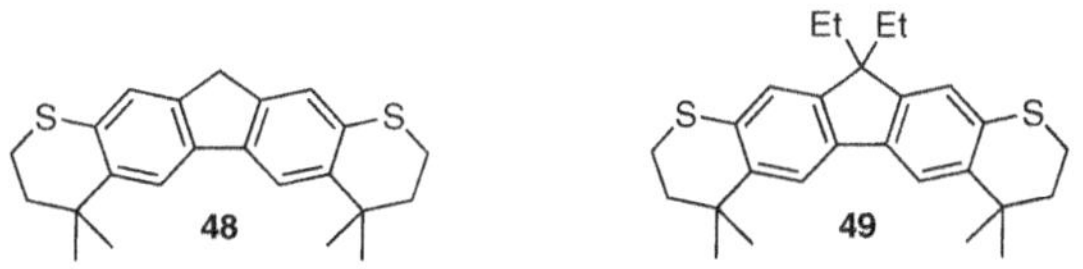

Fig. 8 Thioether-substituted fluorenes **48** and **49** [63]

Table 6 Conductance and first oxidation potential of PAHs **47–49** [63]

Compound	Conductance ($G_0 \times 10^3$)	$E_{1/2}$ (V vs Ag/Ag$^+$)
47	3.6 ± 0.2	0.509
48	3.5 ± 0.3	0.633
48	4.4 ± 0.3	0.627

48 and **49** (Fig. 8) [63]. The essential biphenylene-forming reaction was achieved by Ullman coupling of the 2,2-dibromobiaryl **45**. Acidic cyclization of **46** is completely regioselective at the β position to give **47**, as cyclization at the α position would involve a carbocation cation intermediate with greater cyclobutadiene character.

The conductances of compounds **47–49** were measured using a scanning tunnel microscope (STM) based break-junction technique. Results from the conductance and first oxidation potentials are summarized in Table 6. Surprisingly, although biphenylene **47** is more easily oxidized than the fluorene analogues, decreasing oxidation potential did not seem to correlate with increasing conductivity. These

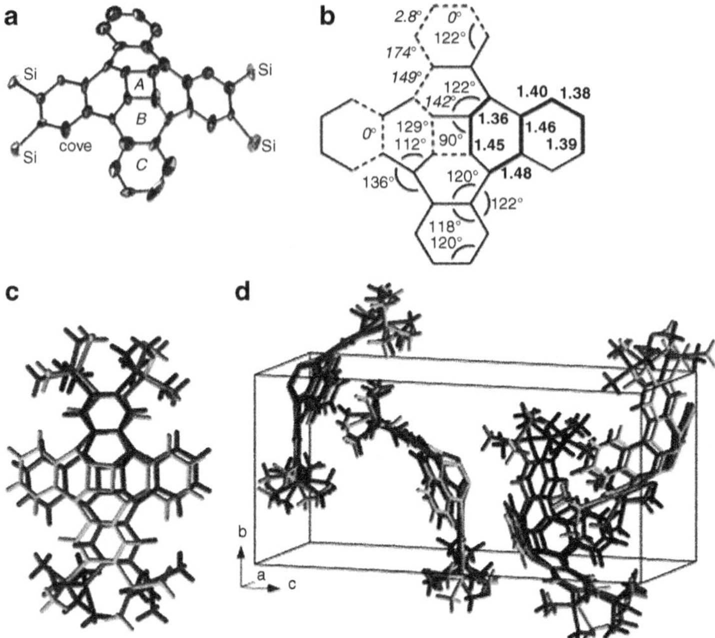

Fig. 9 Structure of **52**: (**a**) X-ray crystal structure of major contributor (methyl groups and H atoms omitted for clarity), (**b**) calculated geometry, (**c**) disorder in the crystal, and (**d**) crystal packing. Reprinted with permission from [66]. Copyright 2009 John Wiley and Sons

results reflect the complexities of materials designs and difficulties in translating molecular properties to solid-state and device properties.

Nonbenzenoid rings in graphitic structures induce curvature, the most classic example being the 12 5-membered rings that allow the spherical shape of C_{60}. The [n]circulenes are a class of PAHs derived from the fusion of benzenoid rings to the corresponding [n]radialene frame [65]. The curvature and bowl-to-bowl inversion barrier of these materials is expected to increase with decreasing size of the central ring. The first synthesis of benzannulated [4]circulene, quandrannulene (**52**) was reported by King and coworkers in 2010 [66]. The four-membered ring is derived from the photodimer of naphthoquinone (**50**). The authors take advantage of the Co-catalyzed cyclotrimerization's well-established ability to form highly-strained rings to form the bridges in going from tetrayne **51** to **52**.

The X-ray crystal structure and calculated geometry (B3LYP/6-311G**) of **52** are shown in Fig. 9. Calculated bond lengths and NICS value support the radialene-structure as depicted in Scheme 14, with ring A having an NICS value of 4.5 ppm. The bowl depth, the distance between the plane containing ring A and the plane formed by the eight atoms that fuse rings B and C, is 1.36 Å, significantly deeper than a depth of 0.87 Å for [5]circulene.

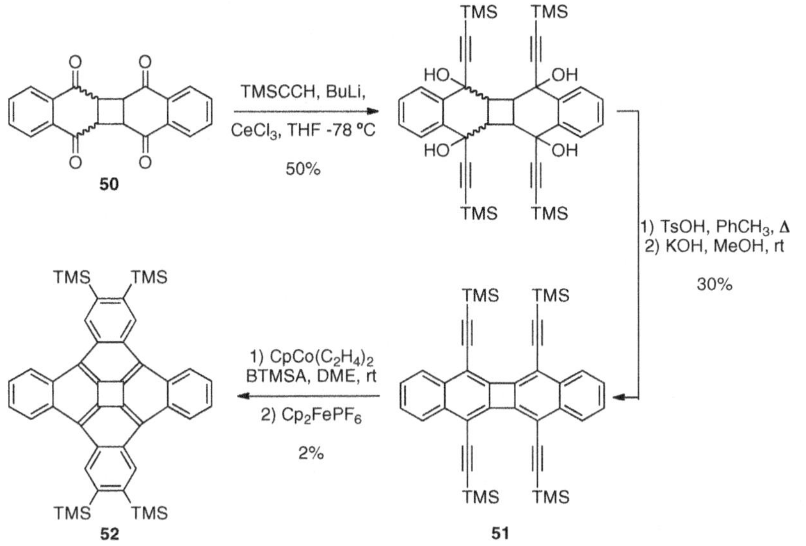

Scheme 14 Synthesis of quadrannulene **52** [66]

2.5 Polymeric and Graphitic [N]Phenylenes

Currently, ladder polymers and 2D or 3D graphitic materials containing phenylenes have mostly been an area of theoretical interest due to synthetic challenges. Inspired by the successful synthesis of the [N]heliphenes up to $N = 9$ [30, 31], calculations on the properties of even longer linear, angular, and helical phenylenes, of $N = 14$, 19, and 24, were reported in 2003 [67]. Notably, in each case the authors found that bond lengths and NICS value converge on the centermost rings (Fig. 10), the zigzag and helical topologies taking on increased bond alternation, while all bonds of the six-membered rings are nearly equal in the linear topology.

Despite the increase in bond alternation, isotropic magnetic susceptibility calculations indicate that the zigzag topology is the most aromatic followed closely by the helical analogue. For zigzag and helical topologies the terminal benzenoid ring is the most aromatic while the penultimate is the least, whereas for the linear case aromaticity decreases from the terminus towards the center. The four-membered rings in the linear topology are more antiaromatic than their zigzag and helical counterparts.

The polyhedral carbon allotrope containing phenylene units and a molecular formula of C_{120} is known as archimedene **53** (Fig. 11) [69–75]. Ab initio calculations were performed on **53** as well as eight related bowl-shaped hydrocarbons in 2005 by Schulman and Disch [76]. The resulting geometries are very consistent with what has been shown experimentally for angular and branched [N]phenylenes: the six-membered rings have significant bond alternation, 1.479 Å and 1.364 Å, and the four-membered rings are nearly square. NICS values for the

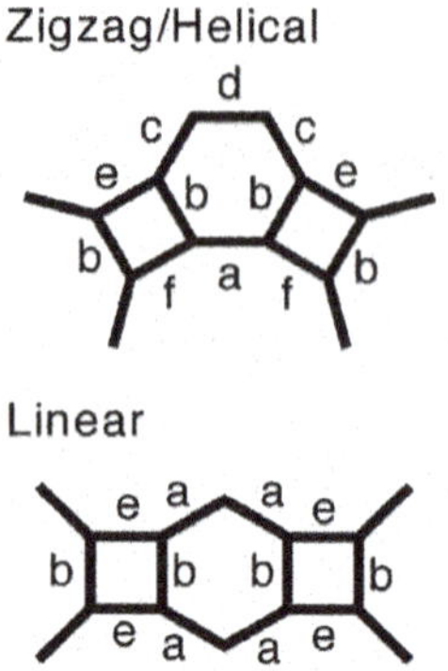

Zigzag/Helical

Linear

Bond	Zigzag	Helical	Linear
a	1.361	1.362	1.395
b	1.452	1.453	1.417
c	1.371	1.371	-
d	1.438	1.435	-
e	1.492	1.492	1.511
f	1.491	1.496	-

Fig. 10 Convergent bond lengths in extended [N]phenylenes [67]

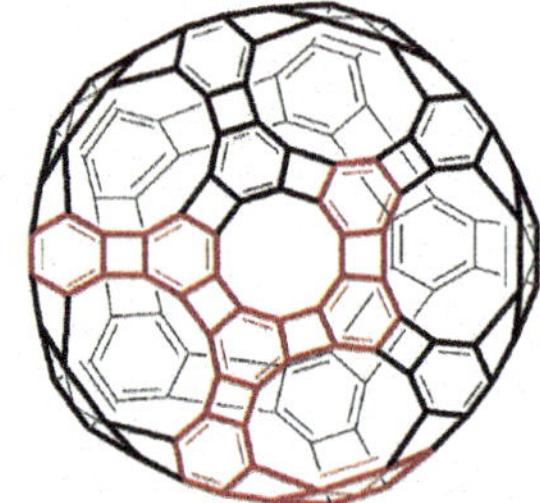

Fig. 11 Archimedene **53**. Reprinted with permission from [68]. Copyright 2003 American Chemical Society

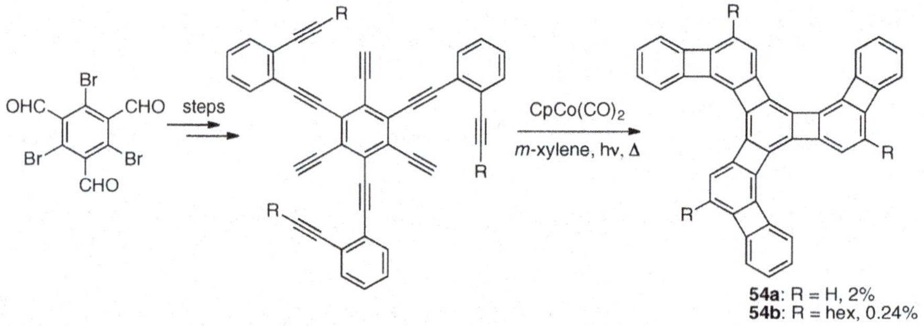

54a: R = H, 2%
54b: R = hex, 0.24%

Scheme 15 Threefold cyclotrimerization to give archimedene fragment **54** [68]

six- and ten-membered rings are slightly negative, −1.9 and −0.2 respectively, while those of the four-membered ring are slightly positive, 3.2 indicating that the system is neither strongly aromatic nor antiaromatic.

Synthetic attempts towards **53** have so far focused on assembling smaller fragments. In 2003, Vollhardt and coworkers reported the synthesis of the C_{3h}-symmetric fragment [7]phenylene **54** (highlighted in red in Fig. 10) by a threefold cyclotrimerization of the appropriate nonayne precursor (Scheme 15) [68].

Fig. 12 Insertion of 2-carbon fragment into C60. Reprinted with permission from [77]. Copyright 2003 American Chemical Society

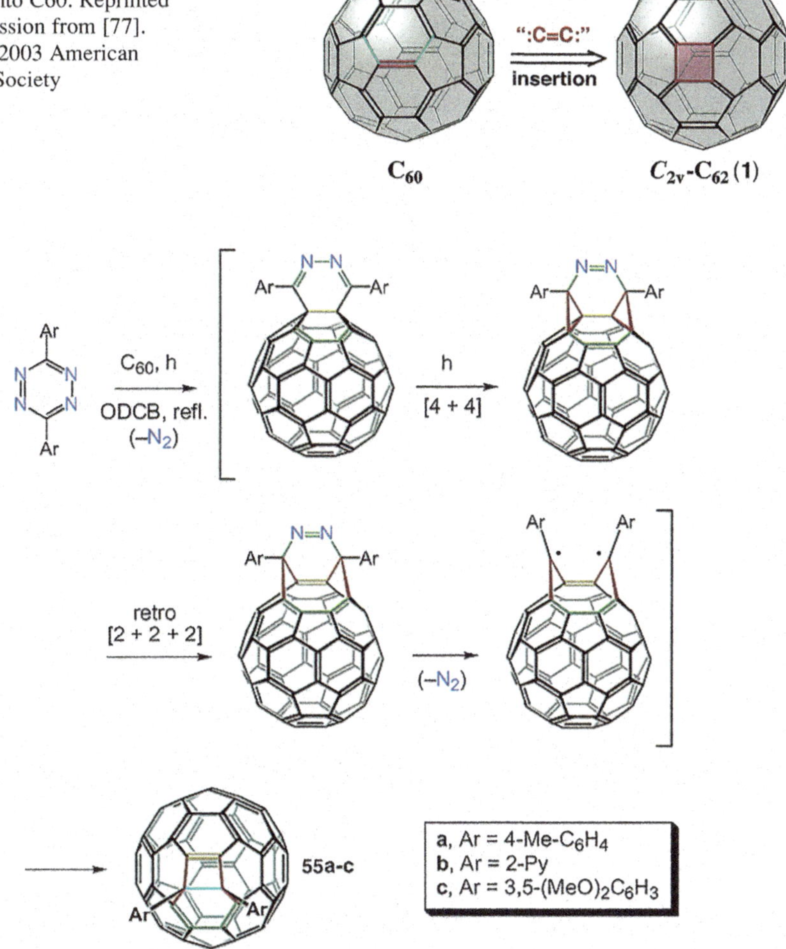

Scheme 16 One-pot synthesis of C_{62} analogue **55**. Reprinted with permission from [77]. Copyright 2003 American Chemical Society

Unfortunately, insolubility prevented some NMR studies and X-ray analysis. However, DFT and NICS calculations of **54a** indicated that the central ring has greater diatropic character than the triangular [7]phenylene with linear arms.

There is also interest in incorporating rings of different sizes into the standard C_{60} skeleton, in the hope of creating new functional materials. Rubin and coworkers proposed a conceptual approach for inserting a two-carbon fragment into fullerene to generate C_{62} in 2000 (Fig. 12) [78]. Attempts to isolate the parent hydrocarbon C_{62} were not successful, likely due to its predicted high reactivity. Therefore the authors turned to a related system that allowed them to access diaryl-C_{62} **55** [77]. The one-pot reaction, commencing with the Diels-Alder reaction of diaryltetrazine

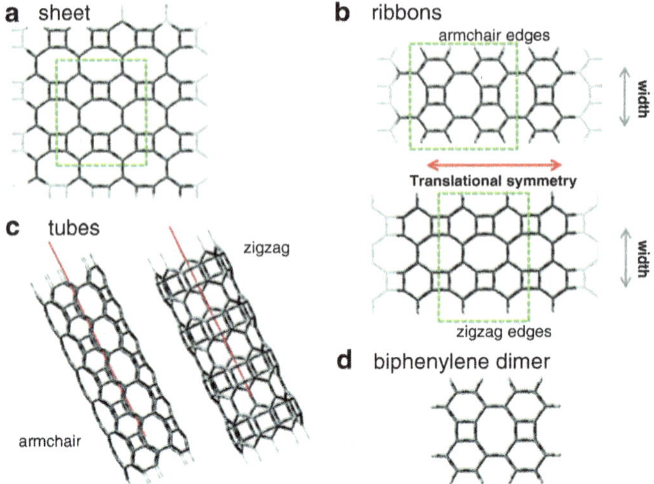

Fig. 13 Structures of phenylene-containing graphitic materials: (**a**) biphenylene sheet, (**b**) biphenylene nanoribbons with armchair and zigzag edge morphology, (**c**) biphenylene nanotubes, and (**d**) biphenylene dimer. Reproduced with permission from [79]. Copyright 2010 American Chemical Society

and C_{60}, involves a sequence of electrocyclic reactions and the eventual extrusion of 2 equiv. nitrogen gas (Scheme 16).

Other graphitic materials containing phenylenes, such as nanoribbons, nanosheets, and nanotubes, can also be imagined and have been subjects of theoretical interest (Fig. 13) [79]. The biphenylene nanoribbons (BNR, Fig. 13b) were dubbed either armchair or zigzag depending on the edge morphology, analogous to graphene nanoribbons (GNR). For the armchair BNR, calculations predict the band gap will decrease monotonically with the width of the ribbon, whereas zigzag BNR are predicted to have metallic character for all but the narrowest ribbon. Spin-polarized calculations predict both edge configurations to be diamagnetic, unlike GNR for which the zigzag morphology is predicted to have interesting magnetic properties [80]. Both armchair and zigzag biphenylene nanotubes are predicted to possess metallic character.

3 Diaryl[*a,e*]pentalenes

Another commonly studied formally antiaromatic moiety is the 8π-electron bicyclic pentalene molecule (Fig. 14a) [81]. Pentalene is locked in a planar geometry, unlike cyclooctatetraene, an 8π-electron system that adopts a boat geometry to reduce antiaromaticity. Unsubstituted pentalene is highly reactive and dimerizes at temperatures above −100 °C [82–84]. The 10π-electron dianion of pentalene, however, is a stable and well-characterized species [85–87].

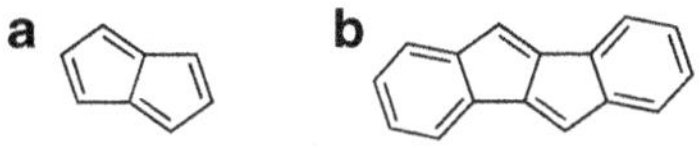

Fig. 14 Structures of (**a**) antiaromatic pentalene [81–87] and (**b**) dibenzo[*a,e*]pentalene [88]

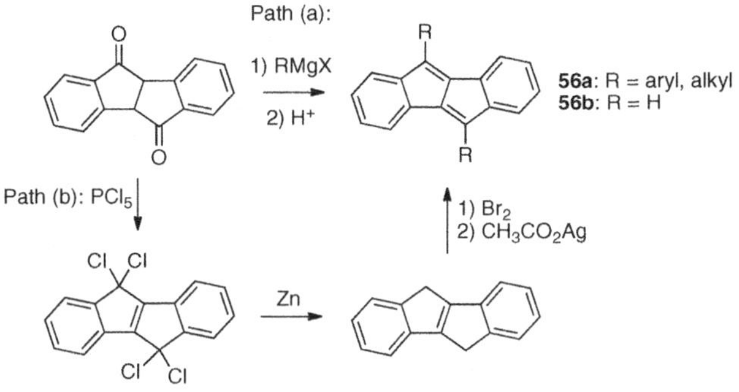

Scheme 17 Original synthetic routes to (*a*) 5,10-disubstituted dibenzo[*a,e*]pentalene **56a** [89], and (*b*) unsubstituted dibenzo[*a,e*]pentalene **56b** [90]

Analogous to the relationship between cyclobutadiene and [N]phenylenes, it is possible to synthesize stable ladder structures by fusing aryl rings to a pentalene core resulting in diaryl[*a,e*]pentalene (DAP, Fig. 14b) [88]. The first example of a 5,10-substituted dibenzo[*a,e*]pentalene (**56a**) was reported in 1912 [89], while the parent hydrocarbon (**56b**) was not synthesized until 1952 (Scheme 17) [90]. Both synthetic routes relied on the diketone shown in Scheme 17 as the starting material and did not provide much versatility beyond substitution at the C5/10 position. Increasing interest in PAHs as materials for optoelectronic applications during the last decade has caused a resurgence in research on the synthesis and properties of DAPs.

3.1 *Recent Synthetic Developments in Diaryl[*a,e*]pentalenes*

Several different syntheses of DAPs have been reported since 2008. Seeking to investigate the reactivity of strained alkynes towards nucleophiles, Orita, Otera, and coworkers discovered an efficient method for synthesizing unsymmetrical dibenzo[*a,e*]pentalenes [91]. A nucleophile, such as an alkyl or aryllithium, adds across one triple bond of 5,6,11,12-tetradehydrodibenzo[*a,e*]cyclooctene (**57**), generating a vinylic anion (Scheme 18a). The anion can then add across the remaining triple bond and is quenched with an electrophile ("E"). Yields were high in each case, varying both the nucleophile and the electrophile, ranging from 57% to 76%. Using iodine as the electrophile resulted in vinyl iodide **58a**, which could then be used in metal-catalyzed cross-coupling reactions to synthesize additional derivatives (Scheme 18b).

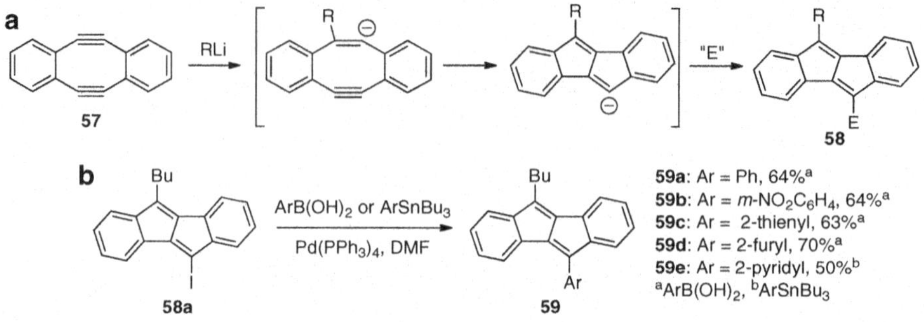

Scheme 18 (**a**) Synthesis of unsymmetrical dibenzo[*a*,*e*]pentalenes. (**b**) Metal-catalyzed cross-coupling of vinyl iodide **58a** [91]

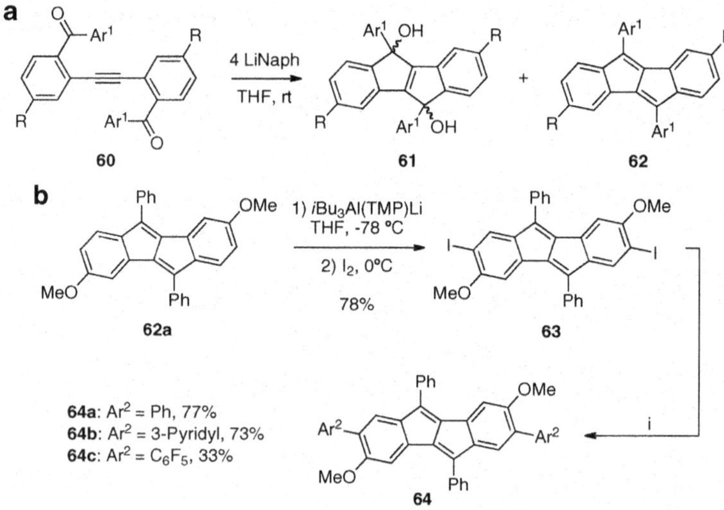

Scheme 19 (**a**) Reductive double cyclization of *o*,*o*′-bis(arylcarbonyl)-diphenylacetylene **60** to give methylene-bridged stilbene **61** and DAP **62**. (**b**) Metal-catalyzed cross-coupling of 2,7-diiodo-substituted DAP **63** [92]. Conditions: for **64a/b** (i) $ArB(OH)_2$, $Pd_2(dba)_3{\bullet}CHCl_3$/S-Phos, K_3PO_4, $PhCH_3/H_2O$ or 1-butanol, for **64c** (i) (1) *n*BuLi, THF −78 °C, (2) C_6F_6, −78 °C to rt

Similarly inspired by the reactivity of phenylacetylenes, Yamaguchi and coworkers reported a new synthesis of DAPs and the closely related methylene-bridged stilbene scaffold [92]. Treatment of *o*,*o*′-bis(arylcarbonyl)-diphenylacetylenes (**60**) with lithium naphthalide caused a two-electron reduction. Calculations indicated that each radical anion resides at one of the carbonyl positions, and thus the reduction is followed by a twofold radical 5-*endo-dig* cyclization onto the acetylene moiety to give the tetracyclic products (Scheme 19a). The authors determined that DAP **62** resulted from the over-reduction of **61** by excess lithium naphthalide. Using the electron-rich **62a** the authors are able to iodinate the DAP at the C2/7 position, and then further extend the DAP through metal-catalyzed cross-coupling (Scheme 19b)

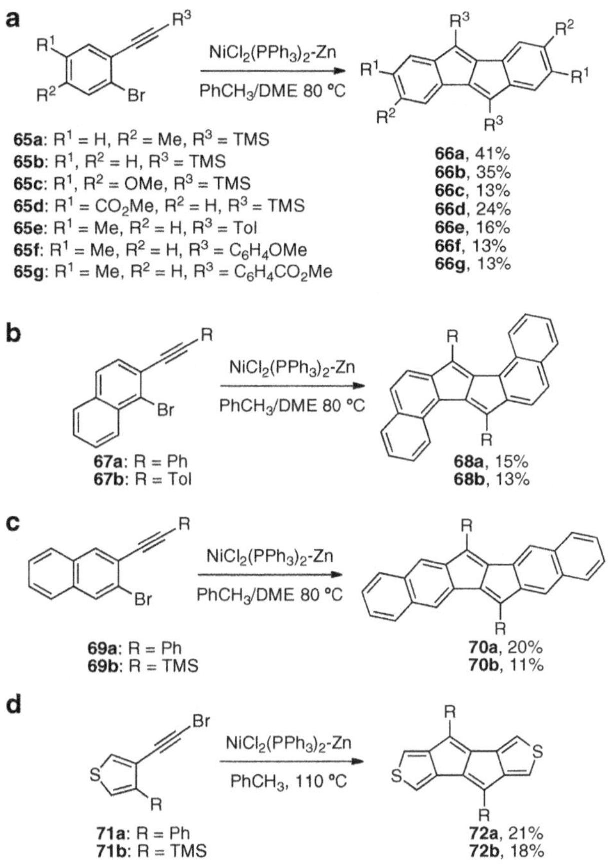

Scheme 20 Nickel-catalyzed synthesis of DAPs from 2-bromo-1-ethynylarenes: (**a**) dibenzopentalenes **66** [94], (**b**) angular **68** and (**c**) linear dinaphthopentalenes **70** [95], (**d**) di-(3,4)-thienopentalene **72** [96]

a

$NiCl_2(PPh_3)_2$-Zn

$PhCH_3$/DME 80 °C

65a: R^1 = H, R^2 = Me, R^3 = TMS
65b: R^1, R^2 = H, R^3 = TMS
65c: R^1, R^2 = OMe, R^3 = TMS
65d: R^1 = CO_2Me, R^2 = H, R^3 = TMS
65e: R^1 = Me, R^2 = H, R^3 = Tol
65f: R^1 = Me, R^2 = H, R^3 = C_6H_4OMe
65g: R^1 = Me, R^2 = H, R^3 = $C_6H_4CO_2Me$

66a, 41%
66b, 35%
66c, 13%
66d, 24%
66e, 16%
66f, 13%
66g, 13%

b

$NiCl_2(PPh_3)_2$-Zn

$PhCH_3$/DME 80 °C

67a: R = Ph
67b: R = Tol

68a, 15%
68b, 13%

c

$NiCl_2(PPh_3)_2$-Zn

$PhCH_3$/DME 80 °C

69a: R = Ph
69b: R = TMS

70a, 20%
70b, 11%

d

$NiCl_2(PPh_3)_2$-Zn

$PhCH_3$, 110 °C

71a: R = Ph
71b: R = TMS

72a, 21%
72b, 18%

Work reported in 1999 found that a palladium catalyst could transform a substituted 2-iodo-1-ethynylbenzene into a DAP [93]. However, the reaction appeared to be an isolated case as substrates of that nature more often form phenylene-ethynylene macrocycles under the given conditions. Despite the lack of generalizability, the synthetic accessibility of starting materials and the ability to form three bonds in one pot made this transformation intriguing for further investigation [94]. Optimized conditions using a variety of 2-bromo-1-ethynylarenes (**65**) and a nickel catalyst in the absence of additives successfully suppressed the competing biaryl coupling to give the desired DAPs (**66**) in moderate yields (Scheme 20a).

This new methodology allowed access to DAPs with a wide variety of substituents. The same conditions were used to synthesize successfully extensions of the PAH to dinaphthopentalenes of different connectivities (**68**/**70**, Scheme 20b, c) [95]. Additionally, it has been shown to provide access to di-(3,4)-thieno[*a*,*e*]pentalene **72** (Scheme 20d) [96].

DAPs **66a**, **68a**, and **70a** were all analyzed by X-ray crystallography and, as with the [N]phenylenes, the degrees of bond alternation can be an indication of

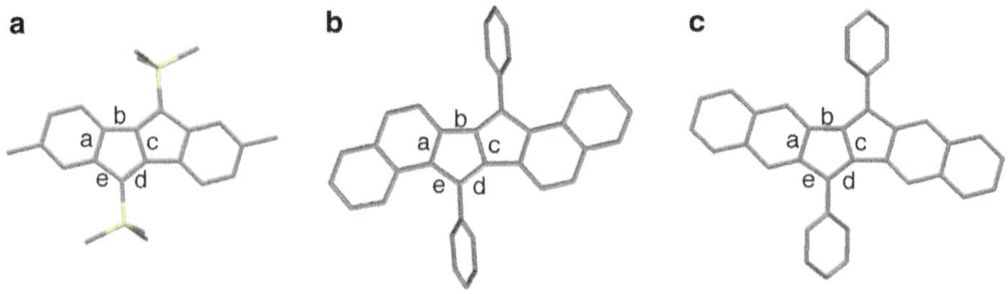

d

Bond	**66a** (Å)	**68a** (Å)	**70a** (Å)
a	1.425	1.407	1.441
b	1.471	1.465	1.475
c	1.471	1.454	1.449
d	1.364	1.363	1.356
e	1.503	1.498	1.488

Fig. 15 X-Ray crystal structures of (**a**) **66a**, (**b**) **68a**, (**c**) **70b**, and (**d**) selected bond lengths. Crystal structure data from [94, 95] was gathered from the Cambridge Structural Database (*CSD*)

aromaticity (Fig. 15). The calculated NICS values for unsubstituted **56b** are NICS $= +5.88$ for the pentalene fragment and NICS $= -6.23$ for the benzene rings, which is approximately one-third of the value for pentalene and one-half of the value for benzene, respectively [94]. The difference in length of the bond fusing the five- and six-membered rings (bond “a”) is notably shorter in **68a**, and longer in **70a**, than in **66a** [95]. In **68a**, the shorter length indicates decreased bond alternation and therefore increased 8π-electron character in the pentalene unit. Considering the resonance structures shown in Scheme 20b, **68a** has decreased antiaromatic character but this structure negates the stabilization of a Clar sextet in the terminal ring. For **70a** it is possible to imagine a resonance structure that benefits from both decreased double bond character of bond “a” and a Clar sextet in the terminal ring as drawn in Scheme 20c.

Independently from the work reported above, Tilley and coworkers developed a related synthetic methodology using palladium as the catalyst [97]. The yields reported using these conditions were generally higher than those reported by Kawase and coworkers (Scheme 21a). Additionally, it was demonstrated that the Pd-catalyzed conditions can be used to efficiently synthesize other pentalene-containing compounds di[2,3]-thieno[*a*,*e*]pentalene **76** and hexaaryl-pentalene **78**.

The following year Tilley and coworkers reported a method for synthesizing PAHs containing multiple pentalene units [98]. The synthesis of **81** is shown in Scheme 22 as an example of the methodology also used to synthesize PAHs **82–84**. Bromostilbene derivatives such as **79** were synthesized by Horner-Wadsworth-Emmons reaction of the appropriately substituted xylene. Heck conditions were then used to couple **79** with diphenylacetylene to give *s*-indacene **80**. Oxidative ring closure using $FeCl_3$ then yielded the desired PAH. This reaction sequence allowed access to PAHs with a variety of topologies (Fig. 16).

Scheme 21 Pd-catalyzed syntheses of (**a**) DAPs **74**, (**b**) dithieno[*a*,*e*]pentalene **76**, and (**c**) hexa-aryl-substituted pentalene **78** [97]

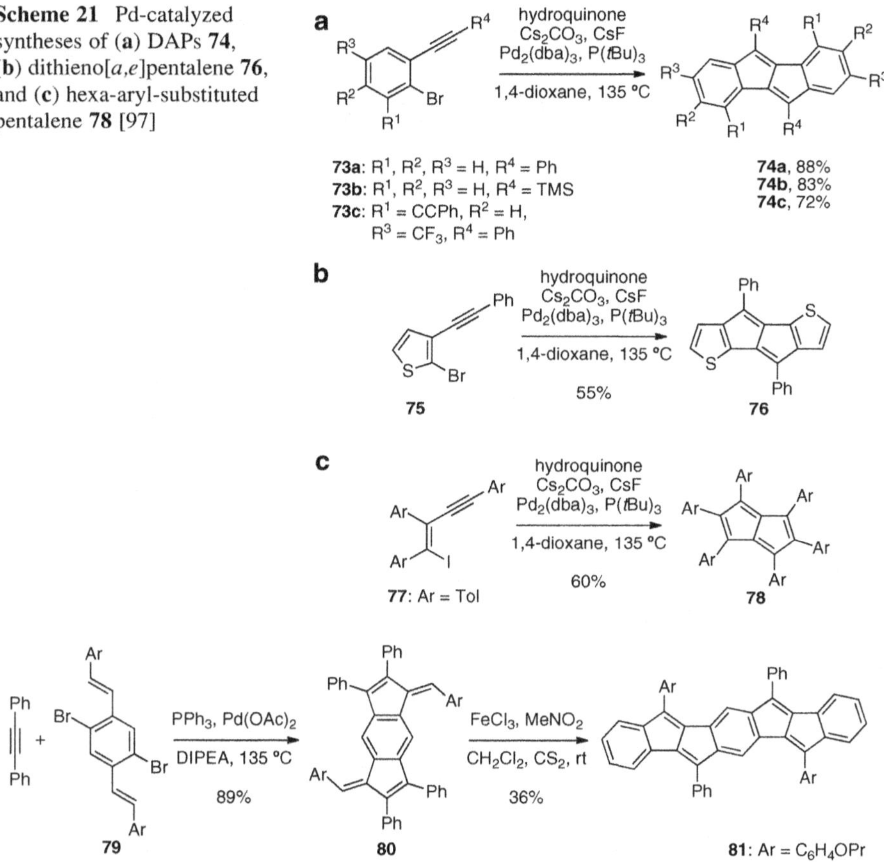

Scheme 22 Synthesis of multiple pentalene-containing PAH **81** [98]

Very recently, a procedure for the synthesis of DAPs using gold catalysis was reported [99]. Rearrangement of 1,2-diethynylarynes **85** allowed the authors to produce DAPs without substituents at the C5/10 position (**86**), including unsubstituted dibenzo[*a*,*e*]pentalene **56b** (Scheme 23a). The conditions were also successful for thiophene-substituted **87**, although the yield was lower as a result of the stability of the product (Scheme 23b). The authors propose a mechanism that passes through a gold vinylidene intermediate.

In 2007, Saito and coworkers found that treatment of TIPS-protected phenylacetylene with lithium followed by treatment with iodine resulted in a 7% yield of DAP **91** and 28% 1,4-diiodo-1,3-butadiene **90** (Scheme 24) [100] The dilithium dibenzopentalenide **89** was eventually characterized by X-ray crystallography and NMR, the first X-ray structure of a DAP dianion (Fig. 17). Unlike the neutral DAP, there is more bond alternation in the six-membered rings of **89** than the five-membered rings.

The surprising formation of **91** inspired the authors to investigate further the reactivity of this DAP derivative. When **91** was treated with methyllithium (MeLi)

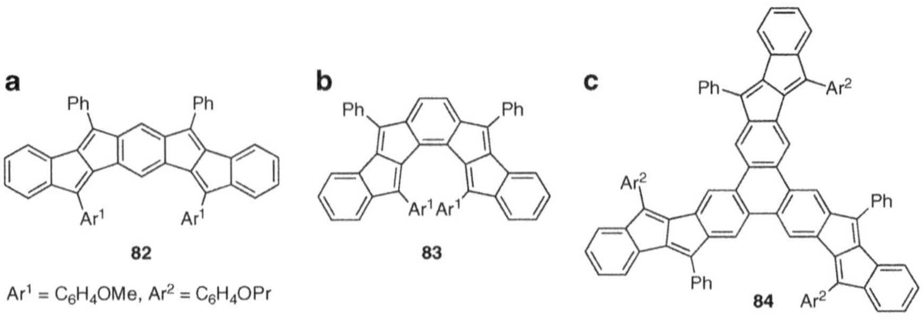
a
Ph
Ph
Ar1
Ar1
82
b
Ph
Ph
Ar1 Ar1
83
c
Ph
Ar2
Ar2
Ph
Ph
84
Ar2
Ar1 = C6H4OMe, Ar2 = C6H4OPr

Fig. 16 Structures of multiple pentalene-containing (**a**) **82**, (**b**) **83**, and (**c**) **84** [98]

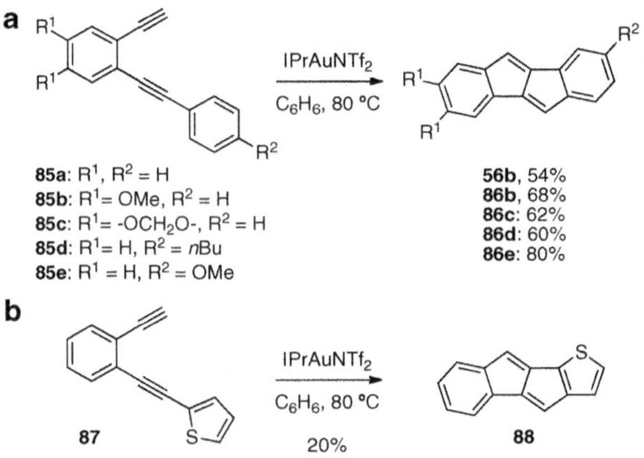
a
R1
R1
R2
IPrAuNTf2
C6H6, 80 °C
R1
R1
R2
85a: R1, R2 = H
85b: R1= OMe, R2 = H
85c: R1= -OCH2O-, R2 = H
85d: R1= H, R2 = nBu
85e: R1 = H, R2 = OMe
56b, 54%
86b, 68%
86c: 62%
86d: 60%
86e: 80%
b
87
IPrAuNTf2
C6H6, 80 °C
20%
88

Scheme 23 Gold-catalyzed rearrangements of (**a**) 1,2-diethynylarenes **85** to DAPs **86** and (**b**) **87** to benzo[*a*]-thieno[*e*]pentalene **88** [99]

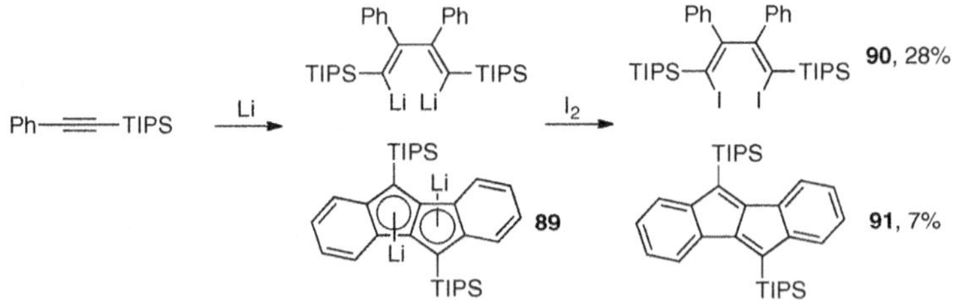
Ph—≡—TIPS
Li
Ph Ph
TIPS
TIPS
Li Li
I2
Ph Ph
TIPS
TIPS
I I
90, 28%
TIPS
Li
Li
TIPS
89
TIPS
TIPS
91, 7%

Scheme 24 Reduction of TIPS-substituted phenylacetylene to give dilithium dibenzopentalenide **89**, and subsequent treatment with I_2 to give DAP **91** [100]

the monoanion **92** formed which could then be trapped with an electrophile to give methylene-bridged stilbene **93** (Scheme 25a) [101]. DAP **91** also reacted with excess bromine and iodine to give 5,10-dihalo-DAPs **94a/b** in good yields (Scheme 25b).

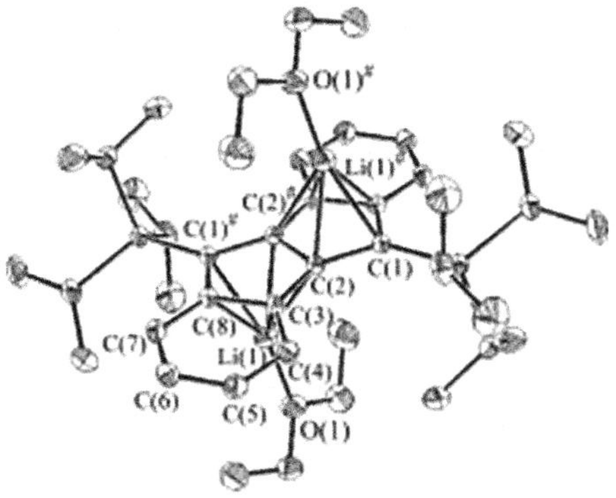

Fig. 17 ORTEP rendering of the X-ray crystal structure of **89**, hydrogens omitted for clarity. Reproduced with permission from [100]. Copyright 2007 John Wiley and Sons

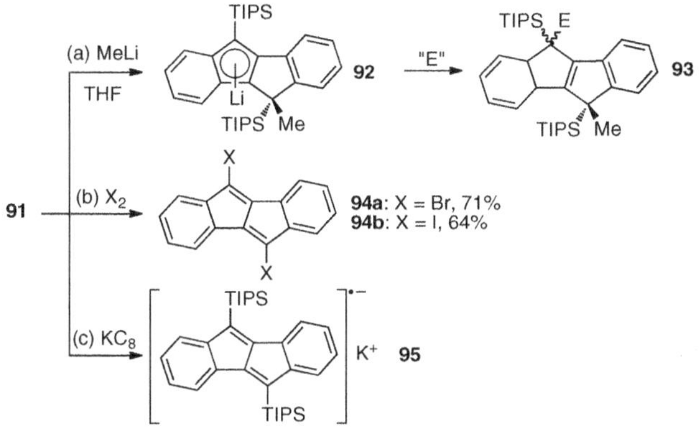

Scheme 25 Reactivity of DAP **91** [101]

Cyclic voltammetry (CV) of **91** [102] contained two reduction peaks, the first of which was reversible and at a more positive potential than that of 5,10-dimethyldibenzo[*a,e*]pentalene [103], −1.48 V and −1.9 V (vs Ag/Ag^+), respectively. Thus the authors attempted a chemical one-electron reduction of **91** by action of potassium graphite (Scheme 25c) [102]. Radical anion **95** was successfully isolated as a potassium salt and characterized by X-ray crystallography.

3.2 Electrochemical and Optical Properties of Diaryl[a,e] pentalenes

As a result of the established stability of the 10π-electron dibenzo[*a,e*]pentalene dianion and the theoretical stability of the dication, these materials are often observed to have amphoteric redox properties. DAPs **64a–c** all undergo a reversible

Table 7 Electrochemical properties of **64a–c** [92]

Compound	$^{red}E_{1/2}$ (V)	$^{ox}E_{1/2}$ (V)
64a	−1.71	0.58
64b	−1.66	–
64c	−1.59	0.80

Potentials (V) vs Fc/Fc^+, 0.1 M $TBAPF_6$ in CH_2Cl_2, 100 mVs

Table 8 Electrochemical properties of **66a–g** [95], **68a**, and **70a** [95]

Compound	$^{red}E_{1,1/2}$ (V)	$^{red}E_{2,1/2}$ (V)	$^{ox}E_{1,1/2}$ (V)	$^{ox}E_{2,1/2}$ (V)
66a	−1.60	−2.17	0.87	1.06
66b	−1.53	−2.17	0.97	1.16
66c	−1.63	−2.21	0.35	0.72
66d	−1.23	−1.85	–	–
66e	−1.62	−2.18	0.87	0.98
66f	−1.66	−2.24	0.81	1.13
66g	−1.38	−1.74	0.94	1.05
68a	−1.52	−2.06	0.55	1.14
70a	−1.79	−2.26	0.73	1.11

Potentials (V) vs Ag/Ag^+, 0.1 M $TBAClO_4$ in DMF, 100 mVs

Table 9 Electrochemical and optical properties of **74a** and **81–84** [98]

Compound	$^{red}E_{1,1/2}$ (V)	$^{red}E_{2,1/2}$ (V)	λ_{max} (nm)
74a	−1.41	−1.96	447
81[a]	−1.33	−1.65	550
82[a]	−1.34	−1.71	534
83	−1.56	−1.89	476
84	−1.64	−2.19	479

Potentials (V) vs Fc/Fc^+, 0.1 M $TBAPF_6$ in CH_2Cl_2, 100 mVs
[a]*tert*-Butyl substituted

one-electron reduction, while **64a** and **64c** both also undergo a reversible one-electron oxidation (Table 7) [92].

Many of the DAPs synthesized by Kawase and coworkers, **66a–g**, exhibit four-stage redox properties (Table 8) [94]. Notably, **66c** undergoes a reversible first oxidation and a pseudoreversible second oxidation at very low potentials. The electron-donating methoxy substituents at the benzene rings increase the energy of the HOMO, an effect also evident in absorption spectra of **66a–d** where **66d** experiences a large bathochromic shift. The opposite effect is seen in the reduction potential of carbonyl-substituted **66d** and **66g**, undergoing reduction at more positive potentials than the other derivatives. DAPs with aryl substituents in 5,10-position, **66e, f**, undergo only a small bathochromic shift of approximately 50 nm compared to **66a–c**.

The extended DAPs **68a** and **70a** showed improved oxidative ability over their shorter counterparts [95]. The first oxidation and reduction waves for **68a** and **70a** were both reversible (Table 8). Angular **68a** was both oxidized and reduced more

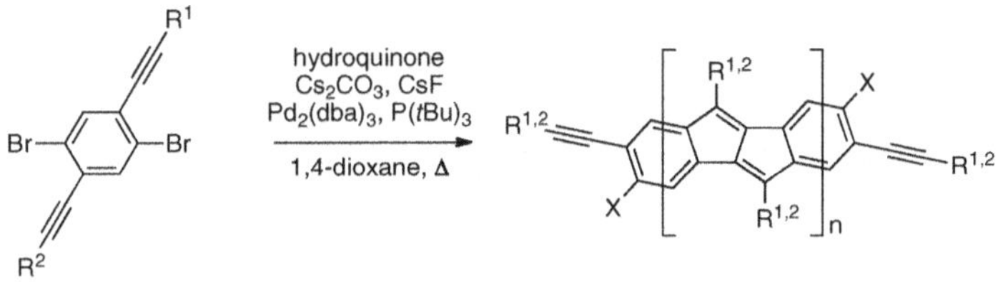

hydroquinone
Cs_2CO_3, CsF
$Pd_2(dba)_3$, $P(tBu)_3$
1,4-dioxane, Δ

96a: R^1, R^2 = $C_6H_4N(C_{16}H_{33})_2$ — **P96a**
96b: R^1 = $C_6H_4N(C_{16}H_{33})_2$, R^2 = $C_6H_4NO_2$ — **P96b**
96c: R^1 = $C_6H_4N(C_{16}H_{33})_2$, R^2 = $OC_{16}H_{33}$ — **P96c**
96d: R^1, R^2 = TIPS — **P96d**

Scheme 26 Polymerization of bis-functionalized monomers **96** to ladder polymers **P96** [105]

Table 10 Yield and molecular weights of ladder polymers **P96a–d** [105]

Polymer	Yield (%)	M_n[a]	PDI[a]	λ_{onset}[b] (nm)	λ_{em}[b] (nm)	$^{red}E_p$ (V)	$^{ox}E_p$ (V)
P96a	92	11,000	1.23	831	453, 500	–	0.56
P96b	89	7,200	1.29	805	478, 509	−1.69	0.52
P96c	90	18,000	1.44	681	457, 549	–	0.56
P96d	68	2,400	1.27	587	452, 526	−2.07	1.45

[a]Molecular weights determined by THF-GPC using a polystyrene standard
[b]Absorption and emission measured in THF

easily than linear **70a**. An amorphous thin film of **70a** exhibited high hole mobilities of 1.8×10^{-3} cm^2 V^{-1} s^{-1}. Bulk heterojunction solar cells constructed from **70a** as the donor and fullerene as the acceptor layer had a power conversion efficiency (PCE) of 0.94% and an open circuit voltage (V_{OC}) of 0.96 V.

The electrochemical and optical properties of 5,10-diphenyl-dibenzo[*a,e*]-pentalene (**74a**) and multiple pentalene-containing **81–84** are summarized in Table 9 [98]. The reductions of dipentalenes **81** and **82** are shifted to slightly more positive potentials than that of monopentalene **74a**. As expected, the absorbance spectra of **81** and **82** also underwent the largest bathochromic shift. This observation is analogous to the trend seen in [N]phenylenes in which linear annelation results in larger decreases in the HOMO-LUMO gap.

3.3 *Diaryl[*a,e*]pentalene Ladder Polymers*

Although conducting polymers containing the DAP framework have been reported previously [104], synthetic access to fully conjugated ladder structures remains a challenge. Traditionally, a single strand polymer precursor is generated first followed by an intramolecular cyclization to form the second strand of the ladder. However, this two-step process allows additional opportunities for defects to occur in the polymer backbone.

In 2010, Wang and Michinobu proposed a one-step synthesis of conjugated ladder polymers based on the Pd-catalyzed methodology developed by Tilley [105]. Symmetrical **96a/d** and unsymmetrical **96b/c** bisfunctionalized monomers were synthesized by sequential Sonogashira couplings of 1,4-diiodo-2,5-dibromobenzene and then submitted to the Tilley conditions (Scheme 26). All of the monomer was consumed after heating for 48 h to give highly soluble polymers **P96a–d** of moderate lengths (Table 10).

The optical and electrochemical properties of polymers **P96a–d** are also summarized in Table 10. As with **66c**, a bathochromic shift is seen in the absorbance spectra of the polymers with the strongest electron-donating substituents, decreasing down the series from **P96a–d**. All of the polymers were weakly fluorescent and exhibited irreversible redox properties that were also tunable depending on the electron-withdrawing or accepting ability of the substituents.

4 Conclusions and Outlook

Much can be learned by studying the interplay of aromaticity and antiaromaticity, and how these dueling theories affect synthetic access to and properties of organic materials. Fully-conjugated ladder structures, both small molecules and polymers, are interesting candidates for optoelectronic applications in that they provide conformationally-rigid systems that promote electron delocalization. The chemistry and properties surrounding the [N]phenylenes are quite well-established. However, the synthesis of polymeric and graphitic materials that reflect those properties still remains a challenge. For the case of diaryl[*a*,*e*]pentalenes, access to a large toolbox of synthetic methodology has allowed researchers the opportunity to develop a diverse array of PAHs containing this framework, as well as extending to the synthesis of ladder polymers. Continued development of synthetic methodology for both systems will allow researchers to confirm experimentally some of the theoretical predictions, and to expand our understanding of the effects of electron delocalization in ladder systems.

Acknowledgements The support of the National Science Foundation DMR-1005810 and the Center for Excitonics, an Energy Frontier Research Center funded by the U.S. Department of Energy, Office of Science, Office of Basic Energy Sciences under Award Number DE-SC0001088, is greatly appreciated.

References

1. Minkin VI, Glukhovtsev MN, Simkin BY (1994) Aromaticity and antiaromaticity. Electronic and structural aspects. Wiley, New York
2. Garratt PJ (1986) Aromaticity. Wiley, New York

3. Hückel EZ (1931) Phys 70:204
4. Dewar MJS (1965) Adv Chem Phys 8:65
5. Breslow R (1973) Acc Chem Res 6:393
6. Maier G (1988) Angew Chem Int Ed 27:309
7. Wilber KB (2001) Chem Rev 101:1317
8. Randić M (2003) Chem Rev 103:3449
9. Jusélius J, Sundholm D (2001) Phys Chem Chem Phys 3:2433
10. Jusélius J, Sundholm D (2008) Phys Chem Chem Phys 10:6630
11. Deniz AA, Peters KS, Snyder GJ (1999) Science 286:1119, and references within
12. Miljanić OŠ, Vollhardt KPC (2006) In: Haley MM, Tykwinski RR (eds) Carbon-rich compounds: from molecules to materials. Wiley-VCH, Weinheim, Chap 4
13. Lothrop WC (1941) J Am Chem Soc 63:1187
14. Shepherd ML (1991) Cyclobutarenes: the chemistry of benzocyclobutene, biphenylene, and related compounds. Elsevier, New York
15. Toda F, Garratt PJ (1992) Chem Rev 92:1685
16. Schleifenbam A, Feeder N, Vollhardt KPC (2001) Tetrahedron Lett 42:7329
17. Randić M, Balaban AT, Plavšić D (2011) Phys Chem Chem Phys 13:20644
18. Vollhardt KPC (1993) Pure Appl Chem 65:153
19. Vollhardt KPC, Mohler DI (1996) In: Halton B (ed) Advances in strain in organic chemistry. JAI Press, London
20. Trinajstić N, Schmalz TG, Nikolić S, Hite GE, Klein DJ, Seitz WA (1991) New J Chem 15:27
21. Gutman I, Tomović Ž (1994) Monatsh Chem 125:887
22. Schulman JM (1996) J Am Chem Soc 118:8470
23. Schulman JM, Disch PL (1997) J Phys Chem A 101:5596
24. Clar E (1964) Polycyclic hydrocarbons. Academic, London
25. Clar E (1971) The aromatic sextet. Wiley, London
26. Havenith RWA, van Lenthe JH, Dijkstra F, Jenneskens LW (2001) J Phys Chem A 105:3838
27. Bong DT-Y, Gentric L, Holmes D, Matzger AJ, Scherhag F, Vollhardt KPC (2002) Chem Commun 3:278
28. Bong DT-Y, Chan EWL, Diercks R, Dosa PI, Haley MM, Matzger AJ, Miljanić OŠ, Vollhardt KPC (2004) Org Lett 6:2249
29. Mohler DL, Kumaraswamy S, Stanger A, Vollhardt KPC (2006) Synlett 2981
30. Han S, Bond AD, Disch RL, Holmes D, Schulman JM, Teat SJ, Vollhardt KPC, Whitener GD (2002) Angew Chem Int Ed 41:3223
31. Han S, Anderson DR, Bond AD, Chu HV, Disch RL, Holmes D, Schulman JM, Teat SJ, Vollhardt KPC, Whitener GD (2002) Angew Chem Int Ed 41:3227
32. Schmidte-Radde RH, Vollhardt KPC (1992) J Am Chem Soc 114:9713
33. Iglesias B, Cobas A, Peréz D, Guitián E, Vollhardt KPC (2004) Org Lett 6:3557
34. Dosche C, Löhmannsröben H-G, Bieser A, Dosa PI, Han S, Iwamoto M, Schleifenbaum A, Vollhardt KPC (2002) Phys Chem Chem Phys 4:2156, and references within
35. Shizuka H, Ogiwara T, Cho S, Morita T (1976) Chem Phys Lett 42:311
36. Otha N, Fujita M, Baba H, Shizuka H (1980) Chem Phys 47:389
37. Elsaesser T, Lärmer F, Kaiser W, Dick B, Niemeyer M, Lüttke W (1988) Chem Phys 126:405
38. Tetrau C, Lavalette D, Land EJ, Peradejordi F (1972) Chem Phys Lett 17:245
39. Hertzberg J, Nickel B (1988) Chem Phys 132:235
40. Swiderek P, Michaud M, Hohlneicher G, Sanche L (1991) Chem Phys Lett 178:289
41. Suarez M, Devadoss C, Schuster GB (1993) J Phys Chem 97:9299
42. Hirthammer M, Vollhardt KPC (1986) J Am Chem Soc 108:2481
43. Blanco L, Helson HE, Hirthammer M, Mestdagh H, Spyroudis S, Vollhardt KPC (1987) Angew Chem Int Ed 26:1246
44. Yatsuhashi T, Akiho T, Nakashima N (2001) J Am Chem Soc 123:10137
45. Wiersum UE, Jenneskens LW (1993) Tetrahedron Lett 34:6615

46. Brown RFC, Choi N, Coulson KJ, Eastwood FW, Wiersum UE, Jenneskens LW (1994) Tetrahedron Lett 35:4405
47. Preda DV, Scott LT (2000) Org Lett 2:1489
48. Dosa PI, Schleifenbaum A, Vollhardt KPC (2001) Org Lett 3:1017
49. Dosa PI, Gu Z, Hager D, Karney WL, Vollhardt KPC (2009) Chem Commun 1967
50. Matzger AJ, Vollhardt KPC (1997) Chem Commun 1415
51. Kumaraswamy S, Jalisatgi SS, Matzger AJ, Miljanić OŠ, Vollhardt KPC (2004) Angew Chem Int Ed 43:3711
52. Mohler DL, Vollhardt KPC, Wolff S (1995) Angew Chem Int Ed 34:563
53. Berris BC, Hovakeemian BGH, Lai Y-H, Mestdagh H, Vollhardt KPC (1985) J Am Chem Soc 107:5670
54. Oprunenko Y, Gloriozov I, Lyssenko K, Malyugina S, Mityuk D, Mstislavsky V, Günther H, von Firks G, Ebener M (2002) J Organomet Chem 656:27
55. Albright TA, Dosa PI, Grosman TN, Khrustalev VN, Oloba OA, Padilla R, Paubelle R, Stanger A, Timofeeva TV, Vollhardt KPC (2009) Angew Chem Int Ed 48:9853
56. Albright TA, Oldenhof S, Oloba OA, Padilla R, Vollhardt KPC (2011) Chem Commun 47:9039
57. Leung M-k, Viswanath MB, Chou P-T, Pu S-C, Lin H-C, Jin B-Y (2005) J Org Chem 70:3560
58. Holmes D, Kumaraswamy S, Matzger AJ, Vollhardt KPC (1999) Chem Eur J 5:3399
59. Taillemite S, Aubert C, Fichou D, Malacria M (2005) Tetrahedron Lett 46:8325
60. Bingel C (1993) Chem Ber 126:1957
61. Nierengart J-F, Gramlich V, Cardullo F, Diederich F (1996) Angew Chem Int Ed 35:2101
62. Dosche C, Mickler W, Lohmannsroben H-G, Agenet N, Vollhardt KPC (2007) J Photochem Photobiol A Chem 188:371
63. Schneebeli S, Kamenetska M, Foss F, Vazquez H, Skouta R, Hybertsen M, Venkataraman L, Breslow R (2010) Org Lett 12:4114
64. Quinn JR, Foss FW, Venkataraman L, Breslow R (2007) J Am Chem Soc 129:12376
65. Christoph H, Grunenberg J, Hopf H, Dix I, Jones PG, Scholtissek M, Maier G (2008) Chem Eur J 14:5604
66. Bharat Bhola R, Bally T, Valente A, Cyrański MK, Dobrzycki L, Spain SM, Rempała P, Chin MR, King BT (2010) Angew Chem Int Ed 49:399
67. Schulman JM, Disch RL (2003) J Phys Chem A 107:5223
68. Bruns D, Miura H, Vollhardt KPC (2003) Org Lett 5:549
69. Haymet ADJ (1985) J Chem Phys Lett 122:421
70. Fowler PW, Woolrich J (1986) Chem Phys Lett 127:78
71. Kovačević K, Graovac A (1987) Int J Quantum Chem Quantum Chem Symp 21:589
72. Coulombeau C, Rassat A (1987) J Chem Phys 84:875
73. Kurita N, Kobayashi K, Kumahora H, Tago K, Ozawa K (1992) Chem Phys Lett 188:181
74. Schulman JM, Disch RL (1996) Chem Phys Lett 262:813
75. De La Vaissière B, Fowler PW, Deza M (2001) J Chem Inf Comput Sci 41:376
76. Schulman JM, Disch RL (2005) J Phys Chem A 109:6947
77. Qian W, Chuang S-C, Amador RB, Jarrosson T, Sander M, Pieniazek S, Khan SI, Rubin Y (2003) J Am Chem Soc 125:2066
78. Qian W, Bartberger MD, Pastor SJ, Houk KN, Wilkins CL, Rubin Y (2000) J Am Chem Soc 122:8333
79. Hudspeth MA, Whitman BW, Barone V, Peralta JE (2010) ACS Nano 4:4565
80. Son Y-W, Cohen ML, Louie SG (2006) Nature 444:347
81. Zywietz TK, Jiao H, Schleyer PR, de Meijere A (1998) J Org Chem 63:3417
82. Hafner K, Dönges R, Goedecke E, Kaiser R (1973) Angew Chem Int Ed 12:337
83. You S, Neuenschwander M (1996) Chimia 50:24
84. Bally T, Chai S, Neuenschwander M, Zhu Z (1997) J Am Chem Soc 119:1869
85. Rosenberger M, Katz TJ (1962) J Am Chem Soc 84:865
86. Katz TJ, Rosenberger M, O'Hara RK (1964) J Am Chem Soc 86:249

87. Stezowski JJ, Hoier H, Wilhelm D, Clark T, Schleyer PvR (1985) J Chem Soc Chem Commun 1263
88. Saito M (2010) Symmetry 2:950
89. Brand K (1912) Ber Detsch Chem Ges 45:3071
90. Blood CT, Linstead RP (1952) J Chem Soc 2263
91. Babu G, Orita A, Otera J (2008) Chem Lett 37:1296
92. Zhang H, Karasawa T, Yamada H, Wakamiya A, Yamaguchi S (2009) Org Lett 11:3076
93. Chakraborty M, Tessier CA, Youngs WJ (1999) J Org Chem 64:2947
94. Kawase T, Konishi A, Hirao Y, Matsumoto K, Kurata H, Kubo T (2009) Chem Eur J 15:2653
95. Kawase T, Fujiwara T, Kitamura C, Konishi A, Hirao Y, Matsumoto K, Kurata H, Kubo T, Shinamura S, Mori H, Miyazaki E, Takimiya K (2010) Angew Chem Int Ed 49:7728
96. Konishi A, Fujiwara T, Ogawa N, Hirao Y, Matsumoto K, Kurata H, Kubo T, Kitamura C, Kawase T (2010) Chem Lett 39:300
97. Levi ZU, Tilley TD (2009) J Am Chem Soc 131:2796
98. Levi ZU, Tilley TD (2010) J Am Chem Soc 132:11012
99. Hashmi ASK, Wieteck M, Braun I, Nösel P, Jongbloed L, Rudolph M, Rominger F (2012) Adv Synth Catal 354:555
100. Saito M, Nakamura M, Tajima T, Yoshioka M (2007) Angew Chem Int Ed 46:1504
101. Saito M, Nakamura M, Tajima T (2008) Chem Eur J 14:6062
102. Saito M, Hashimoto Y, Tajima T, Ishimura K, Nagase S, Minoura M (2012) Chem Asian J 7:480
103. Willner I, Becker JY, Rabinovitz M (1979) J Am Chem Soc 101:395
104. Yang J, Lakshmikantham MV, Cava MP (2000) J Org Chem 65:6739
105. Wang D, Michinobu T (2011) J Polym Sci A Polym Chem 49:72

Top Curr Chem (2014) 350: 177–278
DOI: 10.1007/128_2014_534

Published online: 9 July 2014

Stereochemistry of Bistricyclic Aromatic Enes and Related Polycyclic Systems

P. Ulrich Biedermann and Israel Agranat

Abstract Bistricyclic aromatic enes (BAEs) and related polycyclic systems are a class of molecular materials that display a rich variety of conformations, dynamic stereochemistry and switchable chirality, color, and spectroscopic properties. This is due to the a subtle interplay of the inherent preference for planarity of aromatic systems and the competing necessity of non-planarity due to intramolecular overcrowding in the fjord regions built into the general molecular structure of BAEs. The conformational, dynamic, and spectroscopic properties may be

---

In Memoriam Professor Dr. John J. Stezowski.

This review chapter is dedicated to the memory of Professor Dr. John J. Stezowski (1942–2007), Professor of Chemistry, University of Nebraska in Lincoln (1991–2007), an eminent scientist and teacher, a leader in crystallography, a dedicated adviser, and a valued friend and colleague. Professor Stezowski was born in Chicago. He received the B.Sc. degree in Chemistry from Case Institute of Technology (1964) and the Ph.D. degree in Chemistry from Michigan State University (1969). He was a postdoctoral Research Associate at Cornell University. In 1972, Professor Stezowski moved to the Institut für Organische Chemie der Universität Stuttgart; his Habilitation was completed in 1981. In 1986 he was appointed Universitätsprofessor in Structural Chemistry at the Universität Stuttgart, where he led a highly successful program in organic and protein crystallography. Professor Stezowski was a founder and chair of the IUCr Commission on Small Molecules (aka Commission on Chemical Crystallography). His research interests and activities were focused primarily on structural studies of complex organic molecules, especially supramolecular inclusion complexes with beta-cyclodextrin hosts. Professor Stezowski has been a structural chemist par excellence.

P.U. Biedermann (✉)
Organic Chemistry, Institute of Chemistry, The Hebrew University of Jerusalem,
Philadelphia Bldg. #201-205, Edmond J. Safra Campus, Jerusalem 91904, Israel

Current Address: Max-Planck-Institut für Eisenforschung GmbH, Max-Planck-Str. 1, 40237, Düsseldorf, Germany
e-mail: u.biedermann@mpie.de

I. Agranat (✉)
Organic Chemistry, Institute of Chemistry, The Hebrew University of Jerusalem,
Philadelphia Bldg. #201-205, Edmond J. Safra Campus, Jerusalem 91904, Israel
e-mail: isri.agranat@mail.huji.ac.il

designed and fine-tuned, e.g., by variation of the bridging groups X and Y, the overcrowding in the fjord regions, extensions of the aromatic system, or other modifications of the general BAE structure, based on the fundamental understanding of the structure–property relationships (SPR). The present review provides an analysis of the conformational spaces and the dynamic stereochemistry of overcrowded bistricyclic aromatic enes applying fundamental symmetry considerations. The symmetry analysis presented here allows deeper insight into the conformations, chirality, and the mechanisms of the dynamic stereochemistry, and will be instrumental in future computational studies.

Keywords Conformations · Dynamic stereochemistry · Intramolecular overcrowding · Symmetry · Thermochromism and photochromism

Contents

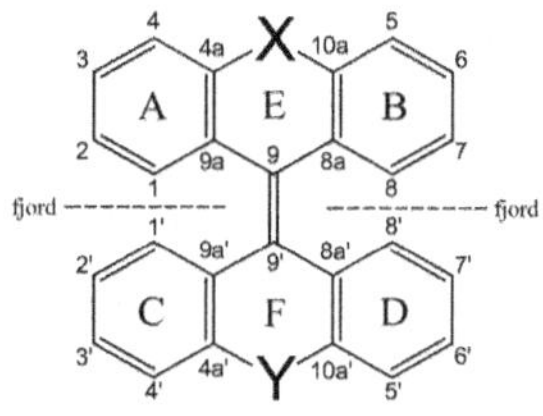

X = Y: homomerous

X ≠ Y: heteromerous

Fig. 1 General structure of bistricyclic aromatic enes (**1**) and atom labeling

1 Introduction

1.1 Presentation of BAEs

Bistricyclic aromatic enes (BAEs **1**) (Fig. 1) [1–3] have fascinated chemists since the bright red hydrocarbon bifluorenylidene (**2**) was first synthesized in 1875 [4] and thermochromism, piezochromism, and photochromism have been revealed in bianthrone (**3**) [5, 6] and dixanthylene (**4**) [7, 8]. BAEs (**1**) are defined by the general structure shown in Fig. 1 and consist of two tricyclic moieties that are connected at the central positions C^9 and $C^{9'}$ by a double bond. The intramolecular overcrowding [9, 10] in the fjord regions on both sides of the central double bond imposes non-planarity on the π-conjugated structures and leads to interesting new molecular properties that may be tuned by variation of the bridges X and Y that close the central rings, by substitutions, or by fusing of additional rings leading to the more general class of polycyclic aromatic enes (PAEs) [2, 11–14]. The color changes in thermo-, piezo-, electro-, and photochromism in BAEs are due to interconversions between different non-planar conformations and triggered by thermal, mechanical (pressure), electrochemical, or optical (UV–vis) stimuli, respectively. More recently, BAEs and related molecules have been developed as functional molecular materials, including molecular switches and molecular motors [15–19]. Furthermore, the proximity of the carbon atoms across fjord regions facilitates cyclizations, which has been exploited in the synthesis of larger polycyclic systems including bowl-shaped fullerene fragments [20–25].

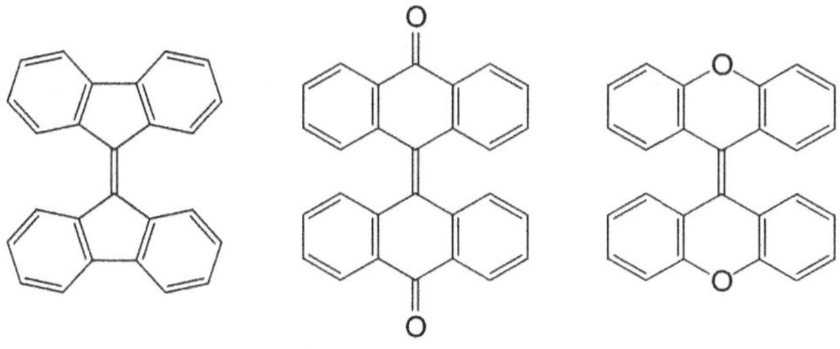

2 **3** **4**

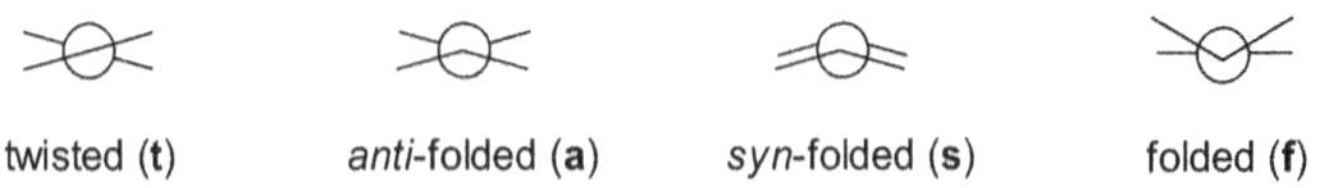

twisted (**t**) *anti*-folded (**a**) *syn*-folded (**s**) folded (**f**)

Fig. 2 Schematic representation of the fundamental BAE conformations

The overall conformations of BAEs may be visualized in a qualitative way as shown in Fig. 2. The most important conformations are the twisted conformation (**t**), the *anti*-folded conformation (**a**), the *syn*-folded conformation (**s**), and the folded conformation (**f**). The folded conformation is found in BAEs with different bridges X $\neq$ Y and characterized by one strongly folded tricyclic moiety, while the other moiety remains (almost) planar. A typical example is the folded fluorenylidene-xanthene (**5**) (see Sect. 2.3). The schematic representations are derived from Newman projections along the $C^9{=}C^{9'}$ double bond. However, the lines represent the conformations of the tricyclic moieties rather than the direction of the bonds attached to the central double bond (as in the original Newman projection). The schematic representations illustrate the degrees of twist and folding in a BAE conformation. Note that this necessarily is a simplified picture, which cannot adequately represent all the important features of the 3D structures.

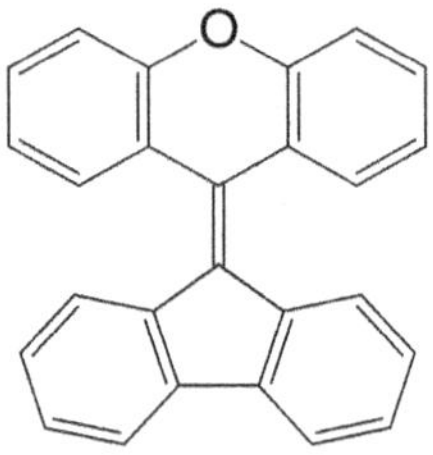

O

5

The structural framework of bistricyclic aromatic enes is eminently suited to stabilize several non-planar conformations. Bistricyclic aromatic enes are an attractive field for the study of non-planar ground state conformations, conformational equilibria, the interrelations of color and conformation, functional molecular materials, and dynamic stereochemistry [1, 2, 26, 27].

BAEs may be also viewed as tetrabenzo[*m*,*n*]fulvalenes (aka tetrabenzofulvalenes) or as doubly bridged tetraarylethylenes with bridges X and Y. They can be classified as *homomerous* bistricyclic enes (**1**, X = Y) and *heteromerous* bistricyclic enes (**1**, X $\neq$ Y) [28]. The topic of bistricyclic aromatic enes has previously been reviewed [1–3, 27, 29, 30].

1.2 Scope of the Review

The central themes of the present review chapter are stereochemistry and symmetry considerations in bistricyclic aromatic enes (BAEs). The focus is on the application of symmetry considerations to the classification and the analysis of molecular conformations of BAEs and to the mechanisms of the dynamic stereochemistry of these fascinating systems.

Section 2 provides an overview of BAE structures and properties. Due to space limitations, only a relatively limited overview of BAEs and related polycyclic systems is presented. The overview is not intended to be comprehensive. The quoted literature published in the last dozen years since our latest 2001 review on BAEs [3] is intended to be selective, emphasizing conceptual understanding. The overview also includes a selection of BAEs for which the crystal and molecular structures have been published. Some variations of the BAE structure **1**, including the removal of the bridges X and/or Y, cyclization across the fjord region, and some related polycyclic aromatic systems, are also mentioned. The term "related polycyclic aromatic systems" is restricted to polycyclic aromatic enes (PAEs) that are benz-annelated BAEs. These BAE variations are treated only briefly.

Section 3 introduces the methodology (general theory) of symmetry analysis in stereochemistry, including a discussion of symmetry operators vs permutation-inversion operators, the molecular symmetry group, transition state symmetry, and practical considerations.

Section 4 addresses the symmetry features of BAEs. Most attention is devoted to symmetry considerations of unsubstituted homomerous BAEs (**1**, X = Y). Due to the highly symmetric topology, the symmetry analysis of the homomerous BAEs gives the most comprehensive results. A mechanism for the dynamic stereochemistry of BAEs in terms of a network of isomerization pathways, based on symmetry analysis and on plausible assumptions of the relative (energetic) stability of the different conformations and transition states, is presented. A brief discussion on symmetry in heteromerous BAEs and (di)substituted BAEs is also included. In cases where the difference between the two bridges (X $\neq$ Y) and/or substitution can be considered a minor perturbation of the conformations, considerably more information may be gained by studying the dynamic stereochemistry based on the molecular symmetry group of the homomerous unsubstituted parent system.

The Chapter concludes with Sect. 5, summarizing the main results and pointing towards some future applications.

2 Overview of BAE Structures and Properties

2.1 Geometrical Parameters

Bistricyclic aromatic enes are overcrowded in the fjord regions (Structure **1**). *Intramolecular overcrowding* is a steric effect shown by aromatic structures in which the (intramolecular) distance between non-bonded atoms is smaller than the sum of the van der Waals radii of the involved atoms [9, 10]. The fjord regions in BAEs resemble the unique rim motif in [5]helicene (dibenzo[*c*,*g*]phenanthrene) [31]. [n]Helicenes are non-planar for $n \geq 4$. Accordingly, cove and fjord regions introduce non-planarity in polycyclic aromatic hydrocarbons (PAHs) [32].

In BAEs, the intramolecular overcrowding requires out-of-plane deformations in order to accommodate the sterically demanding bistricyclic moieties without prohibitively close contacts of non-bonded atoms in the fjord regions on both sides of the central double bond ($C^9 = C^{9'}$). A hypothetical coplanar bistricyclic aromatic ene based on conventional bond lengths [33] and bond angles would maintain very short non-bonded carbon–carbon, carbon–hydrogen, and hydrogen–hydrogen distances at positions 1, 1′, 8, and 8′, leading to a considerable overlap of the van der Waals radii. The associated repulsive interactions may be relieved by deviations from coplanarity and by various bond angle and bond length distortions.

The non-planarity of an ethylene group (a double bond and the four attached atoms) may be described by three parameters [34–36]. In this work, the pure twist $\boldsymbol{\omega}$ and the two pyramidalization angles of the double bond carbons C^9 and $C^{9'}$ will be used. The pure ethylenic twist $\boldsymbol{\omega}$ is defined as the average of the two torsion angles that are smaller than 90° [1, 34–36]:

$$\boldsymbol{\omega} = 1/2\left(\tau\left(C^{9a}\text{-}C^{9}\text{-}C^{9'}\text{-}C^{9a'}\right) + \tau\left(C^{8a}\text{-}C^{9}\text{-}C^{9'}\text{-}C^{8a'}\right)\right) \tag{1}$$

For the pyramidalization, several parameters have been introduced. Their advantages and disadvantages have been reviewed [37, 38]. In the present study, the angles $\chi(C^9)$ and $\chi(C^{9'})$ will be used [1, 34–36, 38]:

$$\boldsymbol{\chi}(C^9) = (\boldsymbol{\tau}(C^{9a}\text{-}C^{9}\text{-}C^{9'}\text{-}C^{8a})\ \mathrm{MOD}\ 360°) - 180°$$
$$\boldsymbol{\chi}(C^{9'}) = (\boldsymbol{\tau}(C^{9a'}\text{-}C^{9}\text{-}C^{9'}\text{-}C^{8a'})\ \mathrm{MOD}\ 360°) - 180° \tag{2}$$

χ is sensitive to small pyramidalization effects and not dependent on local symmetry. The values are readily accessible from computational as well as crystallographic data. In *syn*-pyramidalized double bonds, the pyramidalization angles $\chi(C_9)$ and $\chi(C_{9'})$ have identical signs, whereas in *anti*-pyramidalized double bonds they have opposite signs [1, 34, 35]. Pyramidalization of sp^2-carbons leads to a rehybridization, adding some *s* character to the π-orbital ($s^m p$-hybrid rather than pure *p*-orbital). The rehybridization and the direction of the π-orbital relative to the σ-bonds and to neighboring π-orbitals (Fig. 3) have been extensively studied using the π-orbital axis vector (POAV) theory [37, 39, 40].

The non-planarity of the tricyclic moieties may be described by the torsion angles defining the conformations of the central rings. Alternatively, the folding dihedral A–B (and C–D), defined as the dihedral of the least-squares-planes of the

Fig. 3 Hybridization in *syn*- and *anti*-pyramidalized double bonds

carbon atoms C^1, C^2, C^3, C^4, C^{4a}, C^{9a} and C^5, C^6, C^7, C^8, C^{8a}, C^{10a} of the benzene rings A and B, may be used (Fig. 1).

2.2 Chirality

The non-planarity of the BAE conformations may introduce chirality [28]. The conformations may be described by the following elements of chirality.

Central double bonds in twisted conformations of BAEs are helical and such molecules are chiral [41]. The absolute configuration is defined by the helicity, which is given by the sign of the pure ethylenic twist $\boldsymbol{\omega}$ (Eq. (1)), and may be indicated by stereodescriptors *P* and *M*, following the Klyne–Prelog convention (Fig. 4.) [42].

The helicity of the central ethylene group vanishes in planar conformations ($\boldsymbol{\omega} = 0°$ or $\pm 180°$). Orthogonally twisted conformations ($\boldsymbol{\omega} = \pm 90°$) are also achiral ($D_{2d}$) unless they are appropriately substituted. It should be noted that the helicity describes the inherent chirality of the twisted ethylene group backbone or skeleton irrespective of possible substitution(s). On the other hand, the definition of axial chirality requires at least one substituent on each end of the "axis" and therefore does not apply to unsubstituted BAEs. However, axial chirality may be applied in substituted BAEs. For example, in the above-mentioned case of orthogonally twisted conformations, introduction of one substituent on each tricyclic moiety (or labeling of the atoms) results in a chiral structure; the absolute configuration may be defined using the sign of the torsion angle $\boldsymbol{\tau}(C^{9a}\text{-}C^9\text{-}C^{9'}\text{-}C^{9a'})$ (C^{9a} and $C^{9a'}$ are assumed to be on the substituted sides of the tricyclic moieties) based on axial chirality. In this case the chirality axis passes through the carbon atoms C^9 and $C^{9'}$ of the central double bond (see also Sect. "Orthogonally Twisted Conformations" in Sect. 4.1.3).

In substituted BAEs, folding of the tricyclic moiety introduces another, independent element of chirality, reminiscent of a tripodal unit [41]. The assignment of absolute configuration of a tripodal unit requires two rules, a sequence rule and a conversion rule (assignment rule), as provided by the Cahn–Ingold-Prelog (CIP) system [41, 43]. In the present Chapter, the atom C^9 will be chosen to define the center of the tripod and the atoms C^1, C^8, and $C^{9'}$, to define the position of the 'ligands.' (Using the positions of C^{9a}, C^{8a}, and $C^{9'}$ would define the absolute configuration of the pyramidalization.) The sequence order is $C^1 > C^8, > C^{9'}$. Assuming relative priorities $X > C$ and $R > H$, this sequence is in accord with the CIP rules, which allow ranking of the three tree-graph ligands of C^9 based on the

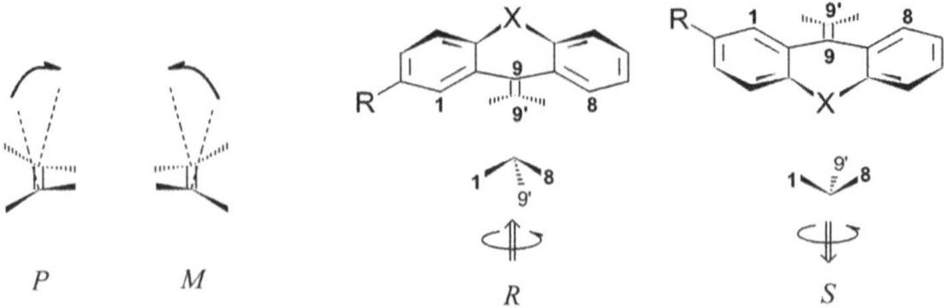

Fig. 4 Chiral elements in BAEs and the corresponding stereodescriptors for absolute configuration

atoms three and four bonds from the center; e.g., for a BAE substituted in the 2-position: C^{9a}\{C[$\underline{X}$(C), C(C,H)], C[C(C,$\underline{R}$), H]\} > C^{8a}\{C[$\underline{X}$(C), C(C,H)], C[C(C, $\underline{H}$), H]\} > $C^{9'}$\{C[$\underline{C}$(Y,C), C(C,H)], C[C(Y,C), C(C,H)]\} [41]. The assignment of the absolute configuration, *R* or *S*, for substituted folded tricyclic moieties is illustrated in Fig. 4. Note that unsubstituted folded moieties are achiral, but nevertheless may be assigned handedness in a labeled atom sense (see below).

Propeller twisted tricyclic moieties are intrinsically chiral, even when unsubstituted. The corresponding helicity may be defined using the sign of a torsion angle across the non-bonding C^{4a}-C^{10a} axis, e.g., $\boldsymbol{\tau}(C^{9a}\text{-}C^{4a}\text{-}C^{10a}\text{-}C^{8a})$. However, the direction of the propeller twist of tricyclic moieties usually depends on the twist of the central double bond and thus is not an independent element of chirality in overcrowded bistricyclic aromatic enes (notable exceptions are perchlorobifluorenylidene [44] and bis-dibenzo[*a,i*]fluorenyllidene [45]). Thus, it is usually unnecessary to designate the propeller twist separately.

The stereodescriptors *E*/*Z* specify the relative positions of the two substituents. The stereodescriptors *syn*- and *anti*- indicate the relative direction of folding of the two moieties. The stereodescriptors *R*/*S* denote the direction of folding of tricyclic moieties. Thus, assigning the absolute configuration *R*/*S* of both moieties includes some redundancy when combined with the *E*/*Z*- and the *anti*-/*syn*-folding classifications. Only the eight combinations shown in Fig. 5 are possible. In the combinations *E*-*anti* and *Z*-*syn*, the folding of the moieties is necessarily heterochiral, i.e., *RS*′ or *SR*′, while in the combinations *Z*-*anti* and *E*-*syn* folding is homochiral, i.e., *RR*′ or *SS*′.

2.3 X-Ray Structures

In bistricyclic aromatic enes, a variety of non-planar conformations have been revealed [1–3, 28]. The molecular architecture gives rise to four basic conformations: twisted (**t**), *anti*-folded (**a**), *syn*-folded (**s**), and folded (**f**). Each conformation represents a compromise between π-delocalization (favoring planarity) and non-planarity due to intramolecular overcrowding. Most of the non-planar

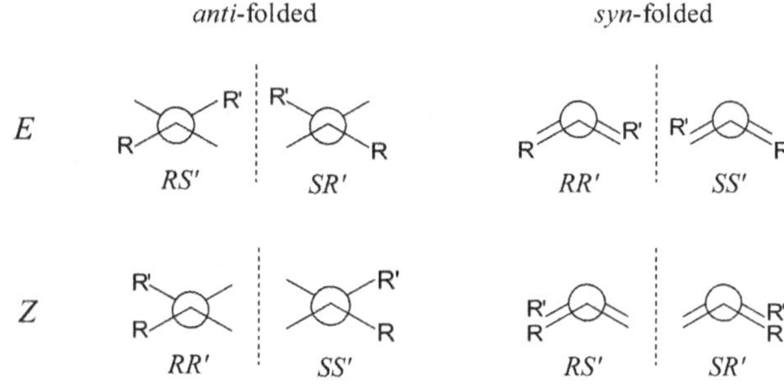

Fig. 5 Interdependence of the descriptors *E*/*Z*, *anti*/*syn*, and *R*/*S*

conformations are still overcrowded, although the intramolecular overcrowding is reduced to a permissible degree [1–3]. The bridges X and Y play an important role in the relative stability of the three conformations, leading in favorable cases to thermochromism [2]. In X-ray crystal structures a combination of twisting and folding is often observed, although usually one mode is by far the dominant one [1, 28]. The molecular structures of leading examples determined by X-ray crystallography are shown in Fig. 6 [46].

The red bifluorenylidene (**2**), the smallest overcrowded bistricyclic aromatic ene (**1**, X,Y: —), adopts a twisted conformation (**t**) in the crystal structure [47, 48]. The two fluorenylidene moieties are nearly planar and the molecule is twisted about the $C^9{=}C^{9'}$ bond by $\boldsymbol{\omega} = 32$–$33°$. The central five-membered rings introduce a bias towards planarity of the tricyclic moieties, while concomitantly rendering them sterically less demanding, especially in twisted conformations [2]. Bifluorenylidenes and their putative constitutional isomers (*E*)- and (*Z*)-biphenalenylidene (**6** and **7**) [49, 50] are the only homomerous BAEs with twisted global minimum conformations [47, 51–54]. The cove regions in *E*-biphenalenylidene (**6**) resemble the rim motif in [4]helicene (benzo[*c*]phenanthrene), while (*Z*)-biphenalenylidene (**7**) has a bay and a fjord region on the two sides of the central double bond. The non-planarity of bifluorenylidene (**2**) was inferred in 1935 on the basis of the dipole moment of its 2,2′-difluoro-derivative [55].

The yellow bianthrone (**3**) with central six-membered rings and bridges X,Y = C=O is *anti*-folded (**a**): the tricyclic moieties are folded in opposite directions and have boat-shaped central rings [56, 57]. According to the first reliable X-ray structure by Apgar and Wasserman quoted in [1], the tricyclic moieties are folded in opposite directions (*anti*) by A–B = C–D = 40°. Recently crystal structure studies at low temperature, at high temperatures, and at high pressures confirmed this structure of the yellow form [58–60]. The results regarding the solid state colored form(s) at high temperature and high pressure are discussed in Sect. 2.5

The colorless (tetrabenzo[7,7′]fulvalene) (**8**) with central seven-membered rings (**1**, X,Y: HC=CH) has two conformational isomers that were characterized by X-ray crystallography: an *anti*-folded and a *syn*-folded polymorph [61]. In the *syn*-folded conformation (**s**) both tricyclic moieties are folded in the same direction and the CH groups in positions 1, 1′, 8, and 8′ are pointing towards each other.

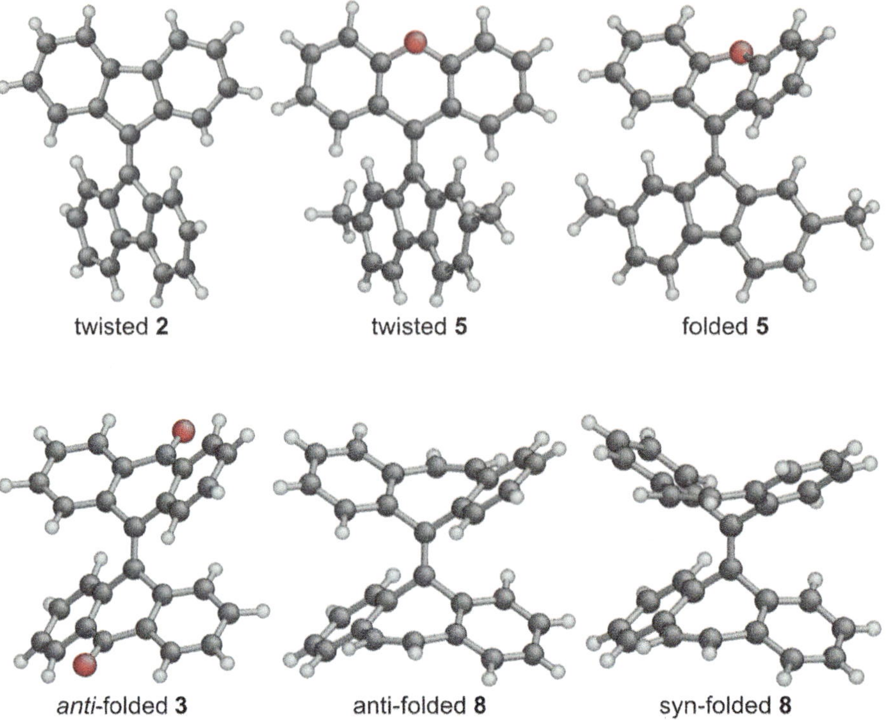

Fig. 6 Molecular structures determined by X-ray crystallography for bifluorenylidene (**2**), the 2′,7′-dimethyl-derivative of fluorenylidene-xanthene (**5**), bianthrone (**3**), and tetrabenzo[7,7′] fulvalene (**8**), typical examples for the twisted, folded, *anti*-folded, and *syn*-folded conformations

The folding dihedrals are very high with A–B (and C–D) = 55.7° in the *anti*-folded form and 62.6° (55.5°) in the *syn*-folded form.

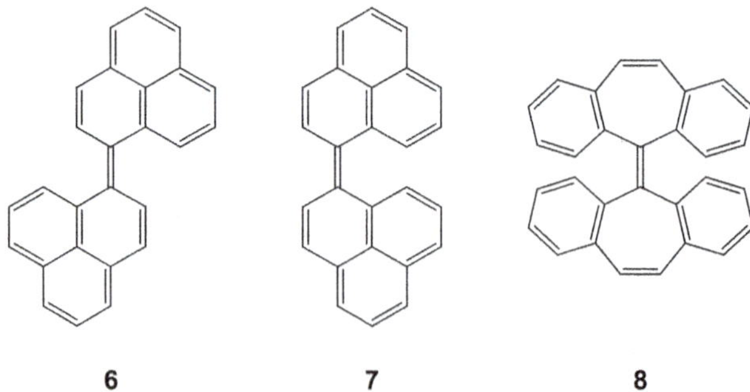

The competition between twisting and folding becomes particularly intense in the case of fluorenylidene-xanthene (**5**) and its derivatives, where twisted (**t**) and

folded (**f**) conformers could be characterized by X-ray crystallography [62, 63]. Crystals with twisted molecules are deep purple, while crystals with folded molecules are yellow. This unambiguously establishes the connection between color and conformation in BAEs [63]. In the twisted conformations, the central double bond has a pure twist of $\omega = 42°$. In the folded conformations, the fluorenylidene moieties are moderately folded (C–D ≅ 20°), while the xanthylidene moieties are folded by A–B = 45°.

Substituents in the fjord regions (positions 1,1′,8 and 8′) substantially increase the twisting in bifluorenylidenes (**1**, X,Y: —) and fluorenylidene-xanthenes (**1**, X: O, Y: —), enhancing the twisting and deepening the color (see Table 1).

The X-ray structures of representative BAEs and those analyzed in our laboratory since the latest review [3] are summarized in Table 1. Further examples of X-ray structures, including details of the conformations, and geometrical parameters can be found in several reviews [1, 2, 28]. Quite a few X-ray structures of BAEs and related compounds have been published recently in the context of applications (see Sect. 2.8).

2.4 BAE Variations

The degree of overcrowding in BAEs (**1**) may be reduced by removing the bridges X and/or Y. The following two variations leading to open BAEs exist: (1) removal of one bridge (X or Y) of **1** to give a BAE-1, i.e., a mono-bridged tetraarylethene, e.g., 9-diphenylmethylene-9*H*-fluorene (**9**, X = –) and 9-diphenylmethylene-9*H*-xanthene (**9**, X = O); (2) removal of the two bridges of **1** to give a BAE-2, i.e., tetraarylethene, e.g., tetraphenylethene (**10**). Removal of the ring constraint (and of π-delocalization if the central ring had aromatic character) facilitates rotations of the aryl groups about the formal single bonds to the central ene group. This conformational flexibility reduces overcrowding in the highly non-planar ground state, while rotation is hindered by steric overlap. The syntheses, molecular and crystal structure, and NMR spectroscopic study of the following naphthologs of BAE-1s have recently been reported: 9-[(di-1-naphthyl)methylene]-9*H*-fluorene (**11**, X = –) and 9-[(di-1-naphthyl)methylene]-9*H*-xanthene (**11**, X=O) [81]. The DFT study of the conformational spaces of **11** and their constitutional isomers, 9-[(di-2-naphthyl)methylene]-9*H*-fluorene (**12**, X = –) and 9-[(di-2-naphthyl) methylene]-9*H*-xanthene (**12**, X=O) was also reported [81]. These naphthologous BAE-1 analogs of BAEs bifluorenylidene (**2**) and dixanthylene (**4**) adopt twisted and folded-twisted conformations, respectively, both in the solid state and in solution, relieving the steric strain by low ethylenic twist angles and high naphthyl twist angles. The conformational preference is governed by the more efficient conjugation between the naphthyl substituents and the twisted central C=C bond and by the overcrowding due to the *peri*-hydrogen atoms in the 1-naphthyl groups. The synthesis and conformational space of the BAE-2 tetrakis(2-naphthyl)ethene (**13**) have recently been reported [82, 83] (see also the discussion of the AIE effect

Table 1 Recent X-ray structures of BAEs including selected PAEs

X	Y	Substitution	Conformation	Color	References
—	—		Twisted	Orange	[47, 48]
—	—	4,5,4′,5′-Tetraaza	Twisted	Orange	[64]
—	—	*Z*-1,1′-Dichloro	Twisted	Red	[65]
—	—	1,1′,3,3′,6,6′,8,8′-Cl	Twisted	Violet	[66]
—	—	Perchloro	Twisted	Blue	[66]
—	—	*E*-Dibenzo[*a,a*′]	Twisted	Violet	[67]
Benzo[def]	Benzo[def]		Twisted	Red	[68, 69]
O	O		*anti*-Folded	Yellow	[70, 71]
CO	CO		*anti*-Folded	Yellow	[56–60]
C=C(CN)$_2$	C=C(CN)$_2$		*anti*-Folded	Yellow	[72]
C(CH$_3$)$_2$	C(CH$_3$)$_2$		*anti*-Folded	Colorless	[73]
S	S	*Z*-2,2′-Tethered	*anti*-Folded	Colorless	[74]
Se	Se		*anti*-Folded	Colorless	[75]
Te	Te		*anti*-Folded	Colorless	[75]
CH=CH	CH=CH		*anti*-Folded	Colorless	[61]
CH=CH	CH=CH		*syn*-Folded	Colorless	[61]
C$_6$H$_4$	C$_6$H$_4$		*syn*-Folded		[76]
CH=CH	CH$_2$-CH$_2$		*anti*-Folded		[76]
O	—		Twisted	Purple	[63]
O	—	2′,7′-Dimethyl	Twisted	Purple	[63]
O	—	2′,7′-Dimethyl	Folded	Yellow	[63]
O	—	Benzo[*b*]	Folded	Yellow	[63]
O	—	1′,3′,6′,8′-Cl-2′,7′-CH$_2$OH	Twisted	Blue	[77]
C=O	—		Folded	Yellow	[78]
C=O	—	1′,8′-Diaza	Folded	Yellow	[78]
C=O	—	Benzo[*b*]	Folded	Yellow	[78]
Se	—		Folded	Colorless	[79]
Te	—		Folded	Colorless	[79]
S	—	1′,8′-Diaza	Folded	Colorless	[80]
Se	—	1′,8′-Diaza	Folded	Colorless	[80]
Te	—	1′,8′-Diaza	Folded	Colorless	[80]

in Sect. 2.8). **13** is considered a naphthologous BAE-2 generated by removal of the two carbonyl bridges from 13-(13-oxo-(6(13*H*)-pentacenylidene)-6(13*H*)-pentacenone (**14**) [84], a naphthalene analog of the BAE bianthrone (**3**).

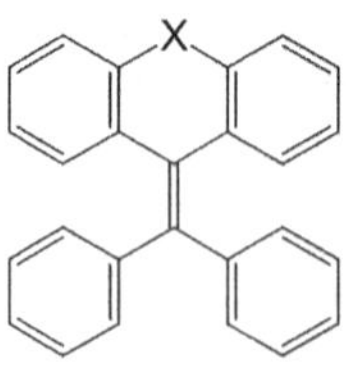

9 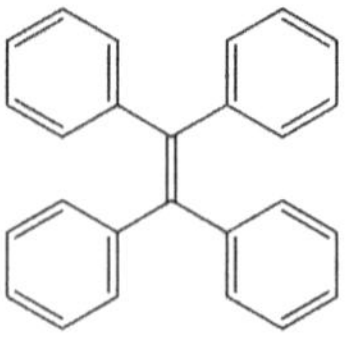 **10** 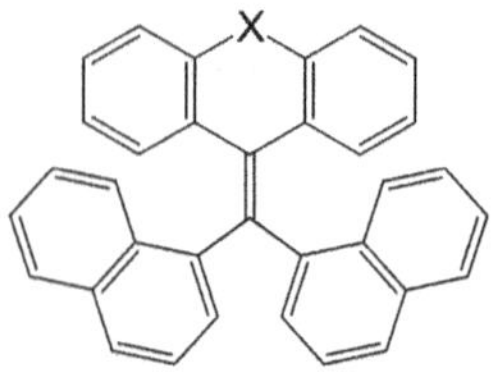 **11**

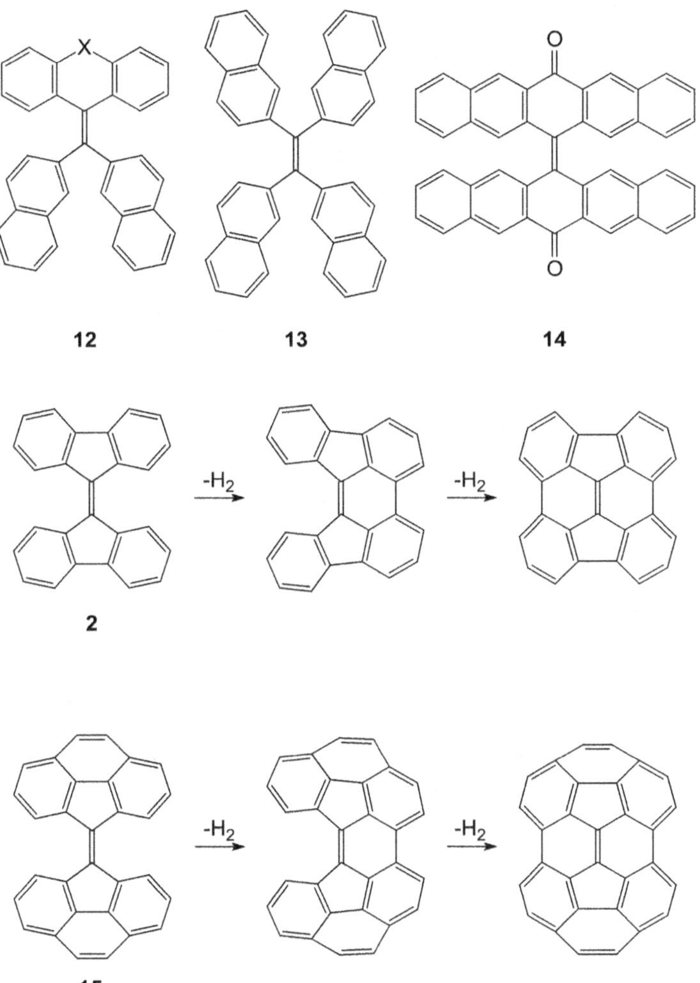

Scheme 1 Formation of bowl-shaped hydrocarbons from **2** and **15**

Another variation of the BAE structure are the cyclizations across the fjord region, which are facilitated by the close proximity of the respective atoms. Photocyclization as well as thermal cyclizations have been reported. Bifluorenylidene (**2**) and the PAE bi-4*H*-cyclopenta[*def*]phenanthren-4-ylidene (**15**) [68, 69], with their central five-membered rings, are fullerene fragments, and potential starting materials for the preparation of bowl shaped hydrocarbons, semifullerenes, and buckyballs (Scheme 1) [20–25, 85–90].

Cyclizations have also been studied in derivatives of bianthrone (**3**) and other BAEs [91–96]. A doubly cyclized bianthrone derivative, hypericin (**16**), is a prominent component in St John's wort (*Hypericum perforatum*, German Johanneskraut), an important remedy for depression [97]. Hypericin also is an antiretroviral agent with potential *anti*-AIDS capabilities [98]. Hypericin has no fjord regions but is nevertheless non-planar and chiral due to the overcrowding in the substituted bay regions. The barrier for enantiomerization is ~98 kJ/mol [99]. Compound **17**, prepared by photocyclization of **14**, has bona fide fjord regions on both sides of a central naphthalene unit and a twisted conformation in the crystal structure [100].

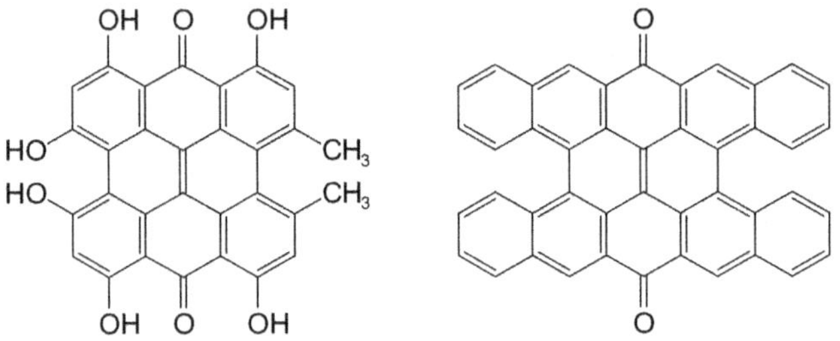

16 **17**

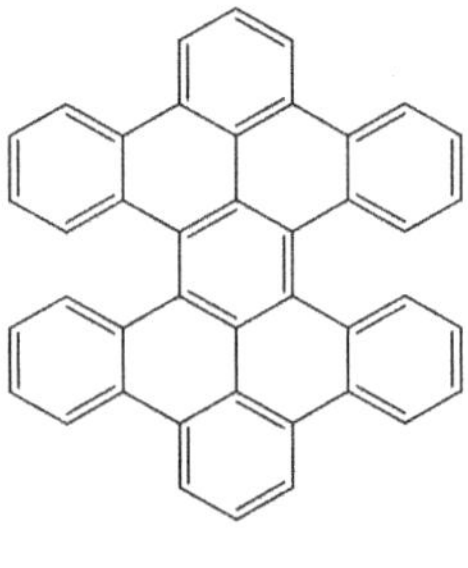

18

The hexabenzoperylene **18** features a red twisted conformer as well as a yellow *anti*-folded conformation [101]. The non-planarity of this benzenoid PAH is due to overcrowded fjord regions that are, however, positioned on both sides of a central benzene ring. The above examples indicate that similar conformational spaces and dynamic stereochemistries as reviewed for BAEs in this Chapter may also be found in large PAHs with double coves and/or fjords [31, 102].

A threefold PAE has been realized in the star-shaped 5,10,15-tri(fluorenylidene) truxene (**19**) [103], which is purple in solution and forms reddish crystals [104]. The three bifluorenylidene units share one central six-membered ring and the overcrowding is enhanced due to the extension of the fjord on one side of each double bond to a [7]helicene-like region. Related star-shaped PAHs with three-fold symmetry are the [3]radialenes (**20**). The [3]radialene with three fluorenylidene moieties (**20**, X= –) is deep-blue in solution and forms reddish-black crystals with a metallic luster [105, 106]. The [3]radialenes with three xanthylidene (**20**, X=O) or thioxanthylidene (**20**, X=S) moieties form black-violet and blue crystals, respectively [106, 107].

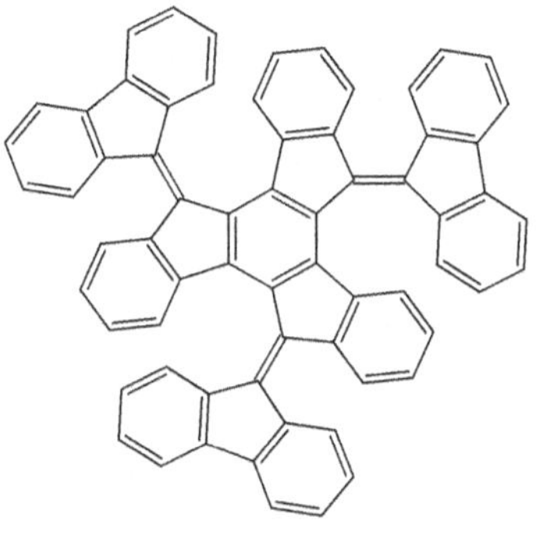

19 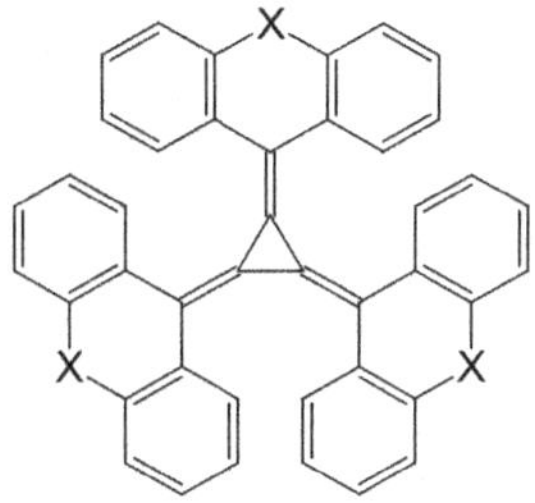 **20**

2.5 *Thermochromism, Piezochromism, Photochromism, and Electrochromism*

Thermochromism is a reversible change of the UV–vis absorption spectrum of a compound in a certain range of temperatures [108]. Temperature-dependent spectroscopic studies of BAE solutions in high boiling point solvents showed that thermochromism is a bona fide equilibrium $\mathbf{A} \underset{}{\overset{kT}{\rightleftarrows}} \mathbf{B}$ between a colorless or yellow ambient temperature form **A** and a blue or green high temperature form **B** with a new absorption band in the visible region (at 600–740 nm) [108–115]. There is no dissociation or association involved as the Bouguer–Lambert–Beer law is observed [108, 109, 112]. The process is completely reversible as long as there is no irreversible side reaction or decomposition due to excessive temperature [112, 116, 117]. The enthalpy ΔH of the transformation $\mathbf{A} \underset{}{\overset{kT}{\rightleftarrows}} \mathbf{B}$ is within the range 2.7–3.9 kcal/mol for bianthrone (**3**) [109, 110, 114, 115, 118, 119], and 4.9–5.6 kcal/mol for dixanthylene (**4**) [110, 118]. The effect of structural modifications – in particular of the bridges X and Y, of substituents, and benzo-annelation – on thermochromism has been discussed [11–14, 115].

Several of the thermochromic BAEs show intensive colors in solid state or solid solution under mechanical pressure [112, 120]. This phenomenon is called

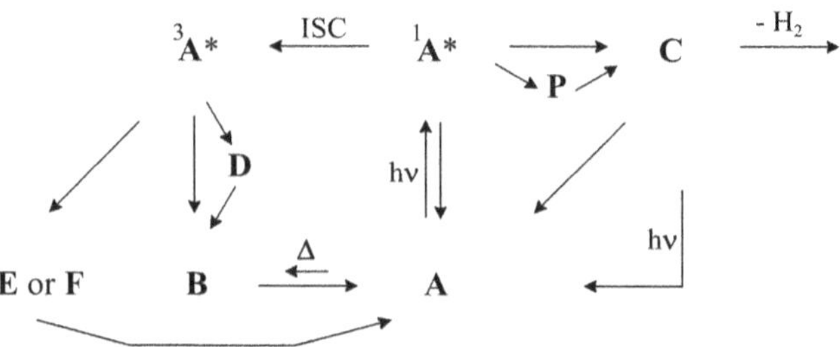

Fig. 7 Generation and decay of the thermochromic and photochromic forms

piezochromism and is reversible after a certain period of time [112]. Piezochromism has been studied under various conditions. Colors were observed for bianthrone (**3**) dissolved in polymethyl methacrylate polymers at pressures of 40–120 kBar at room temperature [120] and when pure crystals or mixtures with KBr, MgO, CaF_2, or $BaSO_4$ are strongly sheared, e.g., by grinding [112, 113, 116, 117, 120, 121]. It is generally assumed that the thermochromic and piezochromic **B** forms are identical [117, 121]. The **B** form has a smaller partial molar volume than the **A** form [120, 122]. A recent UV–vis spectroscopic and crystallographic study of bianthrone crystals in diamond anvil cells revealed a red coloration due to a mixture of two phases [60]. The bulk of the crystal remains in a compressed **A** form with slightly flattened molecules and a broadened and red-shifted UV adsorption. On the surface of the crystals a new phase forms with an adsorption at 600–700 nm, characteristic of the green **B** form. However, the diffraction data was insufficient to determine the crystal structure of the new phase.

Hirschberg discovered photochromism in BAEs [123]. Solutions of bianthrone (**3**) irradiated at low temperatures develop strong colors that persist after irradiation as long as the solutions are kept at low temperatures. Upon warming, the colors fade and disappear [124]. The photochromism of bistricyclic aromatic enes has been extensively reviewed [108, 125–131]. Photochromism in BAEs is a much more complex phenomenon than thermochromism and piezochromism. Several photochromic forms have been observed depending on the molecules studied, the irradiation technique used for excitation, the temperature [117], and the solvent polarity [117, 132] and viscosity [133]. The reversible photochromic forms have been labeled **B**, **C**, **D**, **E**, **F**, and **P** (Fig. 7). They have been characterized by their spectra and conditions of formation and decay [13, 108, 125, 126]. In addition, the singlet and triplet excited states of the normal form **A** and the photochromic forms have been considered. Moreover, the situation may be complicated by various irreversible photochemical reactions such as photocyclization and oxidation.

Photochromism is observed under non-equilibrium conditions and a considerable or even complete conversion to the colored forms may be produced under suitable conditions [108, 118]. It is not observed in crystals or in the glass state [108, 118, 134]. High viscosity (low temperature) reduces the quantum yield [128, 135–138].

The non-planarity of the π-electron system of BAEs leads to a bathochromic shift of the long-wavelength UV–vis absorptions as compared to a hypothetical planar conformation, in particular in the case where the central double bond is affected [108, 127, 139–141]. In a long and controversial debate on the nature of the colored form, which was instrumental in forming today's understanding of the interrelation of color and structure in organic compounds, zwitterionic valence isomers [142] and betain-type mesomeric structures [143–145], biradicals [146–154], thermally populated triplets [109–111, 155, 156], cyclized forms [127, 157, 158] including Woodward and Wasserman's electrocyclized isomers [159], planar [13, 160–165], twisted [108, 112–114, 122, 125, 132, 134, 155, 158, 166–168], and double chair [116, 129, 169] conformations have been discussed as the colored **B** form. Today the puzzle is resolved [63] and it is generally accepted that the room temperature form **A** is the *anti*-folded conformation (**a**) [108, 112, 115, 134, 146, 166, 170–173]. The deeply colored **B** form is the twisted conformation (**t**) [112–114, 122, 125, 134, 158, 166–168, 173]. Each of the two tricyclic moieties are nearly planar; however, the molecules are twisted about the central double bond by ca. 50°–60° [138, 167]. Thermochromism, piezochromism, and the photochromic **B** form have been attributed to this conformational change [63, 113, 122, 141, 166].

A criterion for thermochromism has been derived from ab initio DFT calculations of the *anti*-folded, *syn*-folded, and twisted conformations of thermochromic and non-thermochromic overcrowded bistricyclic aromatic enes (BAEs). The necessary condition is a small energy difference, <7 kcal/mol, between the global minimum *anti*-folded and the twisted conformations [141]. Even a small population of the twisted conformation at elevated temperatures leads to a striking change of the visible color from yellow to green or blue-green. While easily seen with the naked eye and clearly detected by UV–vis spectroscopy, the concentrations remain too low for X-ray crystallography [116, 155]. Therefore, the molecular structures of dixanthylene (**4**) in the yellow β- and blue-green α-form are identical within the limits of experimental error [70, 155]. A recent state-of-the-art crystallographic study of the solid-state thermochromism in bianthrone (**3**) and another BAE-related compound also showed no significant change of conformation with temperature that could explain the color change [59]. Instead, the increased absorption in the UV–vis spectral range at 400–500 K was assigned to dynamic large amplitude vibrations increasing with temperature. It should be noted that the reported spectra [59] do not show the absorption at 600–740 nm characteristic for the thermochromic **B** form.

Fluorenylidene-anthrone (**21**), its benzo[*b*]fluorenylidene derivative **22**, and 1,8-diazafluorenylidene derivative **23** have folded conformations in their yellow or greenish-yellow crystals. However, these compounds give deeply colored purple, blue, or red solutions at room temperature, indicating the presence of the twisted thermochromic **B** form [78]. This has been termed "thermochromism at room temperature." Indeed, DFT calculations showed that the energy differences between twisted and folded conformations are very small, favoring the twisted conformations of **21** and **22** by a small margin, while the folded conformation is preferred in **23**. On the other hand, the thermochromism of bianthrone (**3**) is suppressed by dicyanomethylene bridges in **24** and **25** due to additional overcrowding at the cyano groups in the twisted conformation [72].

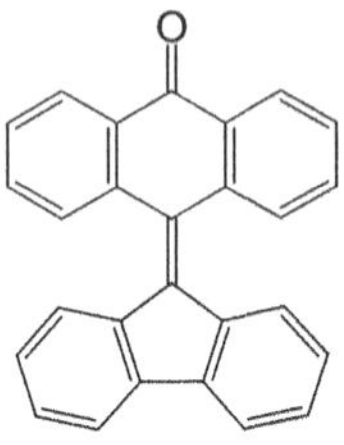

21 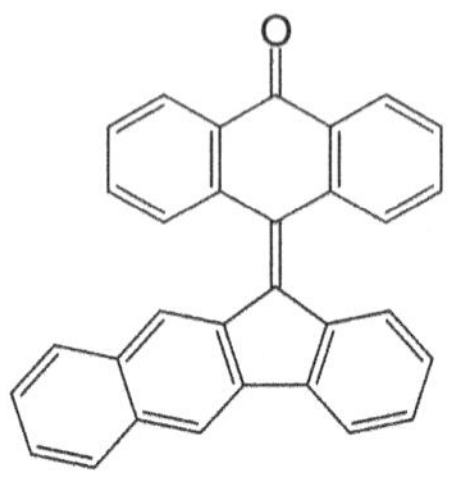 **22** 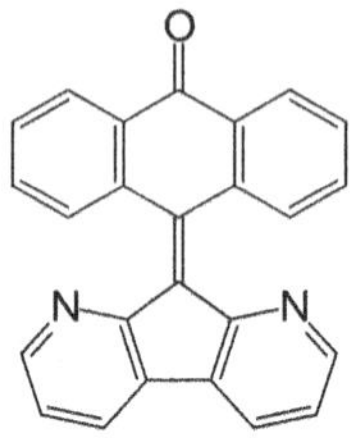 **23**

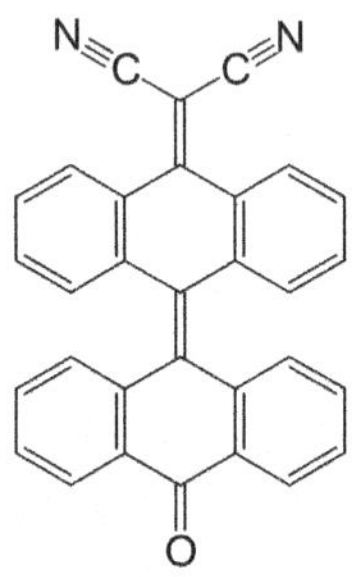

24 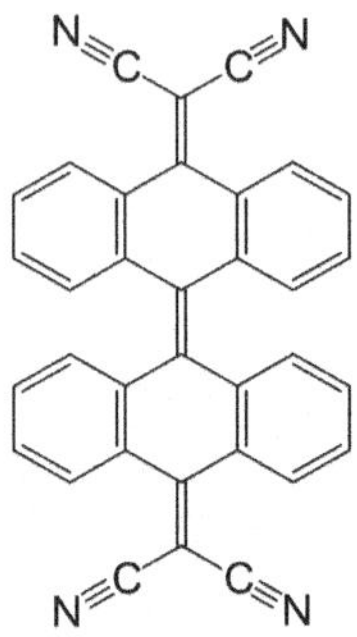 **25** 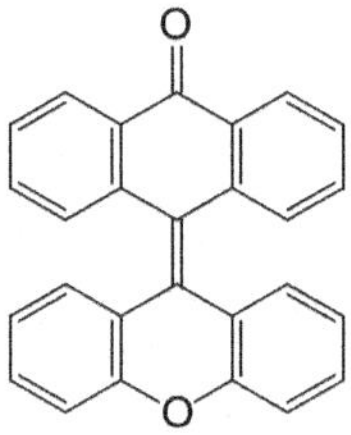 **26** 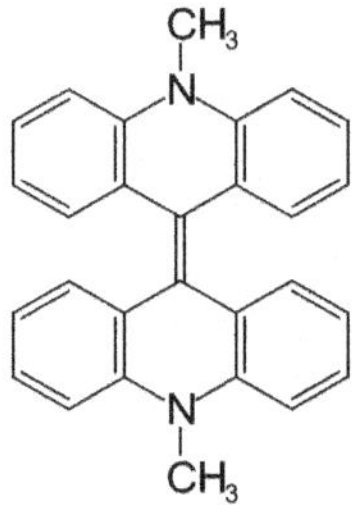 **27**

The conformational transformation into the twisted **B** form may also be triggered by electrochemical reduction or oxidation. Electron transfer leads to a large conformational change because the relative stability of the *anti*-folded and twisted conformations is reversed in the anions of, e.g., bianthrone (**3**) and xanthylidene-anthrone (**26**) [174–178]. The stability of the **B** form in the radical anion and dianion of bianthrone and the reversibility of the reduction/oxidation $\mathbf{B}$ $\mathbf{B}^{\bullet-}$ $\mathbf{B}^{2-}$, furnished another argument in favor of identifying **B** as the twisted conformation [114]. Electrochemical studies of lucigenin, the dication of *N,N′*-dimethyl-biacridane (**27**), indicated analogous conformational changes upon conversion between the cationic and neutral forms [179]. The dianions and dications of bifluorenylidene (**2**) are also highly twisted [50, 180]. Moreover, the electronic and structural properties show that the dianions and the dications of bifluorenylidene (**2**), fluorenylidene-phenalene, and (*E*)-and (*Z*)-biphenalenylidene (**6**, **7**) form a continuum of aromaticity/antiaromaticity [50, 180].

2.6 Dynamic Stereochemistry

BAEs reveal rich dynamic stereochemistry [3, 30, 181–183]. The conformational dynamics of bifluorenylidene has caught attention since the seminal work by Sutherland et al. [184]. In BAEs, four types of dynamic processes were observed:

1. Rapid *E,Z*-isomerizations of twisted and *anti*-folded BAEs as illustrated in Fig. 8.
2. Enantiomerization or conformational inversion, i.e., inversion of the helicity in twisted BAEs (Fig. 9a), or inversions of the boat conformations of the central rings and hence the folding of the tricyclic moieties in *anti*-folded BAEs (Fig. 9b).
3. *syn*-Folded to *anti*-folded conformational isomerization, **s** → **a**, as illustrated in Fig. 9c.
4. Twisted to *anti*-folded conformational isomerization **t** → **a** (Fig. 9d) as observed in the thermal reversion of photochromism **B** → **A** and in electrochemical studies.

It should be noted that enantiomerization and conformational inversion may also be considered in processes (1), (3), and (4).

Dynamic processes with barriers in the range of 6–26 kcal/mol have been determined by dynamic NMR (DNMR) spectroscopy. Even higher barriers have been determined in kinetic equilibration experiments on isolated or enriched isomers. Selected barriers are summarized in Tables 2 and 3. For more examples see the reviews and literature cited therein [3, 27, 30, 181–183].

The low barriers for *E,Z*-isomerizations in BAEs have been attributed to the steric strain inherent in the non-planar conformations destabilizing the ground state and to electronic effects stabilizing a biradical or zwitterionic orthogonally twisted transition state [27, 30, 181–183, 190]. The intrinsic barrier for rotation about an unstrained, isolated C=C bond has been determined as $\boldsymbol{E_a} = 65.9 \pm 0.9$ kcal/mol [191]. In the tetraphenylethylenes, the stabilization of a biradical transition state by delocalization may be similar to BAEs; however, the starting conformations are less overcrowded. Their *E,Z*-isomerization barriers are between 32.9 and 35.3 kcal/mol [192]. Push-pull ethylenes with polar double bonds may have considerably lower *E*, *Z*-isomerization barriers [27, 30].

In homomerous BAEs with bridges X,Y = O, S, and CH_2, identical barriers for conformational inversion and *E,Z*-isomerization were found within experimental error (Tables 2 and 3) [73, 170, 187, 188]. This observation strongly advocates a common highest transition state for both dynamic processes [73, 170] corresponding to the **a** → **t** interconversion, which involves a passage of the hydrogens H^1 and $H^{1'}$ (or H^8 and $H^{8'}$) in the fjord regions [27, 155, 171, 193]. Twisted conformations serve as intermediates in both dynamic processes [170, 171]. The orthogonally twisted biradical transition state for *E,Z*-isomerization of the twisted conformations has a lower energy in the above homomerous BAEs and is thus not rate determining [170, 172, 193]. On the other hand, in the folded

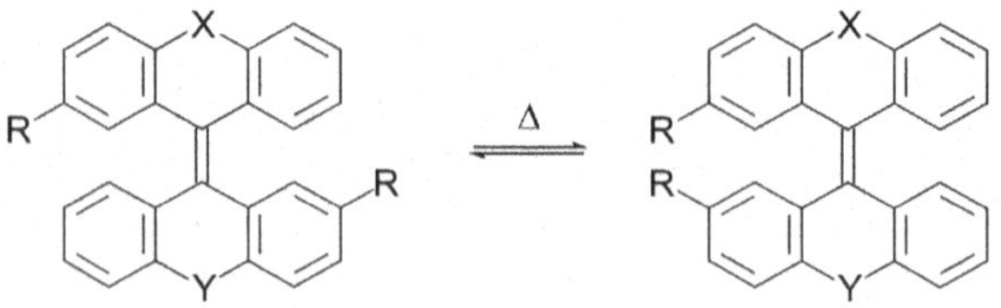

Fig. 8 Thermal *E*,*Z*-isomerization of 2,2′-disubstituted BAEs

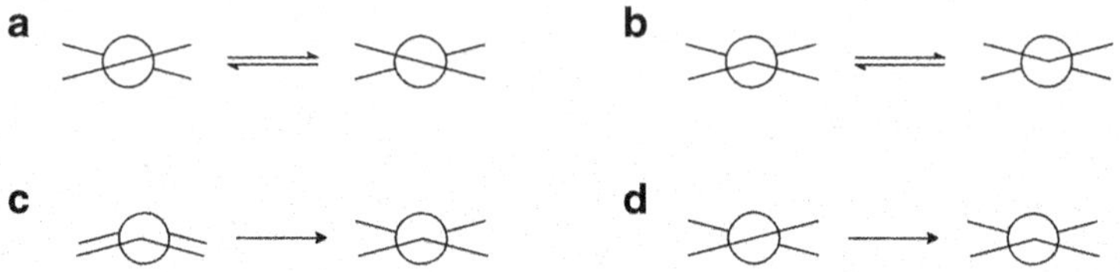

Fig. 9 Conformational isomerization processes in BAEs: (**a**) enantiomerization of twisted BAEs; (**b**) conformational inversion of *anti*-folded BAEs; (**c**) *syn*-folded to *anti*-folded; (**d**) twisted to *anti*-folded conformational isomerizations

Table 2 Selected dynamic NMR results for BAEs

X,Y	Substitution	Process	$\Delta G_c^{\ddagger}$ (kcal/mol)	$\Delta\nu$ (Hz)	T_c (°C)	Solvent	References
—	2-*i*-Pr-	Enantiom.	10.6 ± 0.1	51.3	−55	$CDFCl_2$	[54]
—	2,2′-di-Me-	*E*,*Z*	24.9	2.0	170	DMSO-d_6	[54]
O, —	2-*i*-Pr	Enantiom.	6.3 ± 0.2	58	−139	$CDFCl_2$	[62]
O, —	2-*i*-Pr	*E*,*Z*	19.6 ± 0.2	3.6	84	$CDCl_3$	[62]
O	2,2′-di-Me-	*E*,*Z*	17.5 ± 0.3	2.6	44	$CDBr_3$	[170]
O	2,2′-di-*i*-Pr-	*E*,*Z*	17.9 ± 0.3	3.4	47	$CDCl_3$	[170]
O	2,2′-di-*i*-Pr-	Inversion	17.7 ± 0.5	0.9	36	$CDCl_3$	[170]
C=O	2,2′-di-Me-	*E*,*Z*	20.0 ± 0.5	1.9	84	$CDBr_3$	[115, 171, 185]
C=O	2,2′-di-*i*-PrO-	[a]	21.0		115		[183]
CH_2	2,2′-di-Me-	*E*,*Z*	23.8 ± 0.3	3.8	159	$CDBr_3$	[73]
CH_2	2,2′-di-Me-	Inversion	23.4 ± 0.5	41.2	198	C_4Cl_6	[73]
Se, —	2′-*i*-Pr-	Inversion	14.4	29.4	14	$CDCl_3$	[79]
Se, —	2′-*i*-Pr-	*E*,*Z*	>21.6	17.1	>143	$C_2D_2Cl_4$	[79]
Te, —	2′-*i*-Pr-	Inversion	19.4	20.6	105	$CDCl_3$	[79]
Te, —	2′-*i*-Pr-	*E*,*Z*	>21.6	17.8	>143	$C_2D_2Cl_4$	[79]

[a] *E*,*Z*-Isomerization and conformational inversion

Table 3 Barriers determined by kinetic isomerization experiments in BAEs including thermal decay of photochromism and electrochemical switching

X,Y	Substitution	Process	Barrier (kcal/mol)	T_{Min} (°C)	T_{Max} (°C)	Solvent	Method[a]	References
—	2,2′-di-H_2N-	*E*,*Z*	25	20	80	C_6H_6	[b]	[186]
CHOH		**s** → **a**	12.5				[c]	[125]
C=O		**t** → **a**	15	4	37	DMF	[d]	[174]
NCH_3		**s** → **a**	15				[c]	[125]
NCH_3		**t** → **a**	16.4	−20	23	DMF	[d]	[174]
S		**s** → **a**	18±0.5	−5	80	MCH/2MP	[c]	[137]
S	2-Me-	Racem.	27.3	45	75	*p*-Xylene	Polarimetry	[187, 188]
S	2,2′-di-Me-	*E*,*Z*	27.4				HPLC	[188]
S	2,2′-di-Me-	*Z*-Enantiom.	27.4				HPLC	[188]
CH=CH		**s** → **a**	36.4	174		DMSO-d_6	NMR	[189]
CH=CH, CH_2-CH_2		**s** → **a**	23.6	40		$CDCl_3$	NMR	[76]
CH=CH, C_6H_4		**s** → **a**	36.4	186		DMSO-d_6	NMR	[76]

[a]Method for concentration and/or rate determination
[b]Separation by flash chromatography and photometry
[c]Thermal reversion of the photochromic **E** form
[d]Electrochemical experiments

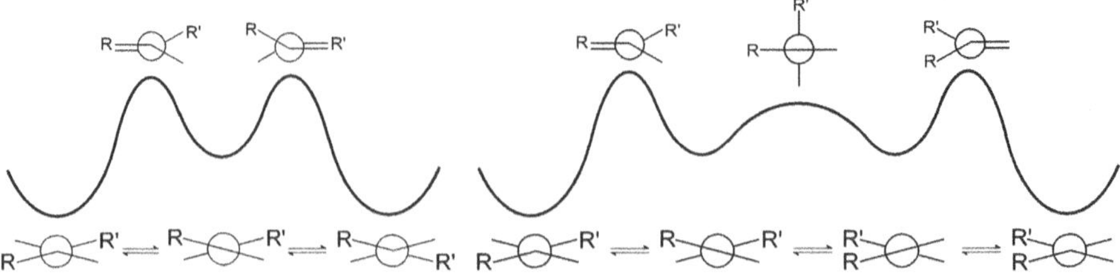

Fig. 10 Mechanisms for conformational inversion and *E,Z*-isomerization in *anti*-folded and folded BAEs

heteromerous BAEs fluorenylidene-selenaxanthene and fluorenylidene-telluraxanthene (**1**, X= — and Y=Se or Te, respectively), low barriers for conformational inversions (edge-passage) were observed, while the barriers for *E,Z*-isomerization (orthogonally twisted transition state) were higher and only a lower limit could be determined [79]. The mechanisms for conformational inversion and *E,Z*-isomerization in *anti*-folded and folded BAEs are illustrated in Fig. 10 [73, 170].

BAEs with twisted global minima **t** like bifluorenylidene and fluorenylidene-xanthene have distinct barriers for enantiomerization and *E,Z*-isomerization via edge-passage and orthogonally twisted transition states, respectively [54, 62].

In BAE **8** with seven-membered central rings (X,Y = CH=CH) a *syn*-folded conformer could be isolated [61, 189, 194–196]. The *syn*-folded to *anti*-folded isomerization barrier **s** $\rightarrow$ **a** was determined to be $\Delta G^{\ddagger} = 36.4$ kcal/mol [189]. Similar high **s** $\rightarrow$ **a** barriers were found for other BAEs with seven-membered central rings and assigned to a seven-membered ring flip [76]. On the other hand, the **s** $\rightarrow$ **a** barriers found for thermal decay of the photochromic **E** forms (*syn*-folded conformation) in BAEs with central six-membered rings and X,Y = S (18.0 kcal/mol), NCH_3 (15.0 kcal/mol), and CHOH (12.5 kcal/mol) (Table 3) are much lower [125, 137].

2.7 Computational Studies

Theoretical methods have been applied to solve the mysteries and puzzles of BAEs and to gain a better understanding as soon as they became available. Pariser–Parr–Pople and Hückel-type calculations were used to estimate spectra, atomic charges, and π-bond orders [127, 139, 156, 197, 198]. Force fields including π-MO approximations quantified the non-planarity and amount of steric strain in the BAE conformations [138, 199–201]. Semiempirical MINDO/3 and PM3 have been used to study the interconversion of *anti*-folded and twisted conformations in BAEs, their anions, cations, and excited state conformations [16, 202, 203]. PM3 has also been used for an extensive study systematically exploring structure–property relationships including trends in the conformational energies, strain energies, folding and twisting of homomerous and heteromerous BAEs by systematically varying the bridges using a series of six different functional groups X,Y and a polycyclic system [2]. Likewise, the conformational space and dynamic stereochemistry of five representative homomerous BAEs has been studied in detail with PM3 [3]. Recently the availability of sufficient computational power and development of efficient algorithms made ab initio electronic structure calculations of BAEs feasible. Density functional theory (DFT) including hybrid density functionals overcame the systematic bias of single-determinant Hartree–Fock against twisted conformations [141]. This allows a theoretical understanding of, e.g., thermochromism, the singlet-triplet gap in the highly twisted bis-dibenzo[*a,i*] fluprenylidene [45], and the prediction of BAE properties based on DFT calculations. Recently experimental studies of BAEs often also include DFT calculations to enhance the understanding of the observed phenomena, e.g., [59, 67, 76, 78, 177, 178, 204]. NMR spectra of the 1,1′-difluoro-derivatives of bifluorenylidene (**2**) show through-space coupling across the fjords, in particular *J*(F,F) in the *Z*-diastereomer and *J*(F,H) in the *E*-diastereomer [205–207]. This highlights the close proximity of the bucking atoms. The *Z*-diastereomer has a higher equilibrium concentration than the *E*-diastereomer, indicating a lower energy. DFT calculations revealed that this is due to a combination of steric strain and electrostatic interactions [205]. The calculated structures are twisted with additional *anti*- and *syn*-pyramidalization of the *Z*- and *E*-diastereomers, respectively, in agreement with symmetry considerations (see below, Fig. 17). DFT calculations also helped to elucidate the mechanism of the unexpected rearrangement in the course of a Peterson olefination, in which, instead of the target – the angularly annelated benzo[*a*]bifluorenylidene (**28**) – its linearly annelated constitutional isomer benzo [*b*]fluorenylidene (**29**) was formed [208].

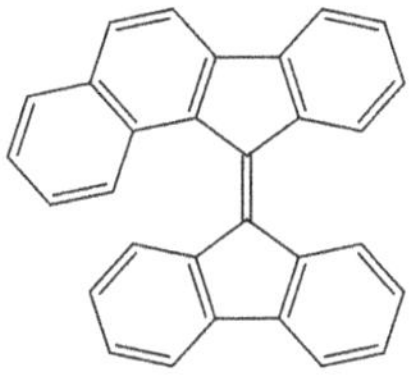

28 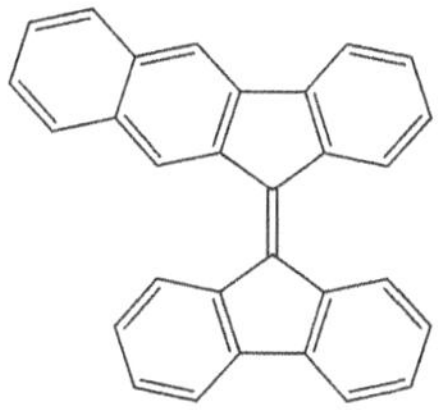 **29**

2.8 *Applications*

The thermochromic, piezochromic, photochromic, and electrochromic properties of BAEs call for applications as molecular switches and functional materials.

Application of BAEs as photochemical memory in a computer was proposed in 1956 [143]. However, until recently organic materials have not been sufficiently reliable as erasable memory [209]. The topic of molecular switches has been reviewed [18, 209–211].

Photo- and electroswitching in BAEs is based on the fact that photoexcitations as well as oxidation or reduction depletes the π-bonding HOMO and/or occupies the π* *anti*-bonding LUMO. In BAEs with *anti*-folded ground state this inverts the equilibrium between *anti*-folded and twisted conformations, leading to a rapid conformational change. Potential energy surfaces for ground and excited states and/or anionic/cationic forms have been reported, e.g., in [16, 76, 202]. After return to the electronic ground state of the neutral molecule, the reverse conformational change requires a thermally activated crossing of the barrier for conformational isomerization. Hence the life time of the twisted conformation depends on the temperature and high barriers are advantageous for stable switches.

Several types of molecular switches derived from BAEs have been developed. A donor substituted bianthrone-derivative (**30**) has been studied as a molecular switch involving conformational changes stabilized by charge transfer [15, 16, 212, 213]. Bianthrone (**3**) as single molecule junction between silver electrodes revealed two states with different conductance and hysteretic ON and OFF switching [214]. Hexabenzo[7,7′]fulvalene (**31**) can be switched by photo- and thermal *anti*/*syn* isomerization, both in very high yields [76].

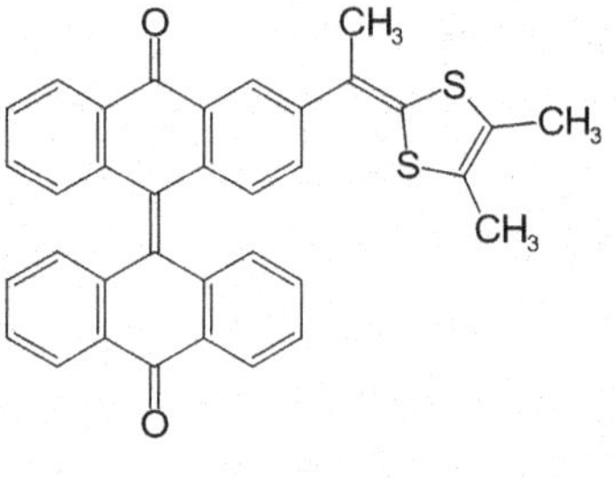

30 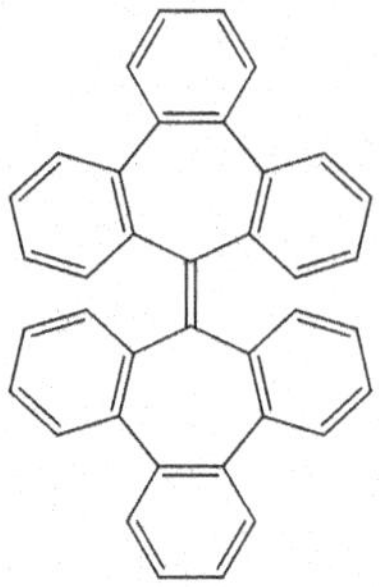 **31** 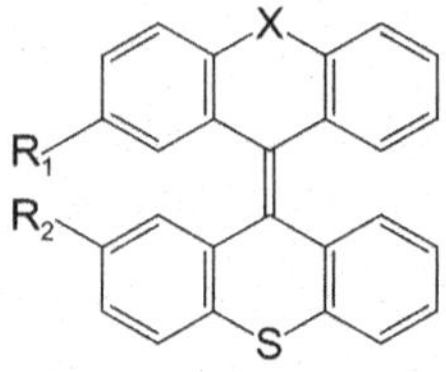 **32**

Feringa et al. took advantage of the inherent chirality of substituted bistricyclic aromatic enes and related overcrowded dissymmetric enes and devised chiral optical molecular switches based on photochemical and thermal *E*,*Z* isomerizations [17, 18, 210, 215]. The systems included the disubstituted BAEs **32** [188, 216]. The benzo[*a*]-annelated polycyclic aromatic ene **33** showed stereospecific thermal and photochemical *P*-*trans* to *M*-*cis* isomerizations [187, 217]. Irradiating an initially racemic sample of **34** with either left or right circular polarized light (CPL) photochemically generated an enantiomeric excess of the (*M*)- or (*P*)-enantiomer, respectively [218]. Linear polarized light led to racemization. By doping a nematic liquid crystal with **34** it was possible to create cholesteric liquid crystal phases of opposite chirality by irradiation with *l*-CPL or *r*-CPL and return to a nematic phase using unpolarized light [218]. This control and amplification of molecular chirality by circular polarized light may be used as a data storage system [218]. The reflection color of liquid crystalline films could also be switched between red and green [219].

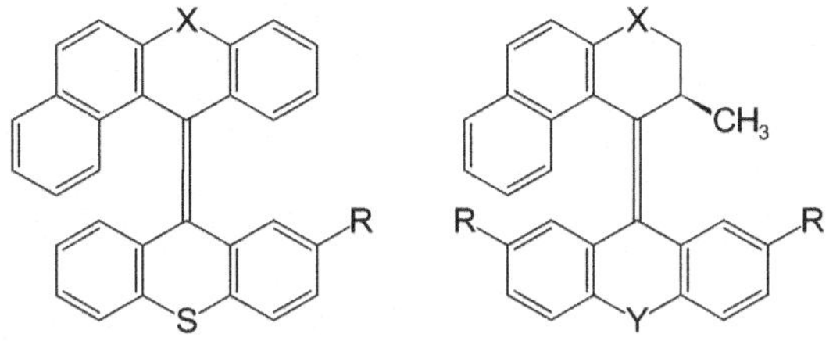

33 X=S, R=CH_3
34 X=O, R=H

35

The structurally related molecules **35**, where one of the aromatic rings is replaced by a methyl group and a stereogenic center is introduced, are light-driven molecular motors [18, 19]. These molecules consume light energy and convert it into a unidirectional rotation of the upper half with respect to the lower half [19]. Introduction of a tetrahedral stereogenic center on the aliphatic carbon of the central ring neighboring the double bond breaks the symmetry. Lifting the equivalence of right-handed and left-handed rotation introduces a bias for unidirectional rotation about the double bond. These motors were optimized [220–223] and it was shown that inverting the absolute configuration at this stereogenic center reverses the direction of rotation [224].

Besides the bi-stable molecular switches, several types of molecules with more than three states have been reported [204, 225, 226] as well as immobilized switches in polymers and on a gold surface [227, 228].

In other applications, BAEs have been introduced into various types of host molecules or modified to become ligands taking advantage of their spectroscopic, structural, and functional properties. A dixanthylene core has been incorporated into a double calix[6]arene [229]. BAEs have been incorporated into crown ethers and their alkali metal binding properties have been studied [230–232]. Metal coordination may also occur for 4,5-diazafluorenylidenes **36** and **37** [64, 233, 234]. The dipyridyl BAE-1 **38** forms a BAE with a particularly large bridge Y and high inversion barrier upon metal complexation [235, 236]. Recently it was shown that the conformation of the fluorenylidene-xanthene **39** may be switched from folded (yellow) to twisted (purple) by encapsulation in a suitable cavity [77].

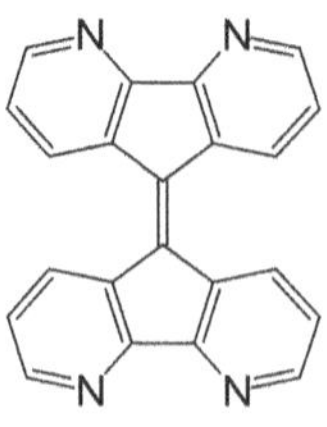

36 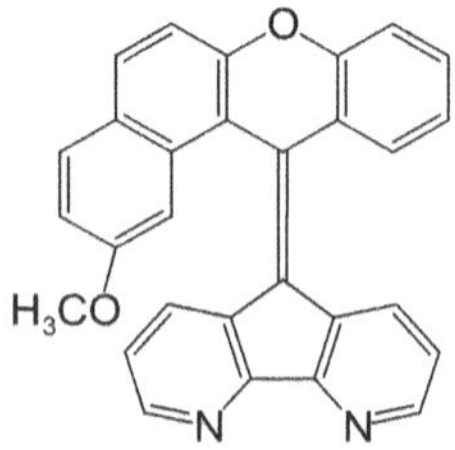 **37** 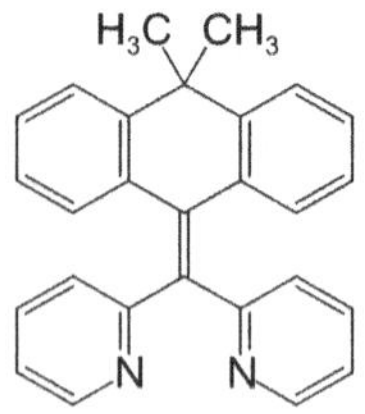 **38**

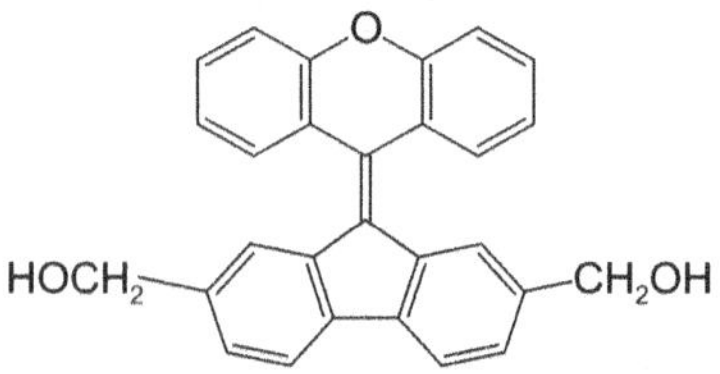

39 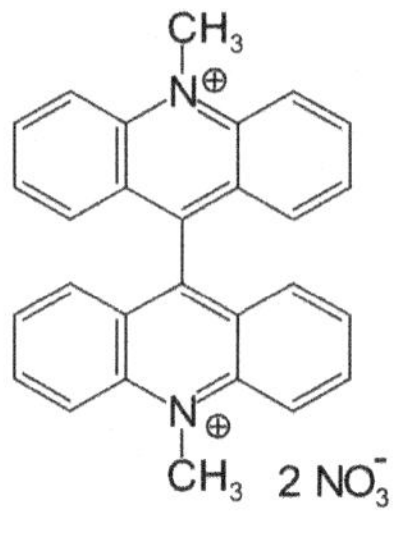 **40**

Lucigenin (**40**), a dicationic salt derived from *N,N′*-dimethyl-biacridane (**27**), was considered to be the most powerful of all synthetic chemiluminescent substances [237–239].

Aggregation-induced emission (AIE) is a novel phenomenon exhibited by propeller-shaped luminogens which are non-emissive in dilute solutions but emit efficiently upon aggregation [240]. AIE luminogens are a class of molecules whose emissions are induced by aggregate formation in contrast to the more commonly observed aggregation-caused quenching (ACQ). The AIE effect has attracted intense interest due to its promising diverse applications in fluorescence sensors, biologic probes, optoelectronic devices, and mechanochromic smart materials [240]. One of the popular AIE luminogens is the twisted, propeller-shaped BAE-2 tetraphenylethene (**10**), described as an ethene stator completely surrounded by phenyl rotors [241]. Its AIE effect has been attributed to the restriction of the intramolecular rotation (RIR) mechanism [71]. A structural evolution in AIE luminogens is the recent synthesis [82, 83], conformational space determination [83], and AIE characterisation [82] of the BAE-2 tetrakis(2-naphthyl)ethene (**13**), in which the four "bulky" naphthalene rings can serve as rotors against the ethene stator to construct the AIE luminogen [82, 83]. Efficient fluorescence of **13** is induced in the aggregate state, where the non-radiative decay pathway is blocked [82]. Recently it has been reported that the locking of the phenyl rings of the BAE-2 tetraphenylethene (**10**) by one and two oxygen bridges, leading to the BAE-1 9-diphenylmethylene-9*H*-xanthene (**9**, X=O) and the thermochromic BAE dixanthylene (**4**), respectively, increased the emission efficiency of the luminogen solutions [71]. The AIE luminogens tetraphenylethene (**10**) and dixanthylene (**4**) differ significantly in their conformational spaces. While **10** is highly twisted about the formal single bonds connecting the phenyl rings to the central ethene group [83, 242, 243], **4** is yellow and adopts an *anti*-folded conformation as the global minimum [70, 71].

Bifluorenylidene and its derivatives have been investigated as a new generation of acceptor compounds in bulk heterojunctions in organic photovoltaic devices [244, 245]. Bifluorenylidene anions are stabilized by steric "twist"-strain relief and gain in aromaticity. Higher open circuit voltage and short circuit currents could be achieved due to the higher LUMO, better overlap with the solar spectrum, and higher electron charge carrier mobility.

Further applications of BAEs and related polycyclic systems will hopefully be forthcoming.

3 Symmetry Analysis in Stereochemistry: General Theory

Symmetry is a powerful tool in analyzing and classifying molecular conformations and the mechanisms of dynamic stereochemistry. In particular, symmetry considerations can be very efficient in analyzing the dynamic stereochemistry of flexible molecules, complexes, and clusters and in predicting possible mechanisms and pathways for isomerizations, conformational rearrangements, and enantiomerizations. Furthermore, prior knowledge about the symmetry is extremely helpful in calculations searching for a transition state (TS).

3.1 Symmetry Operators and Permutation-Inversion Operators

In analyzing the symmetry of molecules, symmetry is usually thought of in terms of rotations about molecule-fixed axes, inversion at the center of mass, reflections at planes, and rotoreflections. This applies to rigid molecules, i.e., molecules that never depart far from their symmetric reference configuration [246]. In this context a reference configuration or *molecular framework* may be visualized as a rigid ball and stick model with labeled atoms [247]. However, e.g., high resolution gas-phase spectroscopy also required symmetry analysis of molecules with large amplitude and/or tunneling motions, i.e., molecules interconverting between several molecular frameworks [246, 247]. Likewise, the symmetry analysis of the dynamic stereochemistry of non-rigid molecules may include systems more complex than a single (reference) conformation and its point group. The dynamic stereochemistry of BAEs with twisted, *anti*-folded, and *syn*-folded conformations is a case in point. For these purposes it is more convenient to describe symmetry based on permutations of identical nuclei and a laboratory-fixed inversion [246, 247]. In the same way as the combination of all proper and improper rotations form the point group, the combination of all allowed permutations and permutation-inversions defines the *molecular symmetry group.*

3.2 *Historical Development of the Molecular Symmetry Group Approach*

The concept of the molecular symmetry group was first introduced by Longuet-Higgins in 1963 [246]. Independently, Pechukas used pairs of rotation/reflection operators and permutation operators to derive symmetry rules for transition states of chemical reactions [248]. The paper by Hougen reviewed some of the spectroscopic applications and also discussed the advantages and disadvantages of the point group treatment and of the permutation-inversion treatment [247]. Rodger and Schipper developed symmetry selection rules for reaction mechanisms and applied them to inorganic complexes [249–251]. Metropoulos translated these results into the permutation-inversion and molecular symmetry group terminology [252]. Bone, Rowlands, Handy, and Stone used the molecular symmetry groups to find the transition states for rearrangements in non-rigid molecular complexes and to deduce the number of such isomerization pathways [253]. In a second paper, Bone used the complete nuclear permutation-inversion (CNPI) group to find the symmetry of transition states and also discussed the interconversion of planar structures via pairs of non-planar labeled-atom enantiomeric transition states [254]. Permutation and inversion operators have also been used in the analysis of permutational isomerization reactions of stereochemically non-rigid trigonal bipyramidal and octahedral molecules where ligand substitution gives rise to constitutional isomers [255–259]. The work by Klemperer and Ruch and Hässelbarth on isomer counting and the classification of modes of rearrangements is reviewed in the book by Brocas et al. [260]. It also includes an introduction into group theory including a discussion of cosets and their properties. Furthermore, the application of permutations in the field of dynamic NMR is discussed. Schaad and Hu have extended the work of Pechukas, deducing a list of allowed point groups of transition states of degenerate reactions for 49 common cases [261]. Bytautas and Klein noted examples where the molecular symmetry group of non-rigid molecules is larger than the Longuet-Higgins permutation/inversion group [262].

3.3 *From the Complete Nuclear Permutation-Inversion Group to Feasible Operations in the Molecular Symmetry Group*

The treatment of symmetry using permutation operators and permutation-inversion operators may be derived from a quantum mechanical viewpoint: a molecule is an aggregate of atomic nuclei and electrons. Its Hamiltonian is invariant under the following types of transformation:

1. Any permutation of the positions and spins of the electrons.
2. Any rotation of the positions and spins of all particles (electrons and nuclei) about any axis through the center of mass.
3. Any overall translation in space.
4. The reversal of all particle momenta and spins (time reversal).
5. The simultaneous inversion of the positions of all particles in the center of mass (parity inversion).
6. Any permutation of the positions and spins of any set of identical nuclei.

The complete (symmetry) group of the Hamiltonian is thus the direct product of several groups [246]. In the context of this Chapter, the operations 5 and 6 are most important since they correspond to symmetry transformations of the nuclear framework of the molecule, i.e., its geometry and dynamic stereochemistry. The spin of the particles may be neglected for the present purpose. Furthermore, the very small (10^{-20} a.u.) parity-violating weak neutral current interaction lifting the exact degeneracy of enantiomers of a chiral molecule is neglected [263, 264].

The number of permutations in the complete nuclear permutation-inversion group is increasing rapidly with the number of identical particles. Fortunately, most of the permutations leaving the Hamiltonian invariant may be disregarded from a dynamic stereochemistry point of view because they involve breaking and forming of bonds. Thus, the corresponding barriers are too high for such processes to be observed under the experimental conditions in consideration. Only *feasible* permutations and permutation-inversions are included in the molecular symmetry group [246, 247, 252].

The concept of feasible permutation-inversions introduces some level of arbitrariness and a dependence on the particular experimental conditions or computational model. On the other hand, the criterion of feasibility is a particularly elegant and transparent way of introducing the necessary chemical information, i.e., the constitution (topology), the conformational flexibility and/or chemical reactivity, into the rigorous mathematical formalism of group theory [247]. For example, the effect of breaking symmetry by substitution may be analyzed in this way. Likewise, one may add or omit certain operators corresponding to a specific conformational isomerization process to the feasible permutation-inversion operators of the molecular symmetry group and thus analyze their effect on the dynamic stereochemistry.

3.4 Transition State Symmetry

The transition state is a characteristic point on the pathway between educts and products in chemical reactions including conformational isomerizations. The transition state is defined as the lowest possible energy maximum (barrier) on a continuous pathway from educts to products (assuming a barrier does exist) [265–268].

The transition state plays a central role in calculating reaction rates [265] and in understanding the mechanism of chemical reactions. It is a stationary point on the potential energy hypersurface with zero first derivatives (gradients). The force constant matrix of a transition state has exactly one negative eigenvalue [265, 269]. The corresponding eigenvector is defined as the transition vector of the reaction [270]. There is only one reaction path through a bona fide transition state [269]. Excluding bifurcations along the pathways and the unlikely case of a monkey saddle, a transition state connects only one educt to one product [269, 270]. Special symmetry rules have been derived for the transition vector and the symmetry of the transition state conformation. Stanton and McIver derived the following four theorems using the physical principle that structures that can be interconverted by symmetry have equivalent energies, group theory, and geometrical arguments [268, 270].

Theorem I. The transition vector cannot belong to a degenerate representation of the transition state point group.

Theorem II. The transition vector must be *anti*-symmetric under a transition state symmetry operation, which converts educts into products.

Theorem III. The transition vector must be symmetric with respect to a symmetry operation, which leaves either educts or products unchanged.

Theorem IV. If the transition vector for the reaction $E_1 \rightarrow P_1$ is symmetric under a symmetry operation $\hat{O}$ which converts educt E_1 into an equivalent educt E_2 and P_1 into P_2, then there exist lower energy transition states for the reactions $E_1 \rightarrow E_2$ and $P_1 \rightarrow P_2$; if the transition vector is *anti*-symmetric under $\hat{O}$ then there exists a lower energy transition state for the reaction $E_1 \rightarrow P_2$.

Pechukas has analyzed the pathways of steepest descent of a potential surface (defined in terms of a mass-scaled coordinate system) from a transition state to the product and the educt [248]. Using the fact that symmetry cannot change along a path of steepest descent and cannot decrease at the endpoints, he showed that "the point group of any point along a reaction pathway, including the transition state can be no larger than the largest subgroup common to both educt and product point groups" [248]. However, educts and products have to be directly linked to the transition state by steepest descent paths without intermediate minima or bifurcations [248]. Moreover, there is one exception: if educts and products are interconverted by a symmetry operation, this may be an additional symmetry element present in the transition state (but not along the steepest descent path) [248]. Schaad and Hu have listed the allowed point groups for transition states in common cases of degenerate reactions [261].

Nourse has further analyzed the case of degenerate isomerizations where educt and product are symmetry related and found that additional symmetry operators can only appear in the transition state of self-inverse degenerate isomerizations [271]. A degenerate isomerization is self-inverse if the coset of permutation-inversion operators leading from the reference isomer to the symmetry equivalent products contains at least one permutation of second order or a permutation of higher (only even) order and its inverse [271, 272].

3.5 Point Group Order and Number of Transition States

Bone et al. have derived two equations for the point group order and the number of versions of the transition state [253]. The first equation relates the point group order $\boldsymbol{h}_{\mathbf{PG}}$ of any conformation including transition states with the number of versions of this conformation $\boldsymbol{n}$ and the order of the molecular symmetry group $\boldsymbol{h}_{\mathbf{MSG}}$ [253]:

$$\boldsymbol{h}_{\mathbf{MSG}} = \boldsymbol{n} \times \boldsymbol{h}_{\mathbf{PG}} \quad (3)$$

Using the permutation operators corresponding to the various types of processes and the versions of the conformations interconverted by them, connectivity networks may be generated. Based on these networks, Bone et al. derived an equation which relates the number of versions of the transition states $\boldsymbol{n}_{\mathbf{TS}}$ with the number of versions of the interconverting minima $\boldsymbol{n}_{\mathbf{Min}}$, the connectivity $\boldsymbol{C}$ at the minima, and the number of equivalent (parallel) pathways $\boldsymbol{p}$ effecting the interconversion of the same two versions of the minimum [253]:

$$2 \times \boldsymbol{n}_{\mathbf{TS}} = \boldsymbol{C} \times \boldsymbol{p} \times \boldsymbol{n}_{\mathbf{Min}} \quad (4)$$

The factor 2 on the right-hand-side of (4) corresponds to the term $\boldsymbol{C} \times \boldsymbol{p}$ on the left-hand-side and reflects the fact that a bona fide TS always connects two and only two minima. Equation (4) does not apply in case of bifurcations along a pathway. Assuming $\boldsymbol{p} = 1$, the number of versions of the transition state $\boldsymbol{n}_{\mathbf{TS}}$ may be inferred, and using (3), the order of its point group $\boldsymbol{h}_{\mathbf{TS}}$. This assumption is equivalent to the principle of maximum symmetry (PMS) of Metropoulos [252]. However, this is only an upper limit for $\boldsymbol{h}_{\mathbf{TS}}$, since there may be more than one pathway ($\boldsymbol{p} > 1$) [251, 253, 273]. Note that a higher number of parallel pathways leads to a lower symmetry transition state. For recent publications on symmetry aspects of transition states in degenerate reactions see [261, 262].

3.6 Searching for Transition States

In summary, the symmetry of a transition state may include the operators of the common subgroup of the educt and product point groups [248]. In the case of a degenerate isomerization, these include all the symmetry operators of the conformation in question. Furthermore, in self-inverse automerizations, the transition state symmetry may also include permutation-inversion operators interchanging the educt and product versions of the conformation. However, note that the combination of all these symmetry operators is the highest possible symmetry of the transition state. All subgroups are also valid candidates. Equations (3) and (4) of Bone et al. [253], and Theorems I and II of McIver and Stanton [268, 270] may serve as tests to eliminate some options. The next step of a theoretical study is to find the stationary point with this point group by symmetry constrained geometry optimization. Note that the optimization may be constrained to the totally symmetric subspace [270]. Furthermore, in the case where the symmetry operator corresponding to the process is part of the point group, a minimum in this subspace has to be found [270]. This will make the computations much more efficient, particularly when the highly symmetric conformations are considered first. Vibrational frequencies should be calculated to verify that the conformation found is a bona fide transition state with one and only one imaginary frequency. It may turn out to be a local energy minimum (i.e., an intermediate conformation with no imaginary frequency) or a higher order saddle point (with more than one imaginary frequency). In the case where a higher order saddle point is found, the symmetry species of the imaginary modes may give useful information. One mode points towards educt and product, while the other mode(s) indicate(s) the point group(s) of the (possible) true transition state(s) and the distortions leading thereto. This information may aid in selecting the next point group to be considered.

In the case where a local minimum was found, this may be an intermediate on the pathway from educt to product. The process in question proceeds in two steps. In this case a new transition state on the way from the original educt to this intermediate product has to be found, and a second transition state on the pathway from the intermediate to the original product. If the original educt and product are symmetry related, the second transition state leading from the intermediate to the original product should be symmetry equivalent to the first transition state.

Even if a bona fide transition state has been found, this may not be the only transition state on this pathway from educt to product. There is the possibility of a multi-step pathway via several intermediates and transition states. This cannot be decided by symmetry considerations. It may be checked by calculating the reaction path with an intrinsic reaction coordinate (IRC) calculation, which will follow the minimum energy pathways from the transition state to the minima corresponding to its immediate educt and product [274, 275]. This technique allows a computational check of the connectivity of transitions states and the corresponding minima, and also the detailed study of the reaction mechanism by providing a continuous series of transient structures along the pathways from the transition state to the corresponding educt and product. Energetic and conformational changes along the reaction path can thus be studied in small steps.

3.7 *Bifurcating Pathways*

Another complication that may occur along a reaction path is the case that a steepest descent path from a transition state **TS1** leads, by symmetry, to another transition

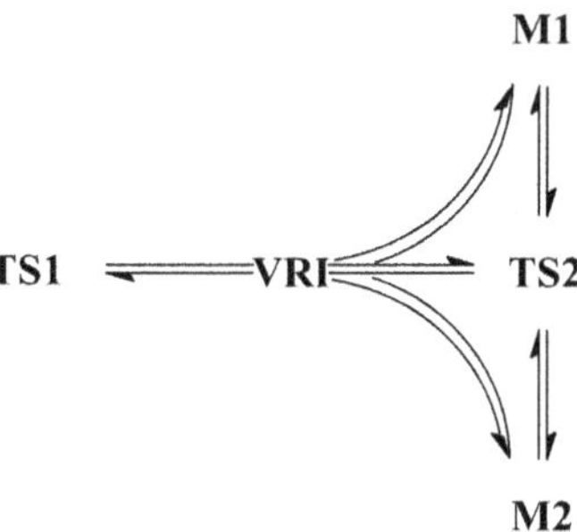

Fig. 11 Scheme of a bifurcating reaction path

state, **TS2**, rather than to a minimum as schematically outlined in Fig. 11. This situation leads to a bifurcation of the reaction paths and occurs, in particular, when the two transition states **TS1** and **TS2** share symmetry elements that are not present in the point groups of the minima **M1** and **M2**, but interchange **M1** and **M2**.

There are two viewpoints describing such a situation. The first opinion assumes steepest descent paths in a mass-weighted coordinate system as a model for the reaction path: a vibration-rotation-less motion of a reacting system is assumed [248, 254, 262, 275–277]. According to this view, the path from **TS1** to **M1** and **M2** proceeds in two steps according to **TS1** ⇌ **TS2** ⇌ **M1** (or **TS1** ⇌ **TS2** ⇌ **M2**) as indicated by the straight reaction arrows in Fig. 11. Note that the products of the second step (the final products of the reactions) may be identified by following the steepest descent paths starting at **TS2**. Symmetry is conserved along these steepest descent paths, and the symmetry rules for reactions apply to each individual step [248, 261, 262, 277]. However, it is accepted that a transition state (**TS2**) may be the end point of a steepest descent path, i.e., the product of a reaction step. Steepest descent paths are useful mathematical tools for studying reactions on potential energy surfaces [261]. However, this picture ignores the following peculiarity: close to the bona fide transition state **TS1**, the reaction system follows the floor of a valley. This is a stable pathway in the sense that there exists a restoring force that tends to drive the trajectories of the system back to the floor. However, at a certain point, the valley floor turns into the crest of a ridge and the symmetry conserving steepest descent path continues along the crest of the ridge down to **TS2**. The reaction path is expected to bifurcate close to the valley-ridge inflection point (VRI), because any small deviation from symmetry (e.g., due to vibration) will cause the system to veer away further and further [278]. Thus the system will head towards the final products **M1** or **M2**, taking a shortcut as indicated by the curved arrows in Fig. 11 and not pass through **TS2**. The reacting molecule does not follow the steepest descent path near and below a valley-ridge inflection point [261, 278]. The valley-ridge inflection point is characterized by a Hessian matrix with a zero eigenvalue, the corresponding eigenvector being orthogonal to the gradient at that point [278]. Note that this is not a stationary point. The gradient does not vanish. In conclusion, steepest descent paths are not good models for bifurcating reaction paths below valley-ridge inflection points [278]. However, steepest descent paths and the corresponding symmetry rules are still very useful tools in studying reaction mechanisms [261]. It may be interesting to note that the transition state **TS2** interconverting **M1** and **M2** has to have a lower energy than **TS1**, due to the steepest descent path leading from **TS1** to **TS2**. This is in accord with the prediction of Theorem IV of Stanton and McIver, which is not based on gradient paths [270].

4 Symmetry of BAEs

4.1 Conformations and Point Groups of BAEs

4.1.1 Topological Classification

Overcrowded bistricyclic aromatic enes (**1**) may be classified as homomerous ($X = Y$) and heteromerous ($X \neq Y$) BAEs, or, according to their substitution pattern, as unsubstituted, monosubstituted, and disubstituted BAEs. These classifications are helpful in analyzing the symmetry of these systems. In this context any modification making one (or more) of the four aromatic rings A, B, C, and D unique will be considered a "substitution." Persubstituted homomerous BAEs (e.g., hexadecachloro-9,9′-bi(9*H*-fluoren-9-ylidene)), and symmetrically tetra-substituted homomerous BAEs (e.g., 2,2′,7,7′-tetramethyl substituted BAE) fall into the same class as unsubstituted BAEs. Analogously, a 2,2′,7-trisubstituted BAE ($R = R' = R''$ and $R = R' \neq R''$; $X = Y$ and $X \neq Y$) would formally fall into the class of monosubstituted BAEs. 2,7-Disubstituted homomerous BAEs have the same symmetry properties as unsubstituted heteromerous BAEs. For symmetry classification, only the more narrowly defined group of 1,1′-, 2,2′-, 3,3′-, and 4,4′-disubstituted BAEs with $R = R'$, and $X = Y$ are considered disubstituted homomerous BAEs. In the class of disubstituted heteromerous BAEs ($X \neq Y$), $R = R'$, $R \neq R'$, and 1,2′-substitution patterns etc. are included, as long as the two substituents are not on the same moiety. In highly substituted BAEs, substituents may be subdivided in equivalent groups (i.e., identical substituents in equivalent positions). For each group the class may be assigned separately. The appropriate class of such a highly substituted BAE will be the "smallest common denominator," i.e., the class corresponding to the common subgroup of all symmetry groups.

The above analysis leads to five classes of overcrowded bistricyclic aromatic enes:

- Unsubstituted homomerous BAEs
- Unsubstituted heteromerous BAEs
- Monosubstituted homomerous BAEs and monosubstituted heteromerous BAEs
- Disubstituted homomerous BAEs
- Disubstituted heteromerous BAEs

This classification has particular relevance for the topological considerations used to derive the feasible permutation-inversion operators and the molecular symmetry group.

Disubstitution (one substituent on each tricyclic moiety) gives rise to *E*- and *Z*-stereoisomers. It should be noted that the systematic nomenclature gives substituents on both moieties the lowest possible number, irrespective of their *E*- or *Z*-relationship. Stereoisomers are distinguished by prefixes (*E*)- and (*Z*)-. Thus atoms 1 and 1′ are *cis* in *Z*-isomers, but *trans* in *E*-isomers. In order to maintain a consistent atom-labeling scheme for unsubstituted BAEs, as well as for *E*- and for *Z*-disubstituted BAEs, this rule will be disregarded in the discussions of the conformations and dynamic stereochemistry below. The labeled general structure shown in Fig. 1 will be considered as the rigid molecular framework (see above) and the *cis* relationship of positions 1 and 1′, etc. in the reference version of the framework will be maintained regardless of the substitution pattern and *E*,*Z*-isomerizations. *E*-isomers thus have their substituents, e.g., at carbon atoms 2 and 7′. Furthermore, the orientation of the molecular framework (before the application of permutations or permutation-inversions) will always be kept fixed as in Fig. 1 with positions 1 and 1′ on the left-hand-side (negative *y*-axis direction) and the unprimed moiety with bridge X at the top (positive *z*-axis direction). Thus, an alternative orientation of the molecule in space (rotated by 180° about the *z*-axis) may be defined by placing substituents at positions 7 and 2′.

4.1.2 Conformational Types

The following classification scheme of the types of conformations of overcrowded bistricyclic aromatic enes may be adopted:

- Planar conformations, **p**
- Orthogonally twisted conformations, $\mathbf{t}_{\perp}$
- Twisted conformations, **t**
- *anti*-Folded conformations, **a**
- *syn*-Folded conformations, **s**
- Folded conformations with a planar (or nearly planar) second moiety, **f**
- Twisted/*anti*-folded conformations, **ta**
- Twisted/*syn*-folded conformations, **ts**
- Twisted/folded conformations, **tf**
- *anti*-Folded/twisted conformations, **at**
- *anti*-Folded conformations with unequal degrees of folding, **au**
- *syn*-Folded/twisted conformations, **st**
- *syn*-Folded conformations with unequal degrees of folding, **su**
- Folded/twisted conformations, **ft**
- Planar central ethylene group with propeller twisted tricyclic moieties, **pt**

The basic conformational types are schematically illustrated in Fig. 12. Here the circle and its center point represent the double bond in a Newman-like projection. The lines represent the tricyclic moieties by indicating the positions 1, 8, 1′, and 8′ in the overcrowded fjord regions.

Fig. 12 Schematic representation of important BAE conformations

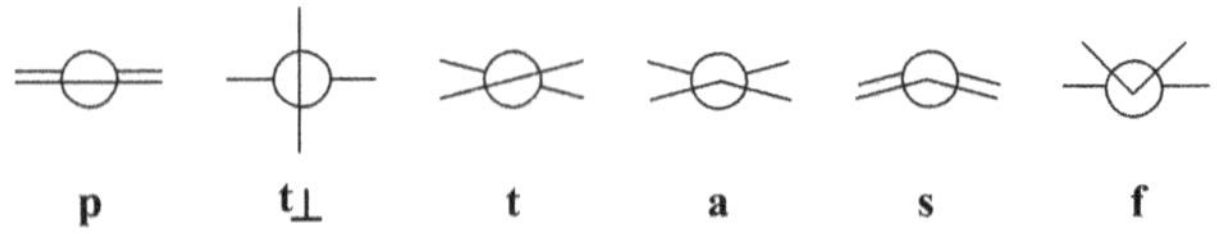

In cases where two out-of-plane modes are combined, the order may be used to designate the dominant mode. Thus, **ta** is closer to a twisted conformation, while **at** is closer to an *anti*-folded conformation. The symbols **au** and **su** describe unequally folded conformations, '**u**', which are close to *anti*- and *syn*-folded conformations, **a** and **s**, respectively. In the same spirit, symbols **tau**, **tsu**, **aut**, and **sut** may be defined. However, in this work the shorter symbols **tf** and **ft** will be preferred for these types of conformations. It should be noted that this classification according to conformation types is an empirical, qualitative approach and is intuitive rather than exact. It may be applied in a strict as well as in an approximate manner. Nevertheless, the conformation type symbols are very useful in describing the conformations of overcrowded bistricyclic aromatic enes.

4.1.3 Conformations and Corresponding Point Groups

An exhaustive list of point groups that may be possible candidates for conformations of overcrowded bistricyclic aromatic enes may be derived from an analysis of the subgroups of the point group D_{4h}, which is isomorphous to the molecular symmetry group of BAEs, G_{16} (see Sect. 4.2.2). The point group D_{4h} includes the symmetry operators of all possible BAE conformations as supergroup. The subgroups of D_{4h} may be found by analyzing the character table and identifying the subgroup associated with each symmetry species. For character tables of D_{4h} and its subgroups see [279]. The results of such an analysis are summarized in Fig. 13 (adapted from [279]). Lines connect subgroups and supergroups. The lines are labeled with the symmetry species of the higher point group leading to the respective lower point group. The point group order h of the groups is given at the left-hand-side. Note that point groups including C_4 axes do not qualify as point groups for any conformations of BAEs since a C_4 axis is not compatible with the topology. Thus, D_{2d} and D_{2h} are the highest possible point groups ($h = 8$) of BAE conformations.

Combining the above classification schemes and descriptors with a symmetry analysis based on the point groups which are subgroups of D_{4h} leads to a detailed classification of the topology, substitution pattern, conformations, symmetry, and chirality of overcrowded bistricyclic aromatic enes. The results are summarized in Table 4.

In Table 4 the unsubstituted homomerous and heteromerous, monosubstituted and disubstituted classes are arranged in columns. Each of the basic confirmations, **p**, $\mathbf{t}_{\perp}$, **t**, **a**, **s**, and **f** (Fig. 12) is listed in a separate block. Conformational symbols and the point group symmetry are given along with the chirality or achirality of the conformation. *E*- and *Z*-isomers, the helicity of the central double bond (*P*/*M*), and the absolute configuration of the (substituted) folded moieties (*R*/*S*) are indicated by appropriate stereodescriptors. For simplicity only one enantiomer (*P*) is given in

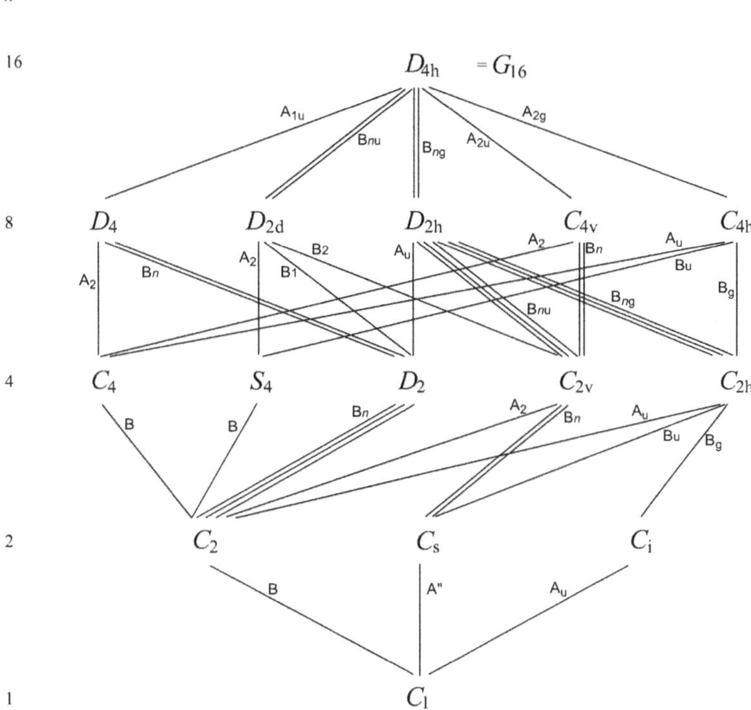

Fig. 13 Subgroups of D_{4h} including the point group order, subgroup–supergroup relations, and symmetry species. The symmetry species of the higher point groups are given next to the lines connecting to the respective subgroups. Subscripts n correspond to $n = 1{,}2$ or $n = 1$–3 as indicated by the number of lines

Table 4. For each of the basic conformations the highest point group is listed first, followed by the subgroups. Corresponding point groups for homomerous, heteromerous, and substituted BAEs are listed on the same line. In cases where the reduced symmetry leads to a different conformational type, this is indicated by a cross reference to the respective point group, which may be found further down in the same column. It should be noted that the listed conformations need not necessarily be minima. Some may serve as intermediates or transition states in the dynamic processes characteristic of BAEs. They could also be higher order saddle points. Furthermore, a specific conformation, (e.g., **t**-D_2) may turn out to be the global minimum in one BAE and a transition state in a dynamic process in a different BAE. In certain cases this may also depend on the computational method employed in the study. The classification presented in Table 4 may also be applied to vibrationally distorted molecules and to molecules in crystal structures, which may be distorted by packing forces.

In cases where the point group symmetry allows twisting, *anti*-folding, or *syn*-folding in addition to the basic conformational mode, this is indicated using symbols for mixed conformations **ta**, **at**, **ts**, **st**, **au**, **su**, and **f**. Note that, from a

Table 4 Conformations, symmetry and chirality of BAEs

Unsubstituted		Monosubstituted	Disubstituted	
R = R′ = H		R ≠ H, R′ = H	R = R	R = R′ or R ≠ R′
Homomerous	Heteromerous	Homomerous or heteromerous	Homomerous	Heteromerous
X = Y	X ≠ Y	X = Y or X ≠ Y	X = Y	X ≠ Y
p-D_{2h} achiral	(C_{2v} (*z*)) [a]	(C_s (*yz*)) [a]	(*E*: C_{2h} (*x*); *Z*: C_{2v} (*y*)) [a]	(*E*,*Z*: C_s (*yz*)) [a]
p-C_{2v} (*z*) achiral	**p**-C_{2v} (*z*) achiral	"	(*E*,*Z*: C_s (*yz*)) [a]	"
p-C_{2h} (*x*) achiral	(C_s (*yz*)) [a]	"	**p**$_E$-C_{2h} (*x*) achiral	"
			(*Z*: C_s (*yz*)) [a]	
p-C_{2v} (*y*) achiral	"	"	(*E*: C_s (*yz*)) [a]	"
			p$_Z$-C_{2v} (*y*) achiral	
p-C_s (*yz*) achiral	**p**-C_s (*yz*) achiral	**p**-C_s (*yz*) achiral	**p**$_E$-C_s (*yz*) achiral	**p**$_E$-C_s (*yz*) achiral
			p$_Z$-C_s (*yz*) achiral	**p**$_Z$-C_s (*yz*) achiral
pt-C_{2h} (*z*) achiral	(C_2 (*z*)) [a]	(C_1) [a]	(*E*: C_1; *Z*: C_s (*xy*)) [a]	(*E*,*Z*: C_1) [a]
t$_\perp$-D_{2d} achiral	(C_{2v} (*d*) [b]) [a]	(C_s (*d*) [c]) [a]	(*E*: C_2 (*x*); *Z*: C_2 (*y*)) [a]	(*E*,*Z*: C_1) [a]
t$_\perp$-C_{2v} (*d*) [b] achiral	**t**$_\perp$-C_{2v} (*d*) [b] achiral	"	(*E*,*Z*: C_1) [a]	"
t$_\perp$-C_s (*d*) [c] achiral	**t**$_\perp$-C_s (*d*) [c] achiral	**t**$_\perp$-C_s (*d*) [c] achiral	"	"
t$_{\perp PM}$-S_4 (*z*) meso [d]	(C_2 (*z*)) [a]	(C_1) [a]	"	"
t$_P$-D_2 *chiral*	(C_2 (*z*)) [a]	(C_1) [a]	(*E*: C_2 (*x*); *Z*: C_2 (*y*)) [a]	(*E*,*Z*: C_1) [a]
t$_P$-C_2 (*z*) *chiral*	**t**$_P$-C_2 (*z*) *chiral*	"	(*E*,*Z*: C_1) [a]	"
ta$_P$-C_2 (*y*) *chiral*	(C_1) [a]	"	(*E*: C_1) [a]	"
			ta$_{Z\text{-}RPR'}$-C_2 (*y*) *chiral*	
ts$_P$-C_2 (*x*) *chiral*	(C_1) [a]	"	**ts**$_{E\text{-}RPR'}$-C_2 (*x*) *chiral*	"
			(*Z*: C_1) [a]	

(continued)

Table 4 (continued)

Unsubstituted		Monosubstituted	Disubstituted	
R = R′ = H		R ≠ H, R′ = H	R = R	R = R′ or R ≠ R′
Homomerous	Heteromerous	Homomerous or heteromerous	Homomerous	Heteromerous
X = Y	X ≠ Y	X = Y or X ≠ Y	X = Y	X ≠ Y
a-C_{2h} (y) achiral	(C_s (yz)) [a]	(C_1) [a]	(E: C_i,; Z: C_2 (y)) [a]	(E,Z: C_1) [a]
a-C_i achiral	"	"	$\mathbf{a}_{E\text{-}RS'}$-$C_i$ meso [d]	"
			(Z: C_1) [a]	
$\mathbf{at}_P$-C_2 (y) *chiral*	(C_1) [a]	"	(E: C_1) [a]	"
			$\mathbf{at}_{Z\text{-}RPR'}$-$C_2$ (y) *chiral*	
au-C_s (xz) achiral	**au**-C_s (xz) achiral	"	(E,Z: C_1) [a]	"
s-C_{2v} (x) achiral	(C_s (yz)) [a]	(C_1) [a]	(E: C_2 (x); Z: C_s (xy)) [a]	(E,Z: C_1) [a]
s-C_s (xy) achiral	"	"	(E: C_1) [a]	"
			$\mathbf{s}_{Z\text{-}RS'}$-$C_s$ (xy) meso [d]	
$\mathbf{st}_P$-C_2 (x) *chiral*	(C_1) [a]	"	$\mathbf{st}_{E\text{-}RPR'}$-$C_2$ (x) *chiral*	"
			(Z: C_1) [a]	
su-C_s (xz) achiral	**su**-C_s (xz) achiral	"	(E,Z: C_1) [a]	"
f-C_s (xz) achiral	**f**-C_s (xz) achiral	(C_1) [a]	(E,Z: C_1) [a]	(E,Z: C_1) [a]
$\mathbf{ft}_P$-C_1 *chiral*	$\mathbf{ft}_P$-C_1 *chiral*	$\mathbf{ft}_{RP}$-C_1 *chiral*	$\mathbf{ft}_{E\text{-}RPR'}$-$C_1$ [e] *chiral*	$\mathbf{ft}_{E\text{-}RPR'}$-$C_1$ [e] *chiral*
			$\mathbf{ft}_{E\text{-}RPS'}$-$C_1$ [f] *chiral*	$\mathbf{ft}_{E\text{-}RPS'}$-$C_1$ [f] *chiral*
			$\mathbf{ft}_{Z\text{-}RPR'}$-$C_1$ [f] *chiral*	$\mathbf{ft}_{Z\text{-}RPR'}$-$C_1$ [f] *chiral*
			$\mathbf{ft}_{Z\text{-}RPS'}$-$C_1$ [e] *chiral*	$\mathbf{ft}_{Z\text{-}RPS'}$-$C_1$ [e] *chiral*

For chiral conformations, only one enantiomer (P) is given. Conformations with C_1 symmetry are represented by **ft** listed at the bottom of the table

[a]See this point group further down in this column

[b]In this orientation of the point group C_{2v} the reflection planes are the diagonal planes $y = x$ and $y = -x$, respectively

[c]The reflection plane is the plane $y = x$

[d]The two moieties have opposite chirality sense

[e]The relative direction of folding of the two moieties is *syn*

[f]The relative direction of folding of the two moieties is *anti*

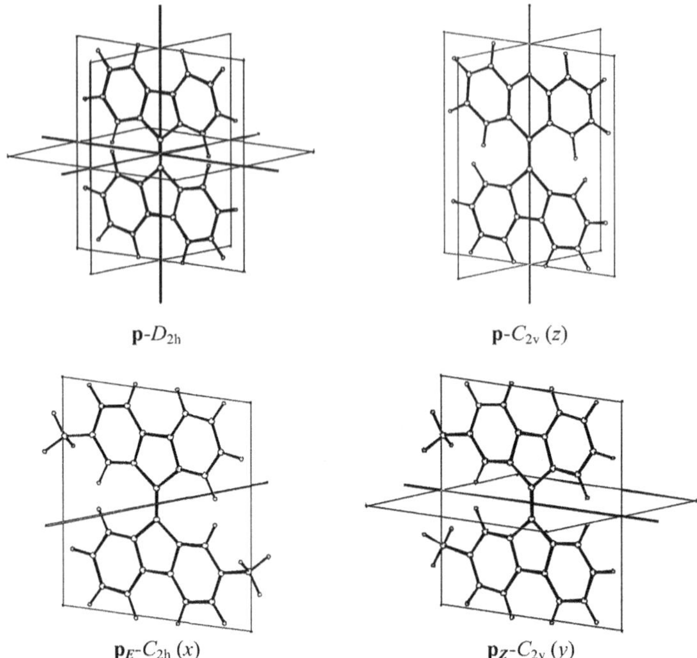

Fig. 14 Three-dimensional projections of the planar conformations of homomerous, heteromerous, *E*-, and *Z*-disubstituted BAEs including their point group symmetry operators

symmetry point of view, the magnitude of the contributions of two modes in a mixed conformation is irrelevant. The point group and symmetry properties are identical. However, the conformations may look significantly different (see below). Any conformation combining three modes (**t**, **a**, and **s**) of out-of-plane deformation has point group C_1. Formally C_1 ($\equiv \{E\}$) is a subgroup of every point group. Thus the corresponding conformations symbolized by **ft**-C_1 (or **tf**-C_1) should be listed as possible sub-groups for each conformation type (except for **p**, which by definition has a minimum symmetry of C_s). However, there is no symmetry information in this fact, and thus **ft**-C_1 is included only once at the bottom of the table rather than repeated for every conformational class.

Planar Conformations

The (hypothetical) planar conformation **p** of a homomerous bistricyclic aromatic ene has point group symmetry D_{2h} or one of its subgroups. Characteristic examples are shown in Fig. 14. The point group C_{2v} (z) corresponds to cases where the two moieties are not equivalent, e.g., in heteromerous BAEs. The point group C_{2v} (y) is characteristic for *Z*-disubstituted BAEs. *E*-Disubstitution leads to point group C_{2h} (x). The point group C_s (yz) corresponds to a loss of all elements of symmetry except for the plane of the molecule. In conformations **pt** with the point group C_{2h}

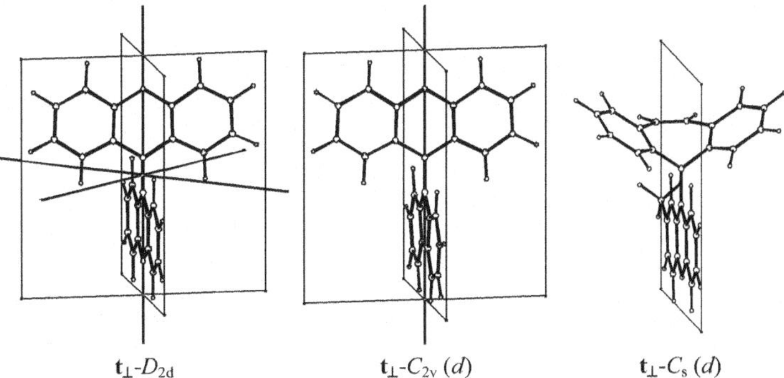

Fig. 15 Three-dimensional projections of the orthogonally twisted conformations of homomerous, heteromerous, and monosubstituted BAEs including their point group symmetry operators

(z) the central ethylene group is confined to planarity. However, the two tricyclic moieties may be twisted out-of-plane like a propeller blade and thus non-planar.

Orthogonally Twisted Conformations

An orthogonally twisted $\mathbf{t}_{\perp}$ homomerous BAE may have D_{2d} symmetry. The point group C_{2v} (d) is characteristic of orthogonally twisted heteromerous BAEs and C_s (d) of monosubstituted BAEs (Fig. 15). Note that in disubstituted BAEs the orthogonally twisted conformation has C_2 symmetry and is chiral. The torsion angles of the central ethylene group are no longer symmetry constrained to ±90°. The symmetry properties of such a conformation are identical to, e.g., $\mathbf{ts}_{\boldsymbol{E\text{-}RPR'}}$-$C_2$ (x) or $\mathbf{ta}_{\boldsymbol{Z\text{-}RPR'}}$-$C_2$ (y) (see above). However, the pure ethylenic twist $\boldsymbol{\omega}$ will be close to 90°. Thus, the definitions of helicity (P/M) based on $\boldsymbol{\omega}$ as well as of E/Z and *syn/anti* will depend on the small deviation from 90°. In this situation it is better to label the two enantiomeric conformations (P) or (M) based on the torsion angle $\boldsymbol{\tau}(C_{9a}\text{-}C_9\text{-}C_{9'}\text{-}C_{9a'})$, which will be either close to +90° or close to −90° and thus clearly distinct (i.e., applying the concept of axial chirality; cf. Sect. 2.2). From a symmetry point of view, $\mathbf{t}_{\perp \boldsymbol{PM}}$-$S_4$ is an interesting case. In this conformation, the two tricyclic moieties are non-planar. They may, e.g., be propeller twisted with positive and negative helicity, respectively, as indicated by the subscript *PM*. As a result, the mirror planes σ_d ($x = y$) and σ_d ($x=-y$) of D_{2d} are lost. However, the two non-planar moieties are still equivalent and symmetry related by S_4, and S_4^3. The torsion angles of the central ethylene group are constrained to ±90°. The conformation is achiral, but both tricyclic moieties are chiral with opposite helicity.

Twisted Conformations

The symmetric twisted conformations of homomerous BAEs are classified as $\mathbf{t}_{\boldsymbol{P}}$-$D_2$ and its enantiomer $\mathbf{t}_{\boldsymbol{M}}$-$D_2$. Non-equivalent moieties in heteromerous BAEs lead to

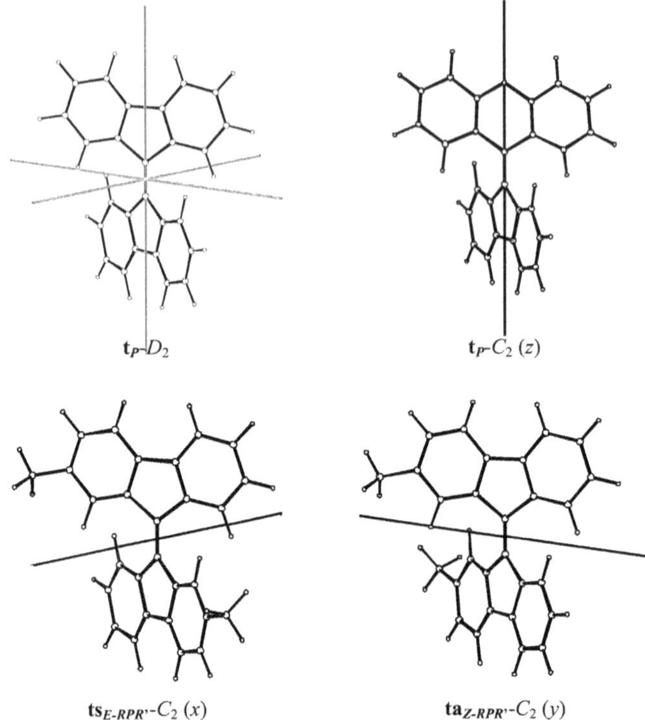

Fig. 16 Three-dimensional projections of the twisted conformations of homomerous, heteromerous, and disubstituted BAEs including their point group symmetry operators

$\mathbf{t}_P$-C_2 (*z*) and $\mathbf{t}_M$-C_2 (*z*) conformations. *E*- or *Z*-disubstitution may introduce *syn*-folding or *anti*-folding, respectively. The various twisted conformations and their point groups are illustrated in Fig. 16.

Note that $\mathbf{ts}_{E\text{-}RPR'}$-$C_2$ (*x*) and $\mathbf{ta}_{Z\text{-}RPR'}$-$C_2$ (*y*) (and their enantiomers $\mathbf{ts}_{E\text{-}SMS'}$-$C_2$ (*x*) and $\mathbf{ta}_{Z\text{-}SMS'}$-$C_2$ (*y*), respectively) correspond to the case where the substituents require additional out-of-plane deformation due to increased overcrowding as in the 1,1'-difluoroderivatives of bifluorenylidene (2) [205]. Reduced out-of-plane deformation at the substituents would lead to $\mathbf{ts}_{E\text{-}SPS'}$-$C_2$ (*x*) and $\mathbf{ta}_{Z\text{-}SPS'}$-$C_2$ (*y*) (and their enantiomers $\mathbf{ts}_{E\text{-}RMR'}$-$C_2$ (*x*) and $\mathbf{ta}_{Z\text{-}RMR'}$-$C_2$ (*y*), respectively) as illustrated in Fig. 17.

anti-Folded Conformations

Symmetric *anti*-folded conformations of homomerous BAEs may have the point group C_{2h} (*y*). Non-equivalent tricyclic moieties in heteromerous BAEs lead to **au**-C_s (*xz*) with unequal degrees of folding. *E*-Disubstitution leads to $\mathbf{a}_{E\text{-}RS'}$-$C_i$. Note that there is no twist of the central double bond due to symmetry constraints. The folded substituted tricyclic moieties are chiral, albeit with opposite absolute

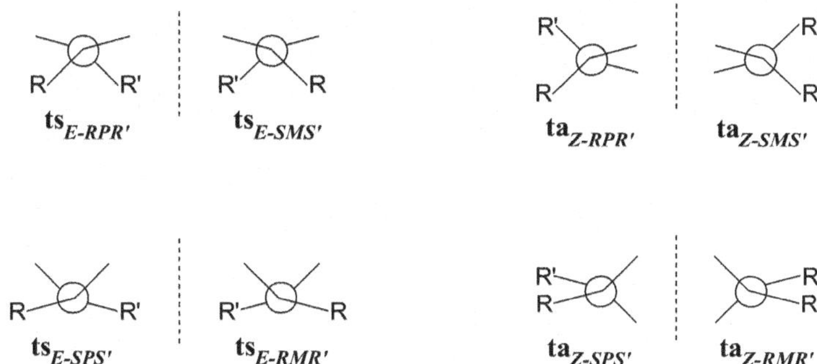

Fig. 17 Schematic representation of the conformations $\mathbf{ts}_{E\text{-}RPR'}$, $\mathbf{ts}_{E\text{-}SMS'}$, $\mathbf{ts}_{E\text{-}SPS'}$, $\mathbf{ts}_{E\text{-}RMR'}$, $\mathbf{ta}_{Z\text{-}RPR'}$, $\mathbf{ta}_{Z\text{-}SMS'}$, $\mathbf{ta}_{Z\text{-}SPS'}$, and $\mathbf{ta}_{Z\text{-}RMR'}$

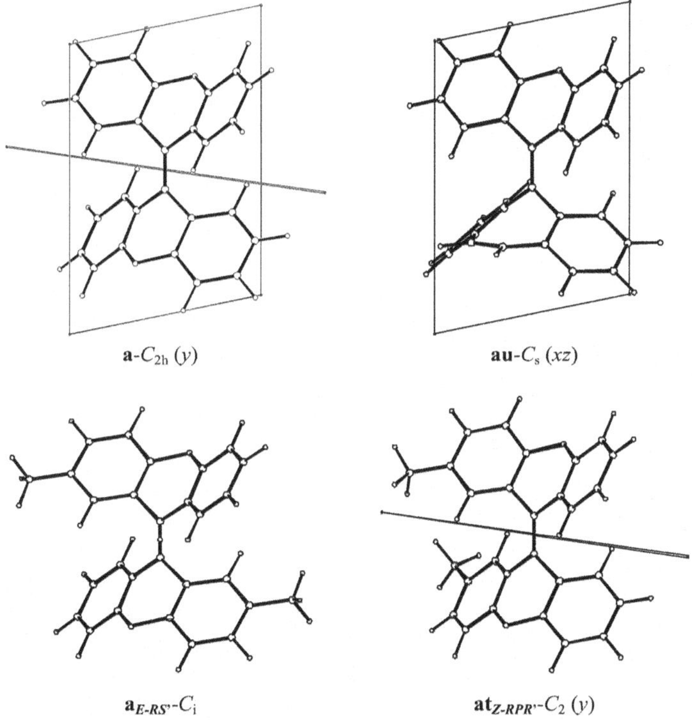

Fig. 18 Three-dimensional projections of the *anti*-folded conformations of homomerous, heteromerous, and disubstituted BAEs including their point group symmetry operators

configuration, as indicated by the subscript RS'. Thus, $\mathbf{a}_{E\text{-}RS'}$-$C_i$ is *meso*. The inverted form, $\mathbf{a}_{E\text{-}SR'}$-$C_i$, is superimposable by rotation and translation. *Z*-Disubstitution leads to $\mathbf{at}_{Z\text{-}RPR'}$-$C_2$ (y), which has the same point group as $\mathbf{ta}_{Z\text{-}RPR'}$-$C_2$ (y) (Fig. 16). Typical examples are shown in Fig. 18.

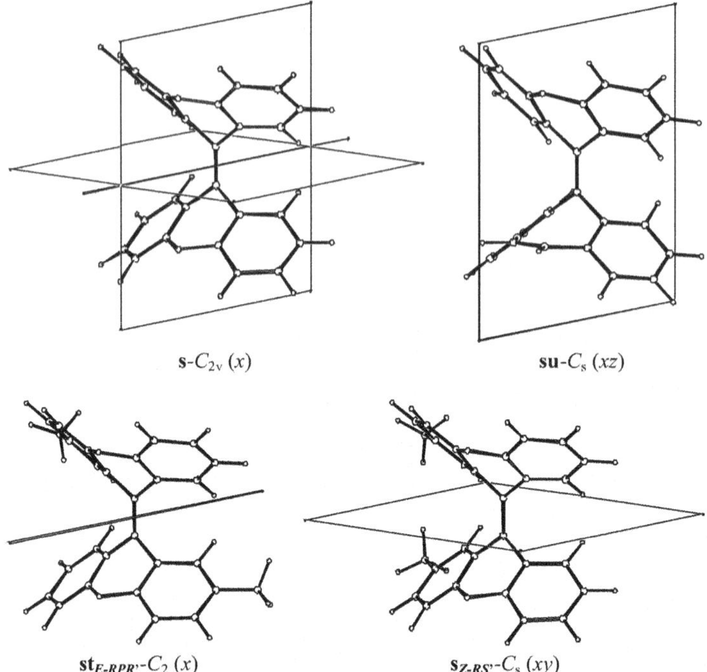

Fig. 19 Three-dimensional projections of the *syn*-folded conformations of homomerous, heteromerous, and disubstituted BAEs including their point group symmetry operators

syn-Folded Conformations

Symmetric *syn*-folded conformations of homomerous BAEs may have point group C_{2v} (x). Non-equivalent tricyclic moieties in heteromerous BAEs lead to **su**-C_s (xz), which has the same point group as **au**-C_s (xz) (Fig. 18) and **f**-C_s (xz). E-Disubstitution leads to **st**$_{E\text{-}RPR'}$-C_2 (x), with the same point group as **ts**$_{E\text{-}RPR'}$-C_2 (x) (Fig. 16). Z-Disubstitution results in **s**$_{Z\text{-}RS'}$-C_s (xy), a *meso* conformation. In the last case, twisting is not compatible with the point group. Typical examples of *syn*-folded conformations are illustrated in Fig. 19.

Folded Conformations

The folded conformation **f**-C_s (xz) may be envisioned as a conformation with one folded and one (nearly) planar moiety, as, e.g., in the folded conformation of fluorenylidene-xanthene (**5**) (Fig. 20). For homomerous and heteromerous BAEs, such a conformation may have C_s (xz) symmetry. Moreover, the conformational type **f**-C_s (xz) may also be used to describe folded conformations with unequal degrees of folding, when it is not known whether the relative direction of folding is *syn* or *anti* (**au**-C_s (xz) or **su**-C_s (xz)). Twisting the central ethylene group reduces

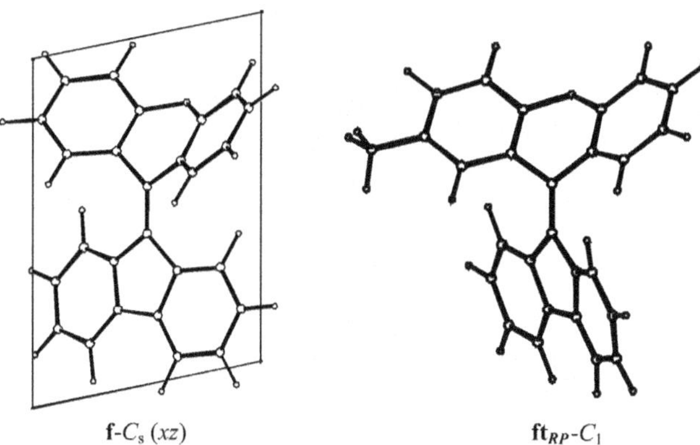

Fig. 20 Three-dimensional projection of the folded conformations of heteromerous and substituted heteromerous BAEs including their symmetry operators

the symmetry to C_1. The conformations **ft**$_{P}$-C_1 and **ft**$_{M}$-C_1 of homomerous and heteromerous BAEs are chiral. Monosubstitution introduces another element of chirality leading to two diastereomeric pairs of enantiomers: **ft**$_{RP}$-C_1 and **ft**$_{SM}$-C_1 vs **ft**$_{RM}$-C_1 and **ft**$_{SP}$-C_1. Adding another substituent on the second moiety results in four diastereomeric pairs of enantiomers: **ft**$_{E\text{-}RPR'}$-C_1 and **ft**$_{E\text{-}SMS'}$-C_1, **ft**$_{E\text{-}RPS'}$-C_1 and **ft**$_{E\text{-}SMR'}$-C_1, **ft**$_{Z\text{-}RPR'}$-C_1 and **ft**$_{Z\text{-}SMS'}$-C_1, and **ft**$_{Z\text{-}RPS'}$-C_1 and **ft**$_{Z\text{-}SMR'}$-C_1. Four structures are *E*-isomers and four are *Z*-isomers. Due to the relationships of *E,Z*-isomerism, *anti*- and *syn*-folding, and the absolute configuration of the substituted moieties (*R*/*S*), the first and the last mentioned enantiomeric pair have *syn*-folded character, while the second and third enantiomeric pair have *anti*-folded character (cf. Fig. 5).

In a global approach to the analysis of the relations between point group symmetry and energy hyper surfaces, Mezey has proved five theorems which allow the prediction of at least one conformation with stationary energy (zero gradients) by comparing the symmetry elements present (or absent) in a closed set of conformations with those of the boundary of the set [280]. In the case of the bistricyclic aromatic enes it may be shown that there is at least one planar and one orthogonally twisted stationary point with the respective highest possible symmetry (Table 4).

4.2 *Symmetry and Dynamic Stereochemistry of BAEs*

In the following, the mechanisms for conformational isomerizations of BAEs will be analyzed in terms of local minima, and transition states. The transition states represent pathways for interconversion between the minima. This analysis leads to the representation of the dynamic stereochemistry of a molecule by the reaction graph, which is a very elegant tool to analyze experimental dynamic stereochemistry and theoretical predictions based on computations.

4.2.1 Notation of Permutation-Inversion Operators

Permutation operators may be written as cycles (123) which should be read as ($1 \leftarrow 2 \leftarrow 3 \leftarrow 1$), i.e., nucleus 1 is replaced by 2, 2 by 3, and 3 by 1 [281]. Likewise, (12)(3) corresponds to ($1 \leftarrow 2 \leftarrow 1$) and ($3 \leftarrow 3$), i.e., 1 and 2 are interchanged, 3 remains in place. Atoms that do not change place (e.g., (3)) may also be omitted for brevity. The identity operator is E and the laboratory-fixed inversion is E^*. Combinations of permutations and the inversion may be written, e.g., (12)* [246, 247, 281].

Using the notation for permutation-inversion operators defined above and the atom labeling in BAEs indicated in Fig. 1, the 180° rotation about the horizontal axis through the center of the overcrowded fjord regions may be written as

$$\left(C^1C^{1'}\right)\left(C^2C^{2'}\right)\left(C^3C^{3'}\right)\left(C^4C^{4'}\right)\left(C^{4a}C^{4a'}\right)\left(C^5C^{5'}\right)\left(C^6C^{6'}\right)\left(C^7C^{7'}\right)\left(C^8C^{8'}\right)\left(C^{8a}C^{8a'}\right)$$
$$\left(C^9C^{9'}\right)\left(C^{9a}C^{9a'}\right)\left(C^{10a}C^{10a'}\right)\left(H^1H^{1'}\right)\left(H^2H^{2'}\right)\left(H^3H^{3'}\right)\left(H^4H^{4'}\right)\left(H^5H^{5'}\right)\left(H^6H^{6'}\right)\left(H^7H^{7'}\right)\left(H^8H^{8'}\right)(XY)$$

This explicit notation of every atom in the molecule is only required when the complete nuclear permutation-inversion group is considered and scrambling of identical atoms is allowed. Considering as feasible only those permutations that do not change the connectivity, i.e., do not break bonds [247], the element symbols may be omitted since C^1 and H^1 necessarily have to undergo analogous permutations in order to conserve the bond C^1-H^1. Furthermore, the (non-hydrogen) atoms of the molecule may be classified according to the topology into sets of atoms with equivalent connectivity: {1, 8, 1′, 8′}, {2, 7, 2′, 7′}, {3, 6, 3′, 6′}, {4, 5, 4′, 5′}, {4a, 10a, 4a′, 10a′}, {8a, 9a, 8a′, 9a′}, {9, 9′}, {X, Y}. Conserving the connectivity of the molecule requires that the atoms in each of the six sets with four elements (members in the set) undergo an analogous permutation. Likewise, the atoms in the two sets with two elements undergo an analogous permutation. Thus, it is sufficient to write only the permutations of one set of four and one set of two, e.g., the sets {1, 8, 1′, 8′} and {9, 9′}, implying that, under the constraint of conserving the connectivity, analogous permutations apply to all the other carbon, hydrogen, and hetero atoms. The above-mentioned 180° rotation may thus be written in shorthand as (11′)(88′)(99′).

Conserving bond connectivity imposes an additional constraint on the feasible permutation operators: if atoms 9 and 9′ are interchanged, all other atoms of the first moiety also have to be interchanged with corresponding atoms of the second moiety, i.e., (99′) necessarily requires either (11′)(88′), or (18′)(81′), or (11′88′), or (18′81′). It should be noted that the above considerations strictly apply only to unsubstituted, homomerous BAEs. For heteromerous BAEs, and substituted BAEs, the different constitution requires some modifications, which reflect the lower symmetry. The corresponding molecular symmetry groups are subgroups of the molecular symmetry group of unsubstituted homomerous BAEs (see below).

4.2.2 The Molecular Symmetry Group of Homomerous BAEs

The molecular symmetry group G_{16} of homomerous BAEs consists of following permutation-inversion operators:

1. E	2. (18)(1′8′)	3. (11′)(88′)(99′)	4. (18′)(81′)(99′)
5. (18)	6. (1′8′)	7. (11′88′)(99′)	8. (18′81′)(99′)
9. E*	10. (18)(1′8′)*	11. (11′)(88′)(99′)*	12. (18′)(81′)(99′)*
13. (18)*	14. (1′8′)*	15. (11′88′)(99′)*	16. (18′81′)(99′)*

A multiplication table and a character table [246] of the permutation-inversion operators of the molecular symmetry group G_{16} are given in Tables 5 and 6, respectively. The molecular symmetry group G_{16} of BAEs is isomorphous to D_{4h} (cf. Sect. 4.1.3 and Fig. 13). Each of the above permutation-inversion operators may be mapped onto a symmetry operator of D_{4h} with identical character such that corresponding multiplication tables and character tables result. (For a multiplication table and character table of D_{4h} see, e.g., [279]). Note that there is no D_{4h} symmetric conformation of a BAE. In particular, there is no C_4 axis in any conformation. It should be noted that the molecular symmetry group includes two operators corresponding to the *E*,*Z*-isomerization: (18) and (1′8′). They correspond to rotating the first or the second tricyclic moiety about $C^9{=}C^{9'}$, respectively. This feasible operation is included in the molecular symmetry group G_{16} in a consistent and natural way.

The permutation-inversion operators of the molecular symmetry group of BAEs were derived from topological considerations and a criterion of feasibility defined by the axiom that no bonds may be broken. Alternatively, the molecular symmetry group may be derived by analyzing the effect of the point group symmetry operators on the labeled atoms and the handedness of the molecular framework, thus correlating symmetry operators with permutation-inversion operators. All operators of the point group(s) of a molecule in its different conformations are combined and operators for known feasible processes (e.g., internal rotations of a methyl group, etc.) are added to give the molecular symmetry group [246, 247, 253]. In both cases a multiplication table should be generated in order to verify that the proposed set of operators is a group [279].

4.2.3 Permutation-Inversion Operators and Corresponding Symmetry Operators of BAEs

For visualization, it may be helpful to define a molecule-fixed coordinate system orienting the molecule in a consistent way in all conformations and to point out the conventional symmetry elements corresponding to the permutation-inversion operators, where appropriate. It should be noted, that the definition of a molecule-fixed coordinate system is not required within the formalism of the molecular symmetry group. Some permutation-inversion operators have no corresponding point group

Table 5 Multiplication table of the molecular symmetry group G_{16}

E	(11′88′)(99′)	(18′81′)(99′)	(18)(1′8′)	(11′)(88′)(99′)	(18′)(81′)(99′)	(1 8)	(1′8′)	*E**	(11′88′)(99′)*	(18′81′)(99′)*	(18)(1′8′)*	(11′)(88′)(99′)*	(18′)(81′)(99′)*	(18)*	(1′8′)*
E	(11′88′)(99′)	(18′81′)(99′)	(18)(1′8′)	(11′)(88′)(99′)	(18′)(81′)(99′)	(1 8)	(1′8′)	*E**	(11′88′)(99′)*	(18′81′)(99′)*	(18)(1′8′)*	(11′)(88′)(99′)*	(18′)(81′)(99′)*	(18)*	(1′8′)*
(11′88′)(99′)	(18)(1′8′)	*E*	(18′81′)(99′)	(1′8′)	(1 8)	(11′)(88′)(99′)	(18′)(81′)(99′)	(11′88′)(99′)*	(18)(1′8′)*	*E**	(18′81′)(99′)*	(1′8′)*	(18)*	(11′)(88′)(99′)*	(18′)(81′)(99′)*
(18′81′)(99′)	*E*	(18)(1′8′)	(11′88′)(99′)	(1 8)	(1′8′)	(18′)(81′)(99′)	(11′)(88′)(99′)	(18′81′)(99′)*	*E**	(18)(1′8′)*	(11′88′)(99′)*	(18)*	(1′8′)*	(18′)(81′)(99′)*	(11′)(88′)(99′)*
(18)(1′8′)	(18′81′)(99′)	(11′88′)(99′)	*E*	(18′)(81′)(99′)	(11′)(88′)(99′)	(1′8′)	(1 8)	(18)(1′8′)*	(18′81′)(99′)*	(11′88′)(99′)*	*E**	(18′)(81′)(99′)*	(11′)(88′)(99′)*	(1′8′)*	(18)*
(11′)(88′)(99′)	(1 8)	(1′8′)	(18′)(81′)(99′)	*E*	(18)(1′8′)	(11′88′)(99′)	(18′81′)(99′)	(11′)(88′)(99′)*	(18)*	(1′8′)*	(18′)(81′)(99′)*	*E**	(18)(1′8′)*	(11′88′)(99′)*	(18′81′)(99′)*
(18′)(81′)(99′)	(1′8′)	(1 8)	(11′)(88′)(99′)	(18)(1′8′)	*E*	(18′81′)(99′)	(11′88′)(99′)	(18′)(81′)(99′)*	(1′8′)*	(18)*	(11′)(88′)(99′)*	(18)(1′8′)*	*E**	(18′81′)(99′)*	(11′88′)(99′)*
(1 8)	(18′)(81′)(99′)	(11′)(88′)(99′)	(1′8′)	(18′81′)(99′)	(11′88′)(99′)	*E*	(18)(1′8′)	(18)*	(18′)(81′)(99′)*	(11′)(88′)(99′)*	(1′8′)*	(18′81′)(99′)*	(11′88′)(99′)*	*E**	(18)(1′8′)*
(1′8′)	(11′)(88′)(99′)	(18′)(81′)(99′)	(1 8)	(11′88′)(99′)	(18′81′)(99′)	(18)(1′8′)	*E*	(1′8′)*	(11′)(88′)(99′)*	(18′)(81′)(99′)*	(18)*	(11′88′)(99′)*	(18′81′)(99′)*	(18)(1′8′)*	*E**
*E**	(11′88′)(99′)*	(18′81′)(99′)*	(18)(1′8′)*	(11′)(88′)(99′)*	(18′)(81′)(99′)*	(18)*	(1′8′)*	*E*	(11′88′)(99′)	(18′81′)(99′)	(18)(1′8′)	(11′)(88′)(99′)	(18′)(81′)(99′)	(1 8)	(1′8′)
(11′88′)(99′)*	(18)(1′8′)*	*E**	(18′81′)(99′)*	(1′8′)*	(18)*	(11′)(88′)(99′)*	(18′)(81′)(99′)*	(11′88′)(99′)	(18)(1′8′)	*E*	(18′81′)(99′)	(1′8′)	(1 8)	(11′)(88′)(99′)	(18′)(81′)(99′)
(18′81′)(99′)*	*E**	(18)(1′8′)*	(11′88′)(99′)*	(18)*	(1′8′)*	(18′)(81′)(99′)*	(11′)(88′)(99′)*	(18′81′)(99′)	*E*	(18)(1′8′)	(11′88′)(99′)	(1 8)	(1′8′)	(18′)(81′)(99′)	(11′)(88′)(99′)
(18)(1′8′)*	(18′81′)(99′)*	(11′88′)(99′)*	*E**	(18′)(81′)(99′)*	(11′)(88′)(99′)*	(1′8′)*	(18)*	(18)(1′8′)	(18′81′)(99′)	(11′88′)(99′)	*E*	(18′)(81′)(99′)	(11′)(88′)(99′)	(1′8′)	(1 8)
(11′)(88′)(99′)*	(18)*	(1′8′)*	(18′)(81′)(99′)*	*E**	(18)(1′8′)*	(11′88′)(99′)*	(18′81′)(99′)*	(11′)(88′)(99′)	(1 8)	(1′8′)	(18′)(81′)(99′)	*E*	(18)(1′8′)	(11′88′)(99′)	(18′81′)(99′)
(18′)(81′)(99′)*	(1′8′)*	(18)*	(11′)(88′)(99′)*	(18)(1′8′)*	*E**	(18′81′)(99′)*	(11′88′)(99′)*	(18′)(81′)(99′)	(1′8′)	(1 8)	(11′)(88′)(99′)	(18)(1′8′)	*E*	(18′81′)(99′)	(11′88′)(99′)
(18)*	(18′)(81′)(99′)*	(11′)(88′)(99′)*	(1′8′)*	(18′81′)(99′)*	(11′88′)(99′)*	*E**	(18)(1′8′)*	(1 8)	(18′)(81′)(99′)	(11′)(88′)(99′)	(1′8′)	(18′81′)(99′)	(11′88′)(99′)	*E*	(18)(1′8′)
(1′8′)*	(11′)(88′)(99′)*	(18′)(81′)(99′)*	(18)*	(11′88′)(99′)*	(18′81′)(99′)*	(18)(1′8′)*	*E**	(1′8′)	(11′)(88′)(99′)	(18′)(81′)(99′)	(1 8)	(11′88′)(99′)	(18′81′)(99′)	(18)(1′8′)	*E*

Table 6 Character table of the molecular symmetry group G_{16}

	E	(18)(1′8′)	(11′88′)(99′) (18′81′)(99′)	(11′)(88′)(99′) (18′)(81′)(99′)	(1 8) (1′8′)	E^*	(18)(1′8′)*	(11′88′)(99′)* (18′81′)(99′)*	(11′)(88′)99′)* (18′)(81′)(99′)*	(1 8)*(1′8′)*	Subgroups [a]
A_{1g}	1	1	1	1	1	1	1	1	1	1	G_{16}
A_{2g}	1	1	1	−1	−1	1	1	1	−1	−1	C_{4h}
B_{1g}	1	1	−1	1	−1	1	1	−1	1	−1	D_{2h}
B_{2g}	1	1	−1	−1	1	1	1	−1	−1	1	D_{2h} (d)
E_g	2	−2	0	0	0	2	−2	0	0	0	
A_{1u}	1	1	1	1	1	−1	−1	−1	−1	−1	D_4
A_{2u}	1	1	1	−1	−1	−1	−1	−1	1	1	C_{4v}
B_{1u}	1	1	−1	1	−1	−1	−1	1	−1	1	D_{2d}
B_{2u}	1	1	−1	−1	1	−1	−1	1	1	−1	D_{2d} (d)
E_u	2	−2	0	0	0	−2	2	0	0	0	

[a]Subgroups of G_{16} corresponding to this symmetry species. D_{2h} and D_{2d} are point groups of the planar and orthogonally twisted conformations, respectively. C_{4h}, C_{2h} (d), D_4, C_{4v}, and D_{2d} (d) may be isomorphic to molecular symmetry groups with less feasible operations

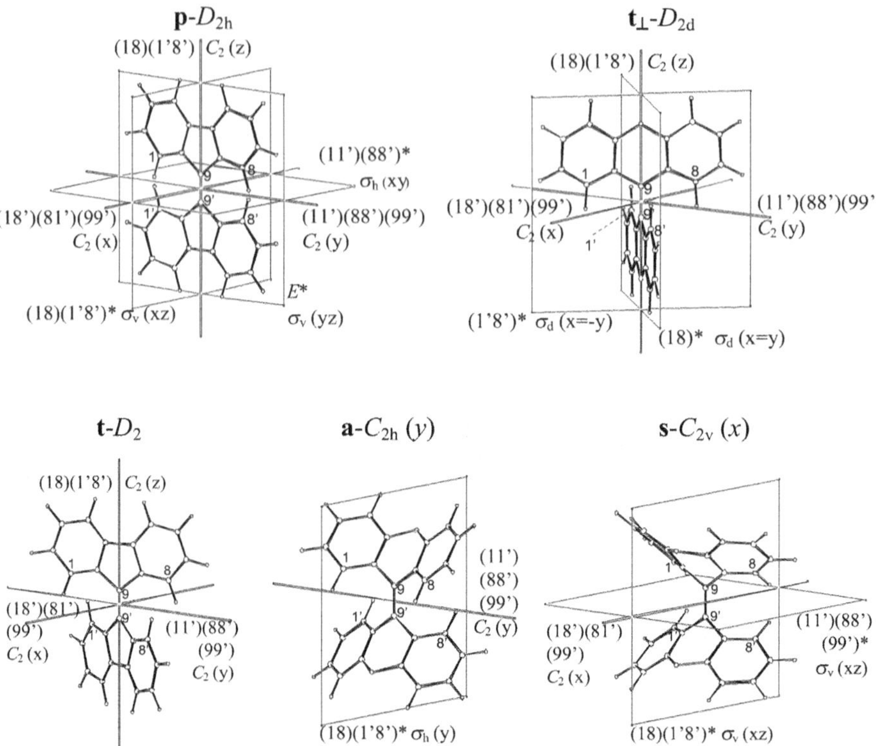

Fig. 21 Rotation axes, reflection planes and corresponding permutation-inversion operators in the conformations **p**-D_{2h}, **t**$_\perp$-D_{2d}, **t**-D_2, **a**-C_{2h} (y), and **s**-C_{2v} (x) of representative BAEs. The rotation-reflection axis S_4 of D_{2d} is oriented along the z-axis; the symmetry center i of D_{2h} and C_{2h} is at the origin

symmetry element at all. Furthermore, the permutation-inversion operators of the molecular symmetry group apply to all conformations (possibly creating new versions of the conformation with different orientation and labeling of the molecular framework; see below), while the point group operators are defined only if they are present in the respective point group of a specific conformation.

In accordance with the conventions customary in vibrational spectroscopy [281], the z-axis may be chosen to go through the atoms C^9 and $C^{9'}$ (and possibly also through X and Y). The y-axis may be chosen to go through the centers of the fjord regions and the x-axis to pass through the center of the double bond and perpendicular to the mean plane of the molecule. Thus, the yz-plane would be the mean plane of the molecule and a projection down the x-axis onto this plane should always give a figure close to the schematic 2D representation shown in Fig. 1, i.e., with the first moiety up in the positive direction of the z-axis and atoms 1 and 1′ on the left side, in the negative direction of the y-axis. Figure 21 illustrates the orientation of the molecule and the permutation-inversion operators corresponding to symmetry axes and reflection planes for the planar D_{2h} symmetric conformation **p** of bifluorenylidene (**2**), the orthogonally twisted D_{2d} symmetric conformation **t**$_\perp$ of dixanthylene (**4**), the twisted D_2 symmetric conformation **t** of bifluorenylidene

Table 7 Permutation-inversion operators of the molecular symmetry group G_{16} of homomerous BAEs and their corresponding rotation, reflection, inversion and rotation-reflection operators

Permutation	Rotation	Permutation-inversion	Reflection, inversion or rotation-reflection
E	E	E^*	$\sigma\ (yz)$
(18)(1′8′)	$C_2\ (z)$	(18)(1′8′)*	$\sigma\ (xz)$
(11′)(88′)(99′)	$C_2\ (y)$	(11′)(88′)(99′)*	$\sigma\ (xy)$
(18′)(81′)(99′)	$C_2\ (x)$	(18′)(81′)(99′)*	i
(18)	*E*,*Z*-Isomerization	(18)*	$\sigma\ (x = y)$
(1′8′)	*E*,*Z*-Isomerization	(1′8′)*	$\sigma\ (x = -y)$
(11′88′)(99′)	—	(11′88′)(99′)*	$S_4^{\ 3}\ (z)$
(18′81′)(99′)	—	(18′81′)(99′)*	$S_4^{\ 1}\ (z)$

(**2**), the *anti*-folded C_{2h} symmetric conformation **a** of dixanthylene (**4**), and the *syn*-folded C_{2v} symmetric conformation of dixanthylene (**4**). The S_4 rotation-reflection axis in $\mathbf{t}_{\perp}$-D_{2d} is the z-axis. The center of symmetry, i, in **p**-D_{2h} and **a**-C_{2h} is at the origin, where the y-axis intersects σ (xz) (and $C^9 = C^{9'}$). The symmetry operators are given in permutation-inversion notation, as well as using point group symmetry operator symbols following the Schönflies notation [282].

The rotations (C_n), reflections (σ), inversion (i), and rotation-reflections ($S_a^{\ nk}$) corresponding to the permutation-inversion operators of the molecular symmetry group G_{16} of homomerous overcrowded bistricyclic aromatic enes are listed in Table 7.

Some comments may be warranted at this point. It may be surprising that the inversion operator, E^*, of the molecular symmetry group does not correspond to the center of symmetry, i. In fact, the treatment of improper rotations, particularly questions related to the effect of i on the molecule-fixed axis system, represents one area of sustained disagreement in the literature [247]. Here, the definitions of Longuet-Higgins [246] and Hougen [247] will be followed: "To yield a mathematically consistent treatment of the symmetry operations, [...] these improper rotations must be associated with the nuclear permutation determined as above [by comparing the original framework to the framework after the improper rotation] followed by (or, equivalently, preceded by) an inversion of the coordinates of all particles in the origin of the laboratory-fixed axis system." [247] For a detailed mathematical treatment see the earlier work by Hougen [283, 284]. Only the second, implicit inversion of the axis system allows superposition of the two frameworks which otherwise would have different handedness. Thus, the inversion at the center of symmetry, i (which is also termed the molecule-fixed inversion [247]), e.g., of a C_{2h} symmetric *anti*-folded conformation, is interchanging atoms 1 and 8′, 8 and 1′, and 9 and 9′, etc., and consequently corresponds to the permutation-inversion operator (18′)(81′)(99′)* (cf. [247]). On the other hand, E^* (the laboratory-fixed inversion [247]) does not permute any atoms and thus should be identified with the σ (yz) operator of D_{2h} in, e.g., the planar conformation, which reflects every atom onto itself (cf. [247]).

The two permutation operators for the *E*,*Z*-isomerization, (18) and (1′8′), moving the first and the second moiety, respectively, do not correspond to any symmetry element in any point group of the BAEs. This is due to the fact that this process converts one form of the molecule (*E*) into another form (*Z*) and thus is a (feasible) chemical reaction interconverting stereoisomers (or automers) rather than a point group symmetry operation. An advantage of the molecular symmetry group formalism is that it allows inclusion of such processes, which may be rapid under certain experimental conditions.

The permutation-inversion operators (18)*, (1′8′)*, (11′88′)(99′)*, and (18′81′)(99′)* correspond to the reflection operators σ_d with the plane $x = y$, $\sigma_{d'}$ with the symmetry plane $x = -y$, and the rotation-reflection operators $S_4{}^3$ (z) and $S_4{}^1$ (z), respectively. They are unique for the D_{2d} symmetric orthogonally twisted conformation.

The permutation operators (11′88′)(99′) and (18′81′)(99′) have no corresponding symmetry operator in any point group of any conformation. They perform a forward and backward cyclic permutation of the positions 1, 1′, 8, and 8′ without inversion, and correspond to the $C_4{}^3$ and $C_4{}^1$ rotation operators in the isomorphous point group D_{4h}. They are generated as products of two permutation operators or two permutation-inversion operators as may be seen from the multiplication table (Table 5). Therefore, they are members of the molecular symmetry group due to the basic postulate of ‘closure’, i.e., the postulate of group theory that for any two operators that are members of a group, their product must also be a member of this group [279].

It may be interesting to note that the permutation operators (18) and (1′8′) corresponding to *E*,*Z*-isomerizations are the product of ‘normal’ symmetry operators, e.g., (18) = (11′88′)(99′)* ⊗ (18′)(81′)(99′)*. Where (11′88′)(99′)* corresponds to $S_4{}^3$ a symmetry operator characteristic of the D_{2d} orthogonally twisted conformation, and (18′)(81′)(99′)* corresponds to *i* in, e.g., the C_{2h} (y) *anti*-folded conformation. Thus, the fact that both an *anti*-folded conformation and an orthogonally twisted conformation are accessible to the molecule leads to the prediction that *E*,*Z*-isomerizations are feasible, based on the group theoretical postulate of ‘closure.’ However, *E*,*Z*-isomerization of an *anti*-folded BAE does not necessarily require (and thus prove) the existence of a D_{2d} symmetric orthogonally twisted conformation as transition state or intermediate. The molecular symmetry group will include the operators (18)*, (1′8′)*, (11′88′)(99′)*, and (18′81′)(99′)*, but these permutation-inversion operators may simply be the product of the *E*,*Z*-operators and some other operator and may not correspond to any symmetry operator in any accessible conformation [246]. (For an illustrative example see the dynamic stereochemistry of tetrabenzo[7,7′]fulvalene (**8**) described in Sect. 5 of [3].)

Likewise, accessibility of a planar D_{2h} conformation for any BAE with twisted (D_2), *anti*-folded (C_{2h} (y)), or *syn*-folded (C_{2v} (x)) global minimum conformation, leads to the prediction of feasible enantiomerizations or conformational inversions, based on the fact that E^* will be an element of the molecular symmetry group. By the same reasoning, group theory predicts feasible enantiomerizations or inversions if conformations with any two of the three point groups D_2, C_{2h} (y), and C_{2v} (x) are accessible to the molecule, i.e., if a molecule can adopt a twisted and an *anti*-folded conformation, or a twisted and a *syn*-folded conformation, or an *anti*-folded and a *syn*-folded conformation. In all three cases, the molecular symmetry group (the smallest group containing all permutation-inversion operators of each two of the above three point groups) is D_{2h} (or higher). However, the conclusion that the planar (D_{2h}) conformation is the transition state of the enantiomerization of a twisted conformation or the transition state for conformational inversion of an *anti*-folded or *syn*-folded conformation is not valid. The planar conformation may not be accessible at all under the given experimental conditions.

Permutation-inversion operators corresponding to feasible isomerizations or inversions of a molecule should not be confused with the reaction mechanism of such a process, i.e., the transition state or a description based on the continuous trajectory of atoms leading from educts to products. They only describe the outcome of the process [255]. For example, dixanthylene (**4**) was found to be *anti*-folded in the solid state, and is thermochromic, implying a thermally populated twisted conformation (the **B** form) (cf. Sects. 2.3 and 2.5) [3]. On the basis of the existence of both an *anti*-folded and a twisted conformation, one may predict that conformational inversion is feasible. This was indeed shown by DNMR [170]. However, this does not prove that the planar conformation is involved as a transition state (or in any other way) in the thermochromic process or in the conformational inversion. Such mechanistic conjectures are not justified.

4.3 Dynamic Stereochemistry of Homomerous BAEs

The molecular symmetry group of unsubstituted homomerous BAEs with feasible *E,Z*-isomerizations is G_{16} (order $\boldsymbol{h}_{\mathbf{MSG}} = 16$). The point groups for planar, orthogonally twisted, twisted, *anti*-folded, *syn*-folded, and various mixed conformations have been defined in the first column of Table 4. For simplicity, in the following discussion only the experimentally observed, highly symmetric conformation types **a**-C_{2h} (y), **s**-C_{2v} (x), and **t**-D_2 will be considered as minima, i.e., as educts and products. The planar conformation is expected not to be a minimum energy conformation in BAEs. The dynamic stereochemistry of the planar conformation and the orthogonally twisted conformation will not be analyzed. (For a discussion of the dynamic stereochemistry of **p**-D_{2h} ethylene (C_2H_4) and $\mathbf{t}_{\perp}$-D_{2d} diborane (B_2H_4) see [254].) However, all conformations will be considered as potential transition states.

In Sects. 4.3.1 to 4.3.3, all degenerate isomerization processes of the three main conformations **a**-C_{2h} (y), **s**-C_{2v} (x) and **t**-D_2 are derived by an analysis of the effect of the permutation-operators of the molecular symmetry group G_{16} on the labeled versions of the respective conformations. In Sects. 4.3.4 to 4.3.7, all conformational isomerization processes interconverting the conformations **a**-C_{2h} (y), **s**-C_{2v} (x), and **t**-D_2 are discussed. For the automerizations and the isomerizations, all possible transition state point groups and conformations for direct one step mechanisms are derived according to the rules of transition state symmetry, and tabulated. The point group order $\boldsymbol{h}_{TS}$, number of versions $\boldsymbol{n}_{TS}$, connectivity $\boldsymbol{C}$, and number of parallel pathways $\boldsymbol{p}$ are determined. The point group symmetry of transient structures along the pathways of steepest descent from the transition state to educt and product and the symmetry species of the transition vector are checked for consistency. Important mechanisms are discussed in detail, showing schematic projections for illustration. In this analysis, it is assumed that the educt and product conformations are bona fide minima and that the transition state derived by symmetry considerations is a true transition state with one and only one imaginary vibrational frequency and directly linked to the educt and product by pathways of steepest descent. The possibility of higher order saddle points or intermediates will be pointed out where relevant. Sect. 4.3 is focused on the symmetry aspects of the dynamic stereochemistry of BAEs and their implications for the mechanisms of automerizations. Energetic aspects were discussed, e.g., in the review [3] reporting semiempirical calculations.

4.3.1 Degenerate Isomerizations of the *anti*-Folded Conformations

The permutation-inversion operators of the molecular symmetry group G_{16} of homomerous BAEs are applied to the *anti*-folded conformation **a**-C_{2h} (y) in Fig. 22. The $\mathbf{a}_{Z\text{-}RR}$ conformation serves as reference conformation (top left in Fig. 22). The respective permutation-inversion operators are given above the schematic projections and the stereochemical descriptors identifying the *E*- or *Z*-configurations and folding direction (*R*/*S*) of the tricyclic moieties are given below.

Using Fig. 22, the permutation-inversion operators may be classified according to their effect on the *anti*-folded conformation **a**-C_{2h} (y) into symmetry operators having no effect on the structure apart from permuting labels of equivalent atoms, operators inverting the folding of one or two moieties, and operators interconverting *E*- and *Z*-configurations. Such a classification is shown in Table 8.

Three types of dynamic processes are predicted by the classification of the permutation-inversion operators:

1. Conformational inversion, i.e., simultaneous inversion of the folding directions of both moieties, e.g., $\mathbf{a}_{Z\text{-}RR'}$ $\mathbf{a}_{Z\text{-}SS'}$
2. *E*,*Z*-isomerization with simultaneous inversion of the first moiety, e.g., $\mathbf{a}_{Z\text{-}RR'}$ $\mathbf{a}_{E\text{-}SR'}$
3. *E*,*Z*-isomerization with simultaneous inversion of the second moiety, e.g., $\mathbf{a}_{Z\text{-}RR'}$ $\mathbf{a}_{E\text{-}RS'}$

Any other process, e.g., inversion of only one moiety or *E*,*Z*-isomerization without simultaneous inversion of a moiety, or with inversion of both moieties, would lead to a different (*syn*-folded) conformation (cf. Sects. 4.3.6 and 4.3.7).

The four permutation-inversion operators listed in Table 8 for each of the three types of dynamic processes are the respective cosets. The cosets may be generated by selecting one of the operators corresponding to the automerization and applying

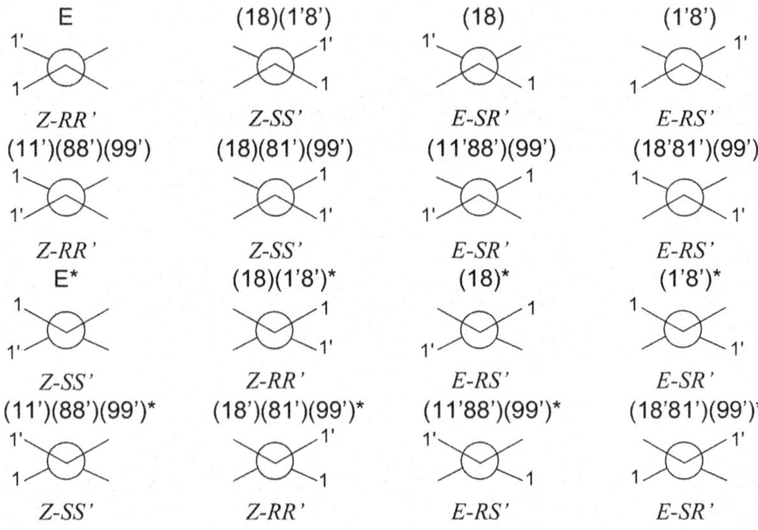

Fig. 22 Effect of the permutation-inversion operators on the *anti*-folded conformation **a**-C_{2h} (y)

Table 8 Effect of permutation-inversion operators on the *anti*-folded conformation **a**-C_{2h} (y)

Permutation-inversion operators	Effect on the conformation **a**-C_{2h} (y)
E, (11′)(88′)(99′), (18)(1′8′)*, (18′)(81′)(99′)*	None [a]
(18)(1′8′), (18′)(81′)(99′), E*, (11′)(88′)(99′)*	Inversion of both moieties
(18), (11′88′)(99′), (1′8′)*, (18′81′)(99′)*	*E,Z*-isomerization and inversion of first moiety
(1′8′), (18′81′)(99′), (18)*, (11′88′)(99′)*	*E,Z*-isomerization and inversion of second moiety

[a] E, (11′)(88′)(99′), (18)(1′8′)*, (18′)(81′)(99′)* correspond to the point group symmetry operators, E, C_2 (y), σ (xz), and i of **a**-C_{2h} (y), respectively

the symmetry operators of the conformation. Each of the cosets contains at least one self-inverse operator. Thus all three dynamic processes are self-inverse and the corresponding transition states may include the permutation-operators of the coset as symmetry operators of their respective point groups [271].

In homomerous BAEs, the two tricyclic moieties are equivalent. Thus, the processes (2) and (3), i.e., *E,Z*-isomerization with inversion of either the first or the second moiety, respectively, cannot be distinguished. They will be discussed together as one dynamic process: *E,Z*-isomerization with inversion of one moiety. Note that the combined mechanism has a connectivity of $\boldsymbol{C} = 2$. Each conformation may be converted into two distinct versions by inverting either the first or the second moiety.

Table 9 Possible point groups and conformations of the transition state for conformational inversion of the *anti*-folded conformation **a**-C_{2h} (y)

Group of permutation-inversion operators	TS	h_{TS}	n_{TS}	C	p	[a]	[b]
{E, (18)(1′8′), (11′)(88′)(99′), (18′)(81′)(99′), E*, (18)(1′8′)*, (11′)(88′)(99′)*, (18′)(81′)(99′)*}	**p**-D_{2h}	8	2	1	1	C_{2h} (y)	B_{2g}
{E, (18)(1′8′), (11′)(88′)(99′), (18′)(81′)(99′)}	**t**-D_2	4	4	1	2	C_2 (y)	B_2
{E, (18)(1′8′), E*, (18)(1′8′)*}	**p**-C_{2v} (z)	4	4	1	2	C_s (xz)	B_1
{E, (11′)(88′)(99′), E*, (11′)(88′)(99′)*}	**p**-C_{2v} (y)	4	4	1	2	C_2 (y)	A_2
{E, (18′)(81′)(99′), (18)(1′8′)*, (11′)(88′)(99′)*}	**s**-C_{2v} (x)	4	4	1	2	C_s (xz)	B_2
{E, (18)(1′8′), (11′)(88′)(99′)*, (18′)(81′)(99′)}	**pt**-C_{2h} (z)	4	4	1	2	C_i	B_g
{E, (11′)(88′)(99′), (18)(1′8′)*, (18′)(81′)(99′)*}	**a**-C_{2h} (y)	4	4	1	2	C_{2h} (y)	A_g
{E, (18′)(81′)(99′), E*, (18′)(81′)(99′)*}	**p**-C_{2h} (x)	4	4	1	2	C_i	B_g
{E, (18)(1′8′)}	**t**-C_2 (z)	2	8	1	4	C_1	B
{E, (11′)(88′)(99′)}	**ta**-C_2 (y)	2	8	1	4	C_2 (y)	A
{E, (18′)(81′)(99′)}	**ts**-C_2 (x)	2	8	1	4	C_1	B
{E, E*}	**p**-C_s (yz)	2	8	1	4	C_1	A"
{E, (18)(1′8′)*}	**f**-C_s (xz)	2	8	1	4	C_s (xz)	A'
{E, (11′)(88′)(99′)*}	**s**-C_s (xy)	2	8	1	4	C_1	A"
{E, (18′)(81′)(99′)*}	**a**-C_i	2	8	1	4	C_i	A_g
{E}	**ft**-C_1	1	16	1	8	C_1	A

[a]Point group symmetry along pathway from transition state to reactant or product, i.e., maximum common subgroup of transition state and reactant or product
[b]Symmetry species of the mode of the transition vector (using the conventional setting of the transition state point group [279])

Conformational Inversion of the *anti*-Folded Conformation

In a conformational inversion of the *anti*-folded conformation the folding direction of both moieties is inverted simultaneously. The E- or Z-configuration of the version is not affected. The symmetry operators of **a**-C_{2h} (y), E, (11′)(88′)(99′), (18)(1′8′)*, (18′)(81′)(99′)*, and the permutation-inversion operators resulting in a conformational inversion, (18)(1′8′), (18′)(81′)(99′), E*, (11′)(88′)(99′)*, combined give the point group D_{2h}. This is the highest possible symmetry for the transition state. Subgroups of D_{2h} may also be considered. All possible groups of permutation-inversion operators and the corresponding transition state conformation and point group are listed in Table 9. The order of the transition state point group $\boldsymbol{h_{TS}}$, the number of versions of this transition state $\boldsymbol{n_{TS}}$ – see (3), the connectivity $\boldsymbol{C}$, and the number of parallel pathways $\boldsymbol{p}$ – see (4), the point group symmetry along the pathways from the transition state to the educt and product, and the symmetry species of the transition vector are also given.

The highest possible symmetry for a transition state is **p**-D_{2h} ($\boldsymbol{h_{TS}} = 8$). There are $\boldsymbol{n_{TS}} = 2$ versions of this conformation corresponding to an E- and a Z-configuration. Equation (4) predicts a connectivity $\boldsymbol{C} = 1$ and only one pathway ($\boldsymbol{p} = 1$). The connectivity of the four versions of the **a**-C_{2h} (y) *anti*-folded conformation via this process and its mechanism are schematically shown in Fig. 23.

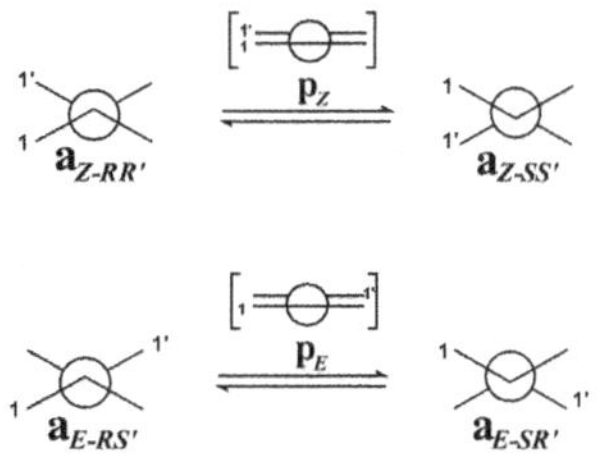

Fig. 23 Schematic mechanism of the inversion of the *anti*-folded conformation **a**-C_{2h} (y) via a planar transition state **p**-D_{2h}

Table 10 Symmetry species of the vibrational modes of the **p**-D_{2h} planar conformation and resulting conformations and point groups

Symmetry species	A_g	B_{1g}	B_{2g}	B_{3g}	A_u	B_{1u}	B_{2u}	B_{3u}
Conformation	**p**-D_{2h}	**pt**-C_{2h} (z)	**a**-C_{2h} (y)	**p**-C_{2h} (x)	**t**-D_2	**p**-C_{2v} (z)	**p**-C_{2v} (y)	**s**-C_{2v} (x)

The mechanism corresponds to a synchronous reduction of the degree of folding of both moieties of the *anti*-folded conformation leading to the planar transition state, followed by folding in the opposite direction to give the inverted version.

Due to the extreme intramolecular overcrowding in a planar **p**-D_{2h} conformation, this is most likely not the lowest energy transition state. It is probably a higher order saddle point with more than one imaginary vibrational frequency. The B_{2g} mode corresponds to the transition vector of the conformational inversion of the *anti*-folded conformation. Additional modes with imaginary frequencies indicate the possibility of transition states (local minima or higher order saddle points) with lower symmetry. The conformations and point group symmetries resulting from deformations of **p**-D_{2h} along the various modes are listed in Table 10 (cf. Fig. 13 or [279]).

More feasible mechanisms for conformational inversion of the *anti*-folded conformation may proceed via the twisted conformation **t**-D_2 or via the *syn*-folded conformation **s**-C_{2v} (x) as transition states. In both cases there are $\boldsymbol{n}_{\mathbf{TS}} = 4$ versions of this transition state with point group order $\boldsymbol{h}_{\mathbf{TS}} = 4$. The connectivities $\boldsymbol{C} = 1$ lead to the conclusion that there must be $\boldsymbol{p} = 2$ parallel pathways. The connectivities and mechanisms are schematically shown in Fig. 24. The versions with *Z*- and *E*-configuration interconvert independently via analogous mechanisms.

In the first inversion mechanism (Fig. 24a) the *anti*-folded conformation is twisted and both moieties are unfolded. Twist may be introduced in two directions, (*P*) and (*M*), leading to the two parallel pathways. Transient structures along the pathway have C_2 (y) symmetry, the common subgroup of C_{2h} (y) and D_2. The transition vector has B_2 symmetry. It should be noted that **t**-D_2 is the midpoint of the reaction paths, and because of its symmetry it has to be a stationary point. However, it may be a local minimum with no imaginary frequency, leading to a two-step process, which will be discussed later (Sect. 4.3.4). Alternatively, it could be a higher order saddle point. This question can only be solved by calculating the transition state and its vibrational frequencies. Additional imaginary frequencies of

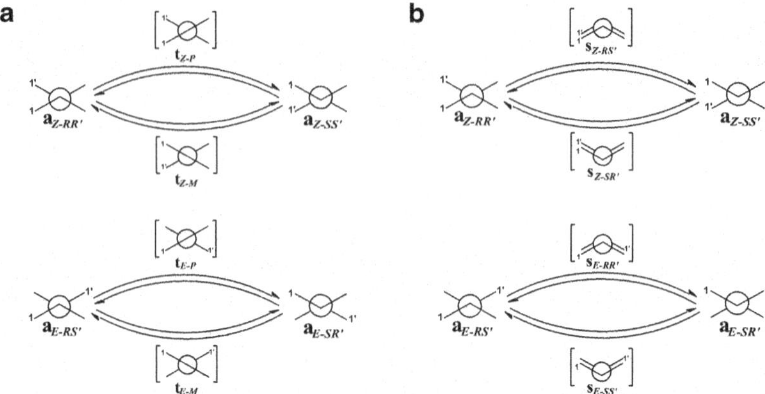

Fig. 24 Schematic mechanisms of the inversion of the *anti*-folded conformation **a**-C_{2h} (y) via (**a**) a twisted transition state **t**-D_2 or (**b**) a *syn*-folded transition state **s**-C_{2v} (x)

B_1 or B_3 symmetry would indicate possible transition states **t**-C_2 (z) or **ts**-C_2 (x), respectively (cf. Fig. 13 and Table 9).

In the second mechanism (Fig. 24b) the two moieties invert one after the other: the *anti*-folded conformation is converted to the *syn*-folded transition state via inversion of one moiety, followed by inversion of the second moiety to give the inverted *anti*-folded conformation. Either the first or the second moiety may invert first, leading to two parallel pathways. In this mechanism the transient structures along the pathways are not twisted and have C_s (xz) symmetry. An imaginary mode with B_2 symmetry corresponds to the transition vector for inversion of the *anti*-folded conformation. In the case where **s**-C_{2v} (x) is a higher order saddle point, additional modes with A_2 and B_1 symmetry indicate possible lower symmetry transition states **st**-C_2 (x) and **s**-C_s (xy), respectively (cf. Fig. 13 and Table 9).

E,*Z*-Isomerization with Simultaneous Inversion of One Tricyclic Moiety of the *anti*-Folded Conformation

Possible point groups for the transition state of the *E*,*Z*-isomerization with simultaneous inversion of one moiety may be constructed by combining the symmetry operators of **a**-C_{2h} (y), E, $(11')(88')(99')$, $(18)(1'8')^*$, and $(18')(81')(99')^*$, with operators corresponding to the process of *E*,*Z*-isomerization with inversion of one moiety: (18), $(1'8')$, $(11'88')(99')$, $(18'81')(99')$, $(18)^*$, $(1'8')^*$, $(11'88')(99')^*$, and $(18'81')(99')^*$, and adding the required operators to ensure closure of the resulting group. Note that, in this case, the smallest group combining all symmetry elements of both sets is G_{16}. The point group D_{2d} combines some operators from both sets plus some additions to give a closed set. Table 11 lists all possible groups of permutation-inversion operators, the corresponding conformations with their point group, order of the point group $\boldsymbol{h}_{\mathrm{TS}}$, number of versions $\boldsymbol{n}_{\mathrm{TS}}$, connectivity $\boldsymbol{C}$, number

Table 11 Possible point groups and conformations of the transition state for *E*,*Z*-isomerization with simultaneous inversion of one moiety of the *anti*-folded conformation **a**-C_{2h} (y)

Group of permutation-inversion operators	TS	h_{TS}	n_{TS}	C	p	[a]	[b]
{E, (18)(1′8′), (11′)(88′)(99′), (18′)(81′)(99′), (18)*, (1′8′)*, (11′88′)(99′)*, (18′81′)(99′)*}	**t**$_{\perp}$-D_{2d}	8	2	2	–[c]	C_2 (y)	–[d]
{E, (18)(1′8′), (11′88′)(99′)*, (18′81′)(99′)*}	**t**$_{\perp}$-S_4	4	4	2	1	C_1	–[d]
{E, (18)(1′8′), (18)*, (1′8′)*}	**t**$_{\perp}$-C_{2v} (d)	4	4	2	1	C_1	–[d]
{E, (11′)(88′)(99′), (18)(1′8′)*, (18′)(81′)(99′)*}	**a**-C_{2h} (y) [e]	4	4	2	1	C_{2h} (y)	A_1
{E, (11′)(88′)(99′)}	**ta**-C_2 (y) [c, f]	2	8	2	2	C_2 (y)	A
{E, (18)(1′8′)*}	**f**-C_s (xz) [e]	2	8	2	2	C_s (xz)	A′
{E, (18)*}	**ft**$_{\perp}$-C_s (d)	2	8	2	2	C_1	A″
{E, (1′8′)*}	**ft**$_{\perp}$-C_s (d')	2	8	2	2	C_1	A″
{E, (18′)(81′)(99′)*}	**a**-C_i [e]	2	8	2	2	C_i	A_g
{E}	**ft**-C_1	1	16	2	4	C_1	A

[a]Point group symmetry along pathway from transition state to reactant or product, i.e., maximum common subgroup of transition state and reactant or product
[b]Symmetry species of the mode of the transition vector (using the conventional setting of the transition state point group [279])
[c]There is no integer $\boldsymbol{p}$ which would satisfy (4)
[d]The point group along the pathway does not correspond to any non-degenerate representation of the transition state point group
[e]For **a**-C_{2h} (y), **ta**-C_2 (y), **f**-C_s (xz), and **a**-C_i there is no pathway for *E*,*Z*-isomerization without leaving this point group
[f]See Sect. 4.3.7

of parallel pathways, point group symmetry of transient structures along the pathways, and the symmetry species of the transition vector.

For the transition state **t**$_{\perp}$-D_{2d}, (4) leads to a contradiction. There is no integer solution $\boldsymbol{p}$ for $\boldsymbol{C} = 2$, $\boldsymbol{n}_{TS} = 2$ and $\boldsymbol{n} = 4$. This implies that **t**$_{\perp}$-D_{2d} may be excluded as a possible transition state. For the conformations **t**$_{\perp}$-D_{2d}, **t**$_{\perp}$-S_4 and **t**$_{\perp}$-C_{2v} (d), the largest subgroups shared with the reactant and product conformation, **a**-C_{2h} (y), are C_2 (y), C_1, and C_1, respectively. These are the highest possible symmetries along the respective pathways. The transition vector of the transition states has to be symmetric with respect to the symmetry operators of the pathway, and *anti*-symmetric with respect to all other symmetry operators of the transition state conformation. However, in the above cases, there is no (non-degenerate) representation with this symmetry species. Thus, the conformations **t**$_{\perp}$-D_{2d}, **t**$_{\perp}$-S_4, and **t**$_{\perp}$-C_{2v} (d) do not qualify as transition states for the *E*,*Z*-isomerization with inversion of one moiety in a direct single step process. They may be higher order saddle points or intermediates (local minima), or transition states of pathways with multiple steps, performing the process in question, or transition states of pathways with a bifurcation.

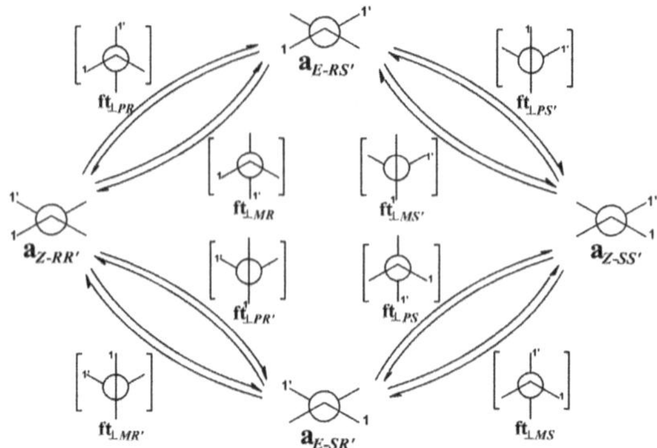

Fig. 25 Schematic mechanism of the *E*,*Z*-isomerization with simultaneous inversion of one moiety of the *anti*-folded conformation **a**-C_{2h} (y) via an orthogonally twisted/folded transition State $\mathbf{ft}_{\perp}$-C_s (d)

For the conformations **a**-C_{2h} (y), **ta**-C_2 (y), **f**-C_s (xz), and **a**-C_i, symmetry constraints do not allow *E*,*Z*-isomerization. The plane σ (xz) in **a**-C_{2h} (y) and **f**-C_s (xz) reflects the left half of each moiety onto the right half. The two halves cannot change sides without coinciding in the plane or breaking the symmetry. Likewise, the center of symmetry i in **a**-C_i relates atoms of different moieties which have a *trans* relationship. An *E*,*Z*-isomerization would necessarily break this *trans* relationship and the C_i symmetry. In **ta**-C_2 (y), the C_2 symmetry axis interrelates atoms *cis* with respect to the central double bond. They would be *trans* after an *E*,*Z*-isomerization. Thus an *E*,*Z*-isomerization cannot be accomplished within the conformational subspaces of C_{2h} (y), C_2 (y), C_s (xz), or C_i symmetry. The conformations **a**-C_{2h} (y), **ta**-C_2 (y), **f**-C_s (xz), and **a**-C_i may thus be excluded as transition states for the *E*,*Z*-isomerization. See, however, Sect. 4.3.7 for a conformational isomerization of the *anti*-folded and *syn*-folded conformations with simultaneous *E*,*Z*-isomerization conserving C_2 symmetry.

The only possible conformations for transition states of a single step *E*,*Z*-isomerization with simultaneous inversion of one moiety without intermediates or bifurcations have low symmetry: $\mathbf{ft}_{\perp}$-C_s (d) or $\mathbf{ft}_{\perp}$-C_s (d') with one planar moiety perpendicular to the second, symmetrically folded moiety, or **ft**-C_1, a transition state (and pathway) without any symmetry.

The transition states $\mathbf{ft}_{\perp}$-C_s (d) and $\mathbf{ft}_{\perp}$-C_s (d') have a point group order $\boldsymbol{h}_{\mathbf{TS}} = 2$; thus, there are eight versions of this conformation. The mechanism for this process with connectivity $\boldsymbol{C} = 2$ and $\boldsymbol{p} = 2$ parallel pathways is schematically outlined in Fig. 25.

Along the pathway from $\mathbf{a}_{Z\text{-}RR'}$ to $\mathbf{a}_{E\text{-}RS'}$, the folding of the first moiety remains more or less constant, while the second moiety unfolds and rotates. At the transition state, the second moiety is planar and bisecting the first moiety. Folding in the opposite direction and continued rotation leads to the product. Interchanging the roles of the two moieties, i.e., inverting and rotating the first moiety while keeping the folding of the second moiety constant, leads to a different product, $\mathbf{a}_{E\text{-}SR'}$ (hence

$C = 2$). The option of rotating clockwise or counterclockwise gives rise to two parallel pathways. The symmetry of transient structures along the pathways from transition states to reactants and products is C_1. The transition vector has A" symmetry. Note that this mechanism allows an inversion of the *anti*-folded conformation in a two-step process. Enantiomeric (labeled) versions of equivalent conformations are on opposite sides of the center of Fig. 25.

4.3.2 Degenerate Isomerizations of the *syn*-Folded Conformation

The effects of the permutation-inversion operators on the versions of the *syn*-folded conformation **s**-C_{2v} (x) are illustrated in Fig. 26, using $\mathbf{s}_{Z\text{-}\boldsymbol{RS'}}$ as reference conformation.

A classification of the permutation-inversion operators according to their effect on the folding directions of the two moieties and the *E*- or *Z*-configuration is summarized in Table 12.

The classification of the permutation-inversion operators shown in Table 12 indicates three types of dynamic processes in the *syn*-folded conformation **s**-C_{2v} (x):

1. Conformational inversion, by simultaneous inversion of the folding directions of both moieties, e.g., $\mathbf{s}_{Z\text{-}\boldsymbol{RS'}}$ $\mathbf{s}_{Z\text{-}\boldsymbol{SR'}}$
2. *E*,*Z*-Isomerization with simultaneous inversion of the first moiety, e.g., $\mathbf{s}_{Z\text{-}\boldsymbol{RS'}}$ $\mathbf{s}_{E\text{-}\boldsymbol{SS'}}$
3. *E*,*Z*-Isomerization with simultaneous inversion of the second moiety, e.g., $\mathbf{s}_{Z\text{-}\boldsymbol{RS'}}$ $\mathbf{s}_{E\text{-}\boldsymbol{RR'}}$

The four permutation-inversion operators listed in Table 12 for each of the three dynamic processes correspond to the respective cosets. Each coset contains at least one self-inverse operator. Thus, all three types of degenerate isomerizations of the *syn*-folded conformation are self-inverse and the corresponding transition states may have point groups containing the symmetry operators of **s**-C_{2v} (x) and operators corresponding to the respective dynamic process.

Since the two tricyclic moieties in homomerous BAEs are equivalent, processes 2 and 3 are indistinguishable and will be discussed together as *E*,*Z*-isomerization with simultaneous inversion of one moiety. The combined process has a connectivity $\boldsymbol{C} = 2$.

Conformational Inversion of the *syn*-Folded Conformation

In the conformational inversion of the *syn*-folded conformation **s**-C_{2v} (x), the folding direction of both moieties is reversed simultaneously. As this process is a self-inverse automerization, the point group of the transition state may include symmetry operators of the point group of the reactant (and product) conformations

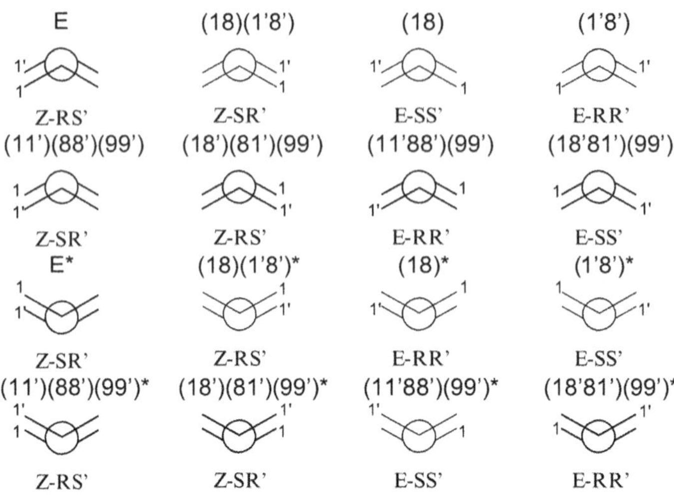

Fig. 26 Effect of the permutation-inversion operators on the *syn*-folded conformation **s**-C_{2v} (x)

Table 12 Effect of permutation-inversion operators on the *syn*-folded conformation **s**-C_{2v} (x)

Permutation-inversion operators	Effect on the conformation **s**-C_{2v} (x)
E, (18′)(81′)(99′), (18)(1′8′)*, (11′)(88′)(99′)*	None[a]
(18)(1′8′), (11′)(88′)(99′), E*, (18′)(81′)(99′)*	Inversion of both moieties
(18), (18′81′)(99′), (1′8′)*, (11′88′)(99′)*	*E*,*Z*-Isomerization and inversion of first moiety
(1′8′), (11′88′)(99′), (18)*, (18′81′)(99′)*	*E*,*Z*-Isomerization and inversion of second moiety

[a] E, (18′)(81′)(99′), (18)(1′8′)*, (11′)(88′)(99′)* correspond to the point group symmetry operators, i.e., E, C_2 (x), σ (xz), and σ (xy) of **s**-C_{2v} (x), respectively

as well as operators interchanging reactant and product. The symmetry operators of **s**-C_{2v} (x) are: E, (18′)(81′)(99′), (18)(1′8′)*, and (11′)(88′)(99′)*. The permutation-inversion operators (18)(1′8′), (11′)(88′)(99′), E*, and (18′)(81′)(99′)* result in conformational inversion. Both sets combined correspond to the point group D_{2h}, which is the highest possible point group for the transition state. All subgroups are also valid possibilities. The groups of permutation-inversion operators, transition state conformations including their point groups, the order of the point group $\boldsymbol{h}_{TS}$, number of versions $\boldsymbol{n}_{TS}$, connectivity $\boldsymbol{C}$, number of parallel pathways, $\boldsymbol{p}$, the point group of transient structures along the pathways, and the symmetry species of the transition vector are listed in Table 13.

The conformation with highest point group order, which, according to the above symmetry considerations, may be a transition state, is **p**-D_{2h}. The order of the point group is $\boldsymbol{h}_{TS} = 8$. There are two versions of this conformation, $\mathbf{p}_Z$ and $\mathbf{p}_E$. The connectivity ($\boldsymbol{C} = 1$), the number of pathways ($\boldsymbol{p} = 1$), and the mechanism are schematically shown in Fig. 27.

Table 13 Possible point groups and conformations of the transition state for conformational inversion of the *syn*-folded conformation **s**-C_{2v} (x)

Group of permutation-inversion operators	TS	h_{TS}	n_{TS}	C	p	[a]	[b]
{E, (18)(1′8′), (11′)(88′)(99′), (18′)(81′)(99′), E*, (18)(1′8′)*, (11′)(88′)(99′)*, (18′)(81′)(99′)*}	**p**-D_{2h}	8	2	1	1	C_{2v} (x)	B_{3u}
{E, (18)(1′8′), (11′)(88′)(99′), (18′)(81′)(99′)}	**t**-D_2	4	4	1	2	C_2 (x)	B_3
{E, (18)(1′8′), E*, (18)(1′8′)*}	**p**-C_{2v} (z)	4	4	1	2	C_s (xz)	B_1
{E, (11′)(88′)(99′), E*, (11′)(88′)(99′)*}	**p**-C_{2v} (y)	4	4	1	2	C_s (xy)	B_1
{E, (18′)(81′)(99′), (18)(1′8′)*, (11′)(88′)(99′)*}	**s**-C_{2v} (x)	4	4	1	2	C_{2v} (x)	A_1
{E, (18)(1′8′), (11′)(88′)(99′)*, (18′)(81′)(99′)*}	**pt**-C_{2h} (z)	4	4	1	2	C_s (xy)	B_u
{E, (11′)(88′)(99′), (18)(1′8′)*, (18′)(81′)(99′)*}	**a**-C_{2h} (y)	4	4	1	2	C_s (xz)	B_u
{E, (18′)(81′)(99′), E*, (18′)(81′)(99′)*}	**p**-C_{2h} (x)	4	4	1	2	C_2 (x)	A_u
{E, (18)(1′8′)}	**t**-C_2 (z)	2	8	1	4	C_1	B
{E, (1′)(88′)(99′)}	**ta**-C_2 (y)	2	8	1	4	C_1	B
{E, (18′)(81′)(99′)}	**ts**-C_2 (x)	2	8	1	4	C_2 (x)	A
{E, E*}	**p**-C_s (yz)	2	8	1	4	C_1	A″
{E, (18)(1′8′)*}	**f**-C_s (xz)	2	8	1	4	C_s (xz)	A′
{E, (11′)(88′)(99′)*}	**s**-C_s (xy)	2	8	1	4	C_s (xy)	A′
{E, (18′)(81′)(99′)*}	**a**-C_i	2	8	1	4	C_1	A_u
{E}	**ft**-C_1	1	16	1	8	C_1	A

[a]Point group symmetry along pathway from transition state to reactant or product, i.e., maximum common subgroup of transition state and reactant or product
[b]Symmetry species of the mode of the transition vector (using the conventional setting of the transition state point group [279])

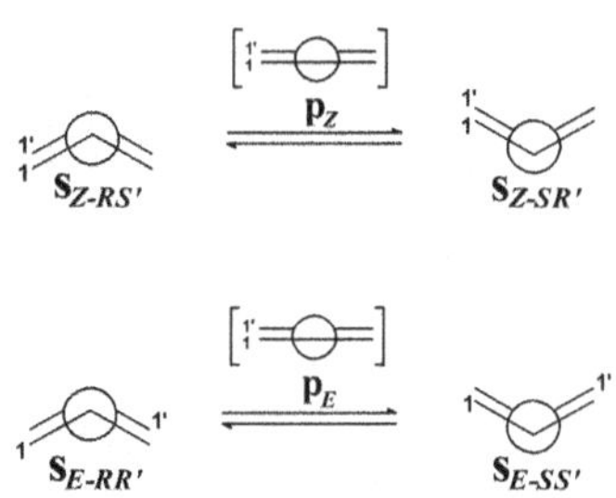

Fig. 27 Schematic mechanism of the inversion of the *syn*-folded conformation **s**-C_{2v} (x) via a planar transition state **p**-D_{2h}

The degree of folding of both moieties is reduced in a synchronous manner, retaining C_{2v} (x) symmetry along the pathway until the planar transition state is reached. Increasing the folding in the opposite direction leads to the inverted conformation. The transition vector leading to products and reactants has B_{3u} symmetry. Note that this planar transition state is the same conformation that was considered as a high symmetry transition state for the inversion of the *anti*-folded conformation (Fig. 23). However, a bona fide transition state can only have steepest descent paths to two minima [248, 270]. Due to the high degree of overcrowding, **p**-D_{2h} is expected to be a higher order saddle point with more than one imaginary frequency and steepest descent paths to several minima, transition states, and/or

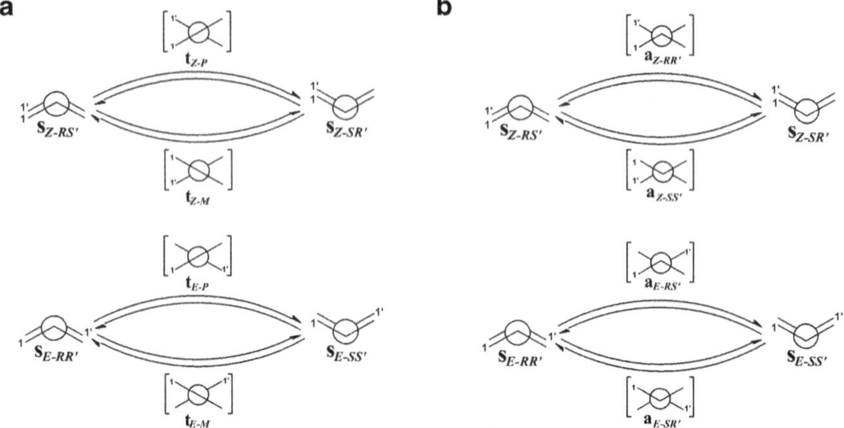

Fig. 28 Schematic mechanisms of the inversion of the *syn*-folded conformation **s**-C_{2v} (x) via (**a**) a twisted transition state **t**-D_2 or (**b**) an *anti*-folded transition state **a**-C_{2h} (y)

higher order saddle points. The conformations corresponding to distortions of **p**-D_{2h} along vibrational modes with imaginary frequency of various symmetry species are indicated in Table 10.

The twisted conformation **t**-D_2 and the *anti*-folded conformation **a**-C_{2h} (y) are less overcrowded candidates for a transition state for inversion of the *syn*-folded conformation. In each case there are $\boldsymbol{n}_{\text{TS}} = 4$ versions of the transition state with point group order $\boldsymbol{h}_{\text{TS}} = 4$. The connectivities $\boldsymbol{C} = 1$ of the two parallel pathways $(\boldsymbol{p} = 2)$ and the schematic mechanisms are shown in Fig. 28. The versions with *E*- and *Z*-configuration interconvert independently via analogous mechanisms.

In the first mechanism (Fig. 28a), the *syn*-folded conformation is twisted and the two moieties are unfolded to reach the twisted transition state. Folding in the opposite direction and untwisting lead to the inverted *syn*-folded conformation. The two parallel pathways are distinguished by positive (*P*) or negative twist (*M*) of intermediate structures. The transient structures along the pathways have C_2 (x) symmetry. The symmetry species of the transition vector is B_3. Note that **t**-D_2 was also considered as a transition state for the inversion of the *anti*-folded conformation (Fig. 24a). The two mechanisms are mutually exclusive and may be distinguished by the symmetry of the vibrational mode of **t**-D_2 that has an imaginary frequency. Distortions of **t**-D_2 along the modes B_1, B_2, and B_3 lead to conformations **t**-C_2 (z), **ta**-C_2 (y), and **ts**-C_2 (x), respectively. Alternatively, the conformation **t**-D_2 may be a local minimum and an intermediate in one or both of the conformational inversion processes (see Sects. 4.3.4 and 4.3.5), or a higher order saddle point.

In the second mechanism (Fig. 28b), the two tricyclic moieties are unfolded and inverted one after the other, leading from a *syn*-folded minimum via an *anti*-folded transition state to the inverted *syn*-folded minimum. Either of the two moieties may be inverted first, giving rise to the two parallel pathways. The transient structures along the pathways have C_s (xz) symmetry. The transition vector has B_u symmetry. Additional imaginary frequencies of vibrational modes with A_u and B_g symmetry (in the case of a higher order saddle point) would suggest conformations **ta**-C_2 (y) and **a**-C_i as possible transition states.

Note the analogy of this mechanism with the inversion of the *anti*-folded conformation via a *syn*-folded transition state (Fig. 24b). Conformations of the same symmetry and type are involved. However, their respective role as minima and transition states along the pathways are interchanged. From a symmetry point of view, both scenarios are possible and cannot be distinguished since they differ only in the relative energy of the conformations and their character as minima or transition states. Alternatively, **a**-C_{2h} (y) may be a local minimum. In this case, the energy profile would be that of a two-step process as discussed in Sect. 4.3.6.

E,*Z*-Isomerization with Simultaneous Inversion of One Tricyclic Moiety of the *syn*-Folded Conformation

Table 14 is constructed by combining the symmetry operators of the **s**-C_{2v} (x) conformation, E, $(18')(81')(99')$, $(18)(1'8')^*$, and $(11')(88')(99')^*$, with operators corresponding to the process of *E*,*Z*-isomerization with inversion of one moiety, (18), $(1'8')$, $(11'88')(99')$, $(18'81')(99')$, $(18)^*$, $(1'8')^*$, $(11'88')(99')^*$, and $(18'81')(99')^*$, and adding the required operators to achieve closure of the resulting set of operators. The possible transition state conformations and their point groups, point group order $\boldsymbol{h}_{TS}$, number of versions $\boldsymbol{n}_{TS}$, connectivity $\boldsymbol{C}$, and number of parallel pathways $\boldsymbol{p}$ are listed in Table 14. The symmetry along the pathways of steepest descent connecting the transition state to the *syn*-folded conformation and the symmetry species of the vibrational mode corresponding to the transition vector is also given. Note that, as in the case of *E*,*Z*-isomerization of the *anti*-folded conformation, the smallest group combining all symmetry elements of both sets is G_{16}. The point group D_{2d} combines some symmetry operators and some operators corresponding to the dynamic process plus some additions to give a closed set.

In the case of $\mathbf{t}_\perp$-D_{2d}, (4) has no integer solution for $\boldsymbol{p}$, indicating that this is not a valid mechanism. For the transition states $\mathbf{t}_\perp$-D_{2d}, $\mathbf{t}_\perp$-S_4, and $\mathbf{t}_\perp$-C_{2v} (d) the point groups along the pathway (C_2 (x), C_1, and C_1, respectively), and hence the symmetry properties of the transition vector do not correspond to any non-degenerate irreducible representation of the point group of the transition state. Thus, these conformations do not qualify as bona fide transition states for the *E*,*Z*-isomerization with simultaneous inversion of one moiety. However, they may be intermediates (local minima), transition states of multi-step pathways with other intermediates on the way, or transition states of pathways with bifurcations.

For the conformations **s**-C_{2v} (x), **ts**-C_2 (x), **f**-C_s (xz), and **s**-C_s (xy), symmetry constraints do not allow an *E*,*Z*-isomerization without leaving the respective point group. Thus, the only options for direct (one step, no bifurcation)

Table 14 Possible point groups and conformations of the transition state for *E,Z*-isomerization with simultaneous inversion of one moiety of the *syn*-folded conformation **s**-C_{2v} (*x*)

Group of permutation-inversion operators	TS	h_{TS}	n_{TS}	C	p	*a*	*b*
{*E*, (18)(1′8′), (11′)(88′)(99′), (18′)(81′)(99′), (18)*, (1′8′)*, (11′88′)(99′)*, (18′81′)(99′)*}	$\mathbf{t}_\perp$-D_{2d}	8	2	2	–[c]	C_2 (*x*)	–[d]
{*E*, (18)(1′8′), (11′88′)(99′)*, (18′81′)(99′)*}	$\mathbf{t}_\perp$-S_4	4	4	2	1	C_1	–[d]
{*E*, (18)(1′8′), (18)*, (1′8′)*}	$\mathbf{t}_\perp$-C_{2v} (*d*)	4	4	2	1	C_1	–[d]
{*E*, (18′)(81′)(99′), (18)(1′8′)*, (11′)(88′)(99′)*}	**s**-C_{2v} (*x*) [e]	4	4	2	1	C_{2v} (*x*)	A_1
{*E*, (18′)(81′)(99′)}	**ts**-C_2 (*x*) [e, f]	2	8	2	2	C_2 (*x*)	A
{*E*, (18)(1′8′)*}	**f**-C_s (*xz*) [e]	2	8	2	2	C_s (*xz*)	A′
{*E*, (11′)(88′)(99′)*}	**s**-C_s (*xy*) [e]	2	8	2	2	C_s (*xy*)	A′
{*E*, (18)*}	$\mathbf{ft}_\perp$-C_s (*d*)	2	8	2	2	C_1	A″
{*E*, (1′8′)*}	$\mathbf{ft}_\perp$-C_s (*d*′)	2	8	2	2	C_1	A″
{*E*}	**ft**-C_1	1	16	2	4	C_1	A

[a]Point group symmetry along pathway from transition state to reactant or product, i.e., maximum common subgroup of transition state and reactant or product
[b]Symmetry species of the mode of the transition vector (using the conventional setting of the transition state point group [279])
[c]There is no integer ***p*** which would satisfy (4)
[d]The point group along the pathway does not correspond to any non-degenerate irreducible representation of the transition state point group
[e]For **s**-C_{2v} (*x*), **ts**-C_2 (*x*), **f**-C_s (*xz*), and **s**-C_s (*xy*) there is no pathway for *E,Z*-isomerization without leaving this point group
[f]See Sect. 4.3.7

E,Z-isomerizations with simultaneous inversion of one moiety have low symmetry: $\mathbf{ft}_\perp$-C_s (*d*) or $\mathbf{ft}_\perp$-C_s (*d*′) and **ft**-C_1. The C_s symmetric transition states $\mathbf{ft}_\perp$-C_s (*d*) and $\mathbf{ft}_\perp$-C_s (*d*′) have a point group order $\boldsymbol{h}_{TS} = 2$ and $\boldsymbol{n}_{TS} = 8$ versions. The connectivity in this process is $\boldsymbol{C} = 2$ with $\boldsymbol{p} = 2$ parallel pathways connecting each pair of minima. The mechanism is schematically shown in Fig. 29.

In each of the automerization reactions one moiety remains basically unchanged, while the other moiety is unfolded, rotated by 180° either clockwise or counterclockwise ($\boldsymbol{p} = 2$), and refolded in the opposite direction. Rotating the first or second moiety gives different products. Hence, the connectivity $\boldsymbol{C} = 2$. The transient structures along the steepest descent paths from the transition states to reactants and products have C_1 symmetry. The transition vector has A" symmetry. This mechanism facilitates an inversion of the *syn*-folded conformation in a two-step process. Enantiomeric (labeled) versions may be found on opposite sides of the center of Fig. 29. Note that the same type of transition state has been considered for the *E,Z*-isomerization of the *anti*-folded conformation (Fig. 25). The low symmetry of the pathways (C_1) does not allow assigning the reactant and product of this transition state unambiguously from the symmetry species of the vibrational mode. Only explicit calculations of the steepest descent path will allow a decision.

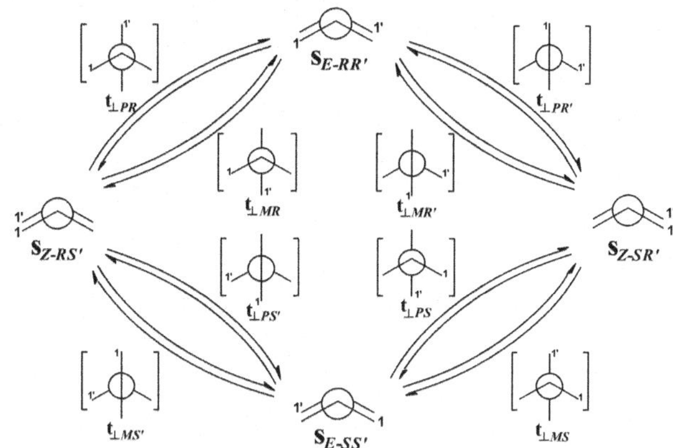

Fig. 29 Schematic mechanism of the *E*,*Z*-isomerization with simultaneous inversion of one moiety of the *syn*-folded conformation **s**-C_{2v} (*x*) via an orthogonally twisted/folded transition state **ft**$_{\perp}$-C_s (*d*)

4.3.3 Degenerate Isomerizations of the Twisted Conformation

The effect of the permutation-inversion operators of the molecular symmetry group G_{16} on the twisted conformation **t**-D_2 is shown in Fig. 30. The conformation $\mathbf{t}_{Z\text{-}P}$ is used as the reference version (top left in Fig. 30). The permutation-inversion operator is given above the schematic projections and stereochemical descriptors referring to *E*- or *Z*-configuration and positive or negative helicity of the central double bond below.

The permutation-inversion operators of the molecular symmetry group G_{16} may be classified according to their effect on the twisted conformation **t**-D_2 as shown in Table 15.

Three types of degenerate isomerization processes may be distinguished according to the classification of permutation-inversion operators:

1. Enantiomerization, i.e., inversion of the helicity of the central double bond, e.g., $\mathbf{t}_{Z\text{-}P}$ $\mathbf{t}_{Z\text{-}M}$
2. *E*,*Z*-isomerization with retention of the helicity of the central double bond, e.g., $\mathbf{t}_{Z\text{-}P}$ $\mathbf{t}_{E\text{-}P}$
3. *E*,*Z*-isomerization with simultaneous enantiomerization, e.g., $\mathbf{t}_{Z\text{-}P}$ $\mathbf{t}_{E\text{-}M}$

The four permutation-inversion operators corresponding to each of the three types of dynamic processes are the respective cosets. Each of the three cosets of operators corresponding to the three different types of degenerate isomerizations contains at least one self-inverse permutation-inversion operator. Thus, all three dynamic processes are self-inverse processes. The corresponding transition states may have point groups including the symmetry operators of **t**-D_2 as well as the permutation-inversion operators of the coset interchanging reactant and product [271].

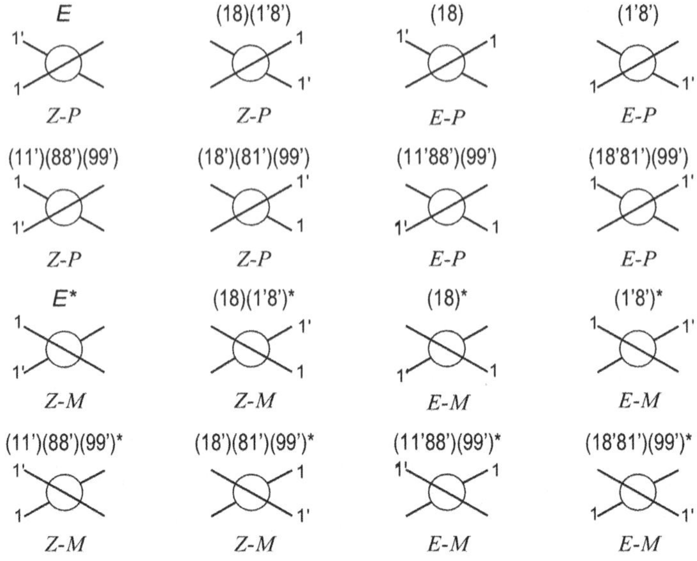

Fig. 30 Effect of the permutation-inversion operators on the twisted conformation **t**-D_2

Table 15 Effect of permutation-inversion operators on the twisted conformation **t**-D_2.

Permutation-inversion operators	Effect on conformation **t**-D_2
E, (18)(1′8′), (11′)(88′)(99′), (18′)(81′)(99′)	None[a]
*E**, (18)(1′8′)*, (11′)(88′)(99′)*, (18′)(81′)(99′)*	Inversion of helicity
(18), (1′8′), (11′88′)(99′), (18′81′)(99′)	*E*,*Z*-Isomerization
(18)*, (1′8′)*, (11′88′)(99′)*, (18′81′)(99′)*	*E*,*Z*-Isomerization and inversion of helicity

[a]*E*, (18)(1′8′), (11′)(88′)(99′), (18′)(81′)(99′) correspond to the point group symmetry operators, *E*, C_2 (*z*), C_2 (*y*), and C_2 (*x*) of **t**-D_2, respectively

Enantiomerization of the Twisted Conformation

The point groups and conformations of possible transition states for the enantiomerization of the twisted conformation **t**-D_2, predicted by the molecular symmetry group are listed in Table 16. The point group order $\boldsymbol{h}_{\mathrm{TS}}$, number of versions $\boldsymbol{n}_{\mathrm{TS}}$, connectivity $\boldsymbol{C}$, number of parallel pathways $\boldsymbol{p}$, the symmetry along the pathways, and the symmetry species of the transition vector are also given. Note that the symmetry operators of **t**-D_2, *E*, (18)(1′8′), (11′)(88′)(99′), and (18′)(81′)(99′) and the operators inverting the helicity of this conformation, *E**, (18)(1′8′)*, (11′)(88′)(99′)*, and (18′)(81′)(99′)* combine to give the point group D_{2h}. All subgroups of D_{2h} are also considered as possible point groups.

The conformation **p**-D_{2h} is the transition state with highest possible symmetry ($h = 8$) for the enantiomerization of **t**-D_2. There are two versions of the transition state, $\mathbf{p}_Z$ and $\mathbf{p}_E$, interconverting the *Z*- and *E*-versions of the twisted conformation,

Table 16 Possible point groups and conformations of the transition state for enantiomerization of the twisted conformation **t**-D_2

Group of permutation-inversion operators	TS	h_{TS}	n_{TS}	C	p	[a]	[b]
{E, (18)(1′8′), (11′)(88′)(99′), (18′)(81′)(99′), E*, (18)(1′8′)*, (11′)(88′)(99′)*, (18′)(81′)(99′)*}	**p**-D_{2h}	8	2	1	1	D_2	A_u
{E, (18)(1′8′), (11′)(88′)(99′), (18′)(81′)(99′)}	**t**-D_2	4	4	1	2	D_2	A
{E, (18)(1′8′), E*, (18)(1′8′)*}	**p**-C_{2v} (z)	4	4	1	2	C_2 (z)	A_2
{E, (11′)(88′)(99′), E*, (11′)(88′)(99′)*}	**p**-C_{2v} (y)	4	4	1	2	C_2 (y)	A_2
{E, (18′)(81′)(99′), (18)(1′8′)*, (11′)(88′)(99′)*}	**s**-C_{2v} (x)	4	4	1	2	C_2 (x)	A_2
{E, (18)(1′8′), (11′)(88′)(99′)*, (18′)(81′)(99′)*}	**pt**-C_{2h} (z)	4	4	1	2	C_2 (z)	A_u
{E, (11′)(88′)(99′), (18)(1′8′)*, (18′)(81′)(99′)*}	**a**-C_{2h} (y)	4	4	1	2	C_2 (y)	A_u
{E, (18′)(81′)(99′), E*, (18′)(81′)(99′)*}	**p**-C_{2h} (x)	4	4	1	2	C_2 (x)	A_u
{E, (18)(1′8′)}	**t**-C_2 (z)	2	8	1	4	C_2 (z)	A
{E, (11′)(88′)(99′)}	**ta**-C_2 (y)	2	8	1	4	C_2 (y)	A
{E, (18′)(81′)(99′)}	**ts**-C_2 (x)	2	8	1	4	C_2 (x)	A
{E, E*}	**p**-C_s (yz)	2	8	1	4	C_1	A″
{E, (18)(1′8′)*}	**f**-C_s (xz)	2	8	1	4	C_1	A″
{E, (11′)(88′)(99′)*}	**s**-C_s (xy)	2	8	1	4	C_1	A″
{E, (18′)(81′)(99′)*}	**a**-C_i	2	8	1	4	C_1	A_u
{E}	**ft**-C_1	1	16	1	8	C_1	A

[a]Point group symmetry along pathway from transition state to reactant or product, i.e., maximum common subgroup of transition state and reactant or product
[b]Symmetry species of the mode of the transition vector (using the conventional setting of the transition state point group [279])

respectively, as shown in the schematic mechanism in Fig. 31. The connectivity is $\boldsymbol{C} = 1$, and there is only one pathway ($\boldsymbol{p} = 1$).

Starting from the twisted conformation, the twist (and hence the non-planarity) of the structure is reduced to reach the planar transition state. Increasing the twist in the opposite direction leads to the other enantiomer. All transient structures along the pathway have D_2 symmetry as the initial and final conformation. The transition vector of **p**-D_{2h} leading to twisted conformations has A_u symmetry. Note that the planar conformation has also been considered as a transition state of the inversion of the *anti*-folded conformation **a**-C_{2h} (y) and/or the inversion of the *syn*-folded conformation **s**-C_{2v} (x). However, the corresponding transition vectors would have different symmetries (Table 10). The conformation **p**-D_{2h} can be a transition state only in one of the three processes. Due to the extreme overcrowding, **p**-D_{2h} is most likely a higher order saddle point and the enantiomerization process will have a lower symmetry transition state.

The *anti*-folded conformation **a**-C_{2h} (y) and the *syn*-folded conformation **s**-C_{2v} (x) are possible transition states with lower symmetry ($\boldsymbol{h}_{TS} = 4$). There are $\boldsymbol{n}_{TS} = 4$ versions of these transition states. They may interconvert the Z- and E-versions of **t**-D_2 in mechanisms with connectivities $\boldsymbol{C} = 1$ and two parallel pathways ($\boldsymbol{p} = 2$), as shown in Fig. 32.

In the first mechanism (Fig. 32a), the two tricyclic moieties of the twisted conformation are folded in opposite directions (*anti*) and the twist of the central

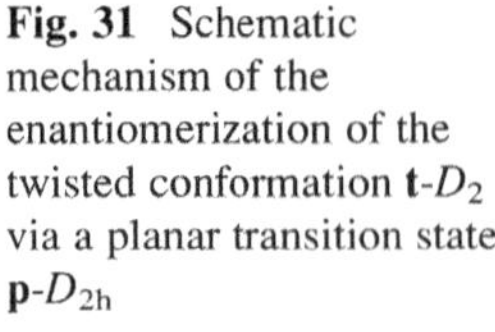
Fig. 31 Schematic mechanism of the enantiomerization of the twisted conformation **t**-D_2 via a planar transition state **p**-D_{2h}

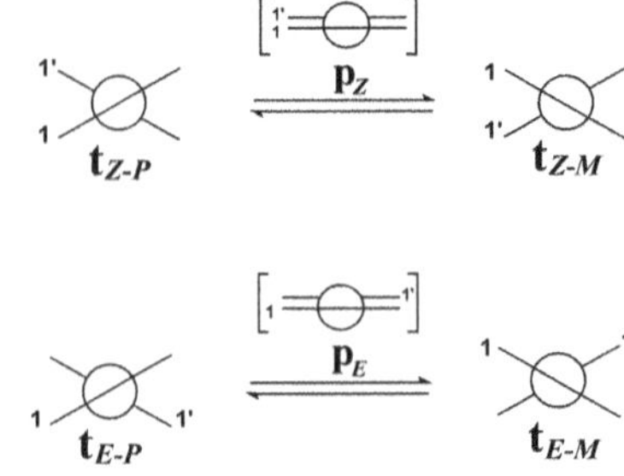

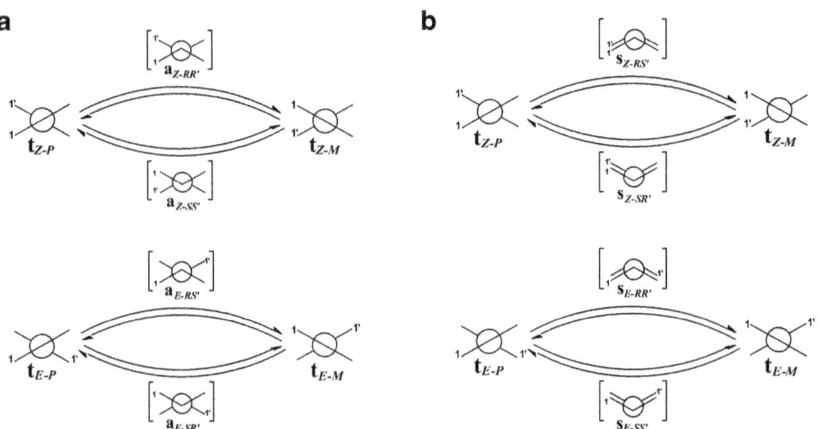

Fig. 32 Schematic mechanisms of the enantiomerization of the twisted conformation **t**-D_2 via (**a**) an *anti*-folded transition state **a**-C_{2h} (y) or (**b**) a *syn*-folded transition state **s**-C_{2v} (x)

double bond is reduced until the transition state, **a**-C_{2h} (y), is reached. Twisting in the opposite direction and unfolding leads to the enantiomeric twisted conformation. The two parallel pathways have transition states with opposite folding directions of the tricyclic moieties. Transient structures along the steepest descent paths have C_2 (y) symmetry. The transition vector of **a**-C_{2h} (y) leading to the twisted conformations has A_u symmetry. Note that **a**-C_{2h} (y) has also been considered as a transition state for the inversion of the *syn*-folded conformation (Fig. 28b). However, this would require a transition vector with B_u symmetry. Alternatively, **a**-C_{2h} (y) may also be a second-order saddle point or a local minimum and intermediate in one or both of these two dynamic processes (see Sects. 4.3.4 and 4.3.5). Note also the analogy to the mechanism of conformational inversion of the *anti*-folded conformation via a twisted transition state (Fig. 24a), in which the roles of minima and transition states are interchanged.

In the second enantiomerization mechanism (Fig. 32b), the two moieties of the twisted conformation are folded in the same direction (*syn*) while the twist of the central double bond is reversed. The two parallel pathways differ in the folding directions of the two moieties. The transient structures along the pathways have C_2 (x) symmetry. A mode with imaginary frequency of A_2 symmetry corresponds to the transition vector leading to the twisted conformations. Note that **s**-C_{2v} (x) has also been considered as transition state for the conformational inversion of the *anti*-folded conformation (transition vector B_2, Fig. 24b). Also note the analogy of this mechanism with the conformational inversion of the *syn*-folded conformation via a twisted transition state (Fig. 28a). Only the roles of minima and transition states are interchanged.

E,Z-Isomerization with Simultaneous Inversion of Helicity of the Twisted Conformation

The possible point groups of the transition states for the *E,Z*-isomerization with simultaneous inversion of helicity may be generated by combining the symmetry operators of the **t**-D_2 conformation, E, $(18)(1'8')$, $(11')(88')(99')$, and $(18')(81')(99')$ with the operators $(18)^*$, $(1'8')^*$, $(11'88')(99')^*$, and $(18'81')(99')^*$ interchanging reactants and products. Note that the two sets of operators combined give the point group D_{2d}. This is the highest possible point group for a transition state for *E,Z*-isomerization with simultaneous inversion of helicity of the **t**-D_2 conformation.

Table 17 enumerates all possible groups of operators, the transition state conformations and point groups, the point group order $\boldsymbol{h}_{\mathbf{TS}}$, number of versions $\boldsymbol{n}_{\mathbf{TS}}$, connectivity $\boldsymbol{C}$, number of parallel pathways $\boldsymbol{p}$, the symmetry of transient structures along the pathways, and the symmetry species of the transition vector.

The transition state with the highest point group order ($\boldsymbol{h}_{\mathbf{TS}} = 8$) is $\mathbf{t}_{\perp}$-D_{2d}. There are $\boldsymbol{n}_{\mathbf{TS}} = 2$ versions of this transition state: $\mathbf{t}_{\perp \boldsymbol{P}}$ and $\mathbf{t}_{\perp \boldsymbol{M}}$. Each of the two "enantiomeric" labeled versions of the transition state interconverts two versions of the twisted conformation as shown in the schematic mechanism in Fig. 33. The connectivity is $\boldsymbol{C} = 1$ with one pathway each ($\boldsymbol{p} = 1$).

Starting from the twisted conformation $\mathbf{t}_{\boldsymbol{Z}\text{-}\boldsymbol{P}}$, the twist of the central double bond is increased to 90° at the transition state. Further twisting leads to the product, $\mathbf{t}_{\boldsymbol{E}\text{-}\boldsymbol{M}}$. Transient structures along the pathway have D_2 symmetry. The transition vector of $\mathbf{t}_{\perp}$-D_{2d} has B_1 symmetry. In the case of a higher order saddle point, deformations along imaginary modes with A_2 and B_2 symmetry lead to possible lower symmetry transition states $\mathbf{t}_{\perp}$-S_4 and $\mathbf{t}_{\perp}$-C_{2v} (d). In $\mathbf{t}_{\perp}$-S_4 the two tricyclic moieties are non-planar (e.g., propeller twisted), while in $\mathbf{t}_{\perp}$-C_{2v} (d) the two moieties are planar and mutually orthogonal, but no longer symmetry equivalent.

E,Z-Isomerization of the Twisted Conformation Without Inversion of Helicity

The point groups of transition states for the *E,Z*-isomerization of the twisted conformation without simultaneous inversion of the helicity are constructed by

Table 17 Possible point groups and conformations of the transition state for *E,Z*-isomerization with simultaneous inversion of helicity of the twisted conformation **t**-D_2

Group of permutation-inversion operators	TS	h_{TS}	n_{TS}	*C*	*p*	*a*	*b*
{*E*, (18)(1′8′), (11′)(88′)(99′), (18′)(81′)(99′), (18)*, (1′8′)*, (11′88′)(99′)*, (18′81′)(99′)*}	$\mathbf{t}_{\perp}$-D_{2d}	8	2	1	1	D_2	B_1
{*E*, (18)(1′8′), (11′88′)(99′)*, (18′81′)(99′)*}	$\mathbf{t}_{\perp}$-S_4	4	4	1	2	C_2 (*z*)	B
{*E*, (18)(1′8′), (18)*, (1′8′)*}	$\mathbf{t}_{\perp}$-C_{2v} (*d*)	4	4	1	2	C_2 (*z*)	A_2
{*E*, (18)(1′8′), (11′)(88′)(99′), (18′)(81′)(99′)}	**t**-D_2	4	4	1	2	D_2	A
{*E*, (18)(1′8′)}	**t**-C_2 (*z*)	2	8	1	4	C_2 (*z*)	A
{*E*, (11′)(88′)(99′)}	**ta**-C_2 (*y*)	2	8	1	4	C_2 (*y*)	A
{*E*, (18′)(81′)(99′)}	**ts**-C_2 (*x*)	2	8	1	4	C_2 (*x*)	A
{*E*, (18)*}	$\mathbf{ft}_{\perp}$-C_s (*d*)	2	8	1	4	C_1	A"
{*E*, (1′8′)*}	$\mathbf{ft}_{\perp}$-C_s (*d′*)	2	8	1	4	C_1	A"
{*E*}	**ft**-C_1	1	16	1	8	C_1	A

[a]Point group symmetry along pathway from transition state to reactant or product, i.e., maximum common subgroup of transition state and reactant or product
[b]Symmetry species of the mode of the transition vector (using the conventional setting of the transition state point group [279])

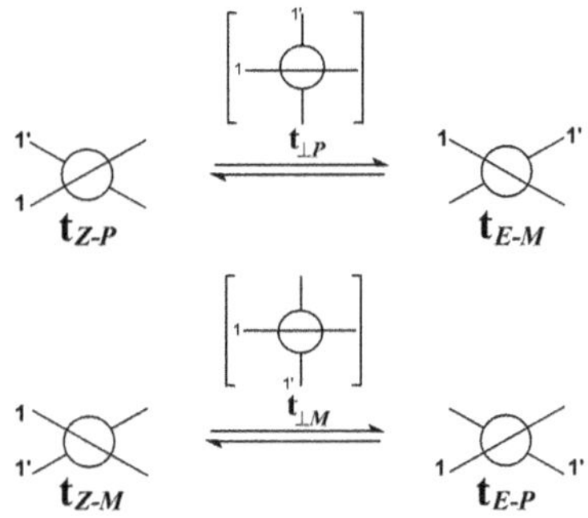

Fig. 33 Schematic mechanism of the *E,Z*-isomerization with simultaneous enantiomerization of the twisted conformation **t**-D_2 via the orthogonally twisted transition state $\mathbf{t}_{\perp}$-D_{2d}

combining symmetry operators of the **t**-D_2 conformation, *E*, (18)(1′8′), (11′)(88′)(99′), and (18′)(81′)(99′), with operators corresponding to the process of *E,Z*-isomerization, (18), (1′8′), (11′88′)(99′), and (18′81′)(99′). Table 18 lists all transition state conformations predicted by the molecular symmetry group theory, including the symmetries of transient conformations along the steepest descent paths and the symmetry species of the vibrational mode of the transition state, which corresponds to the transition vector.

Combining the symmetry operators of **t**-D_2 with the permutation operators corresponding to this dynamic process results in a group that is isomorphous to D_4. However, there is no conformation with D_4 symmetry. The permutation operators (18), (1′8′), (11′88′)(99′), and (18′81′)(99′) do not appear as symmetry operators in any conformation of an overcrowded bistricyclic aromatic ene (cf. Table 7). This is an unusual situation: the permutation-inversion operators corresponding to

Table 18 Possible point groups and conformations of the transition state for *E*,*Z*-isomerization of the twisted conformation **t**-D_2

Group of permutation-inversion operators	TS	h_{TS}	n_{TS}	C	p	[a]	[b]
{*E*, (18)(1′8′), (11′88′)(99′)*, (18′81′)(99′)*}	**t**-D_2	4	4	1	2	D_2	A
{*E*, (18)(1′8′)}	**t**-C_2 (*z*)	2	8	1	4	C_2 (*z*)	A
{*E*, (11′)(88′)(99′)}	**ta**-C_2 (*y*)	2	8	1	4	C_2 (*y*)	A
{*E*, (18′)(81′)(99′)}	**ts**-C_2 (*x*)	2	8	1	4	C_2 (*x*)	A
{*E*}	**ft**-C_1	1	16	1	8	C_1	A

[a]Point group symmetry along pathway from transition state to reactant or product, i.e., maximum common subgroup of transition state and reactant or product
[b]Symmetry species of the mode of the transition vector (using the conventional setting of the transition state point group [279])

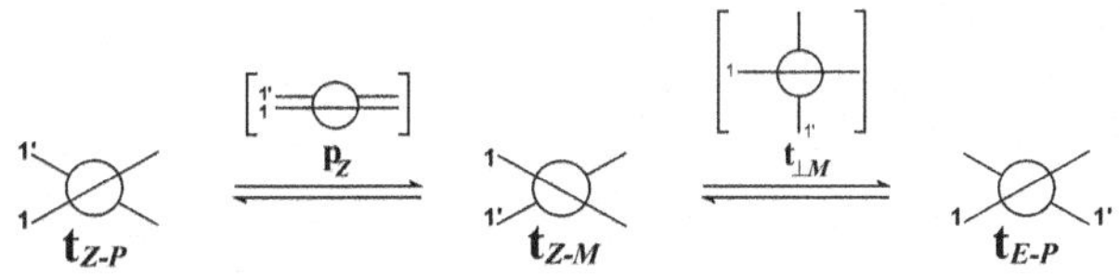

Fig. 34 Counterclockwise rotation of the second moiety of $\mathbf{t}_{Z\text{-}P}$ to give $\mathbf{t}_{E\text{-}P}$

this self-inverse automerization process cannot be symmetry elements of the transition state. The point group of the reactant and product, D_2, and its subgroups are the only possible candidates for a transition state of the *E*,*Z*-isomerization without inversion of helicity.

The *E*,*Z*-isomerization of a twisted conformation without inversion of the helicity of the central double bond requires a 180° rotation of one moiety about the central double bond. This is not likely to occur in one step. In fact, the existence of a stable twisted conformation may be seen as due to a compromise between two energetically unfavorable conformations: an orthogonally twisted conformation, which leads to loss of π-overlap across the central double bond, and a planar (or nearly planar) conformation during passage of the overcrowded regions, which involves high steric strain. A rotation of one moiety by 180° forces the molecule to pass through both unfavorable conformations. This implies a double barrier for a 180° rotation about the double bond (cf. [30]). Thus, the dynamic process of *E*,*Z*-isomerization without inversion of helicity is more likely to proceed in two consecutive steps: an inversion of helicity of the twisted conformation, followed by an *E*,*Z*-isomerization with simultaneous inversion of helicity as discussed in Sects. "Enantiomerization of the Twisted Conformation" and "*E*,*Z*-Isomerization with Simultaneous Inversion of Helicity of the Twisted Conformation", respectively. This is illustrated by the example of a counterclockwise rotation of the second moiety of $\mathbf{t}_{Z\text{-}P}$ to give $\mathbf{t}_{E\text{-}P}$ shown in Fig. 34. Note that the net effect of two consecutive inversions of helicity is retention of helicity.

4.3.4 Interconversions of the Twisted and *anti*-Folded Conformations

In conformational isomerization reactions, the starting geometry and the geometry of the product correspond to different conformations and thus cannot be related by any symmetry operations or permutations of the atom labels. Thus the only symmetry elements available for the transition state, as well as for the pathway connecting it to the reactant and the product, are those common to reactant and product conformations [248]. The transition vector is necessarily totally symmetric with respect to the point group of the transition state [270]. The transition state is not characterized by a higher order point group. All structures along the steepest descent paths from the transition state to the reactant and product conformations have the same point group. Thus, geometry optimizations of the transition state of a conformational isomerization will always require a search for a transition state. A symmetry constrained energy minimization will not converge to the transition state [270].

The permutation-inversion operators corresponding to the point group symmetry operators of the twisted conformation **t**-D_2 and the *anti*-folded conformation **a**-C_{2h} (y) are listed in Table 19 together with their largest common subgroup.

The only common symmetry operators are: E, and $(11')(88')(99')$, leading to the possible transition states listed in Table 20. The highest possible point group symmetry of the transition state for interconversion of twisted and *anti*-folded conformations is C_2 (y). A C_2 axis along y, i.e., through the center of the overcrowded regions, allows twisting of the central double bond and folding of the tricyclic moieties in opposite directions (*anti*). The transition state may be classified as a **ta**-C_2 (y) or **at**-C_2 (y) conformation (Table 4). According to the Hammond principle, twisting is expected to be more pronounced if the twisted conformation has a higher energy than the *anti*-folded conformation. *Anti*-folding will be more pronounced in the transition state if the twisted conformation is more stable. However, these arguments go beyond pure symmetry considerations. In the following analysis, the transition state will be labeled **ta**-C_2 (y), irrespective of the question whether twisting or folding is more dominant.

In addition to the C_2 pathway, a C_1 pathway via a **ft**-C_1 transition state may be considered.

The transition state **ta**-C_2 (y) has a point group order $\boldsymbol{h}_{\mathrm{TS}} = 2$. Thus, there are $\boldsymbol{n}_{\mathrm{TS}} = 8$ versions of this transition state in the molecular symmetry group G_{16}. Four versions have a Z-configuration and convert the two Z-versions of **t**-D_2 ($\mathbf{t}_{Z\text{-}P}$ and $\mathbf{t}_{Z\text{-}M}$) into the two Z-versions of **a**-C_{2h} (y) ($\mathbf{a}_{Z\text{-}RR'}$ and $\mathbf{a}_{Z\text{-}SS'}$). The other four versions have an E-configuration and interconvert the corresponding E-versions of the twisted and *anti*-folded conformations. The mechanism is shown schematically in Fig. 35.

The **ft**-C_1 transition state combines a twisted central ethylene bond with unequal folding of the two moieties in opposite directions (*anti*). The lower symmetry C_1 transition state ($\boldsymbol{h}_{\mathrm{TS}} = 1$) has $\boldsymbol{n}_{\mathrm{TS}} = 16$ versions, leading to a doubling of all pathways as shown in Fig. 36. The versions with E- and Z-configurations

Table 19 Symmetry analysis of the twisted to *anti*-folded conformational isomerization

	Permutation-inversion operators
Symmetry operators of t-D_2	E, (18)(1′8′), (11′)(88′)(99′), (18′)(81′)(99′)
Symmetry operators of a-C_{2h} (y)	E, (11′)(88′)(99′), (18)(1′8′)*, (18′)(81′)(99′)*
Largest common subgroup	E, (11′)(88′)(99′)

Table 20 Possible point groups and conformations of the transition state for interconversion of the **t**-D_2 twisted and **a**-C_{2h} (y) *anti*-folded conformations

Group of permutation-inversion operators	TS	h_{TS}	n_{TS}	a	b
{E, (11′)(88′)(99′)}	**ta**-C_2 (y)	2	8	C_2 (y)	A
{E}	**ft**-C_1	1	16	C_1	A

[a]Point group symmetry along pathway from transition state to reactant or product, i.e., maximum common subgroup of transition state and reactant or product
[b]Symmetry species of the mode of the transition vector (using the conventional setting of the transition state point group [279])

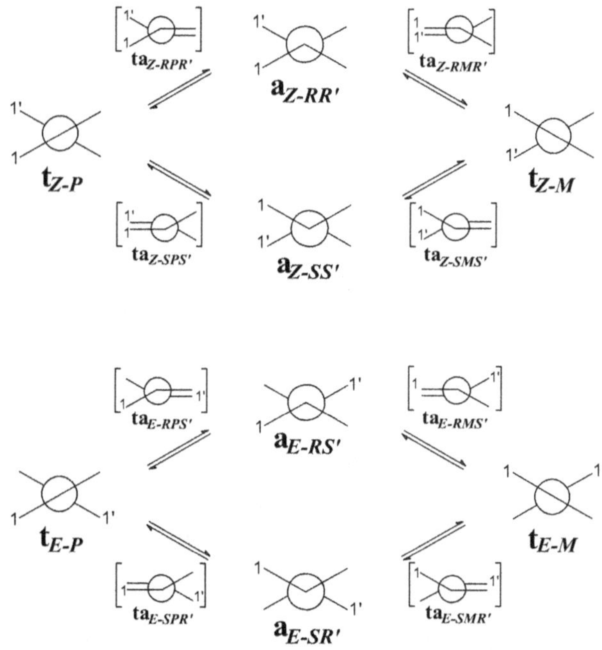

Fig. 35 Schematic mechanism for the interconversion of the **t**-D_2 twisted and **a**-C_{2h} (y) *anti*-folded conformations via a transition state **ta**-C_2 (y)

interconvert independently via analogous mechanisms. Fig. 36 may be derived from Fig. 35 by replacing each transition state and the corresponding pathway by two. In one transition state the first tricyclic moiety is more folded than the second, while in the other transition state the second moiety is more folded than the first.

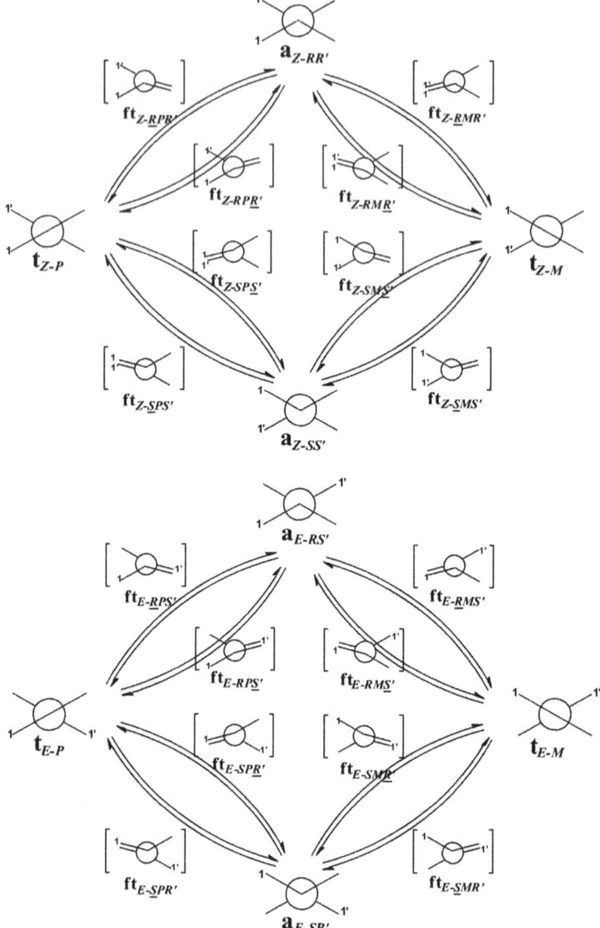

Fig. 36 Schematic mechanism for the interconversion of the **t**-D_2 twisted and **a**-C_{2h} (y) *anti*-folded conformations via a transition state **ft**-C_1

In the labels identifying the versions of the transition state, underlining indicates the more folded moiety. The two versions of the transition state corresponding to parallel pathways, i.e., interconverting the same versions of the twisted and the *anti*-folded conformation, differ only in the degree of folding of the two moieties. In Fig. 36, the first moiety is more folded than the second along the 'outer' pathways. The folding degrees are interchanged in the corresponding 'inner' pathways.

In the mechanism via **ta**-C_2 (y) transition states, the two moieties move in a synchronous manner, while in the mechanism via **ft**-C_1 they fold or unfold independently, one after the other. Note that both mechanisms, via **ta**-C_2 (y) transition states and via **ft**-C_1 transition states interconvert enantiomeric versions (in a labeled atom sense) in a two-step mechanism. Thus they may also serve as an enantiomerization mechanism for the **t**-D_2 conformations via **a**-C_{2h} (y) intermediates, or as conformational inversion mechanism for the **a**-C_{2h} (y) conformations via

t-D_2 intermediates. In Figs. 35 and 36, enantiomeric versions of the same conformation are on opposite sides. *E,Z*-isomerizations are not facilitated by the above two conformational isomerization mechanisms.

4.3.5 Interconversion of the Twisted and *syn*-Folded Conformations

The permutation-inversion operators of the twisted conformation **t**-D_2, the *syn*-folded conformation **s**-C_{2v} (x), and the largest common subgroup are shown in Table 21.

The permutation-inversion operator corresponding to the point groups of the twisted and the *syn*-folded conformation, D_2 and C_{2v} (x) have the common permutation-inversion operators E and $(18')(81')(99')$, corresponding to the point group C_2 (x). The two possibilities for a transition state for the conformational isomerization of the twisted and *syn*-folded conformations are **st**-C_2 (x) (or **ts**-C_2 (x)) and **ft**-C_1, as shown in Table 22.

Conformations with a C_2 axis parallel to x may combine twisting and *syn*-folding: **ts**-C_2 (x) or **st**-C_2 (x) depending on the dominant out-of-plane deformation mode (Table 4). For simplicity, only **st**-C_2 (x) is mentioned in this discussion (it may be replaced by **ts**-C_2 (x) in each case). There are $\boldsymbol{n}_{\mathbf{TS}} = 8$ versions of the **st**-C_2 (x) transition state, four with *Z*-configuration (in a labeled atom sense), and four with *E*-configuration. These four *Z*- and *E*-transition states interconvert the corresponding *Z*- and *E*-versions of **t**-D_2 and of **s**-C_{2v} (x) as schematically outlined in Fig. 37.

The lower symmetry **ft**-C_1 transition state has no symmetry and thus the two moieties may be folded to a different degree. There are $\boldsymbol{n}_{\mathbf{TS}} = 16$ versions of this transition state, 8 *Z*-versions, and 8 *E*-versions. The mechanism for the interconversions via the C_1 transition state may be derived from Fig. 37 by replacing each transition state by two distinct transition states with different degrees of folding of the two moieties. This breaks the C_2 (x) symmetry along the pathways and leads to doubling of all pathways as shown in Fig. 38.

In the labels identifying the versions of the **ft**-C_1 transition states, underlining indicates the more folded moiety. Each two versions corresponding to parallel pathways differ only in the relative degrees of folding of the two moieties.

Comparing the two mechanisms via the **st**-C_2 (x) and via the **ft**-C_1 transition states, in the former, the two moieties move in a C_2 symmetric, synchronous way, while in the latter, the two moieties move independently, giving rise to parallel pathways because either the first or the second moiety may show a higher degree of folding.

In addition to interconverting twisted and *syn*-folded conformations, both mechanisms may also serve as a two-step enantiomerization mechanism for the twisted

Table 21 Symmetry analysis of the twisted to *syn*-folded conformational isomerization

	Permutation-inversion operators
Symmetry operators of **t**-D_2	E, (18)(1′8′), (11′)(88′)(99′), (18′)(81′)(99′)
Symmetry operators of **s**-C_{2v} (x)	E, (18′)(81′)(99′), (18)(1′8′)*, (11′)(88′)(99′)*
Largest common subgroup	E, (18′)(81′)(99′)

Table 22 Possible point groups and conformations of the transition state for interconversion of the **t**-D_2 twisted and the **s**-C_{2v} (x) *syn*-folded conformations

Group of permutation-inversion operators	TS	h_{TS}	n_{TS}	[a]	[b]
{E, (18′)(81′)(99′)}	**st**-C_2 (x)	2	8	C_2 (x)	A
{E}	**ft**-C_1	1	16	C_1	A

[a]Point group symmetry along pathway from transition state to reactant or product, i.e., maximum common subgroup of transition state and reactant or product
[b]Symmetry species of the mode of the transition vector (using the conventional setting of the transition state point group [279])

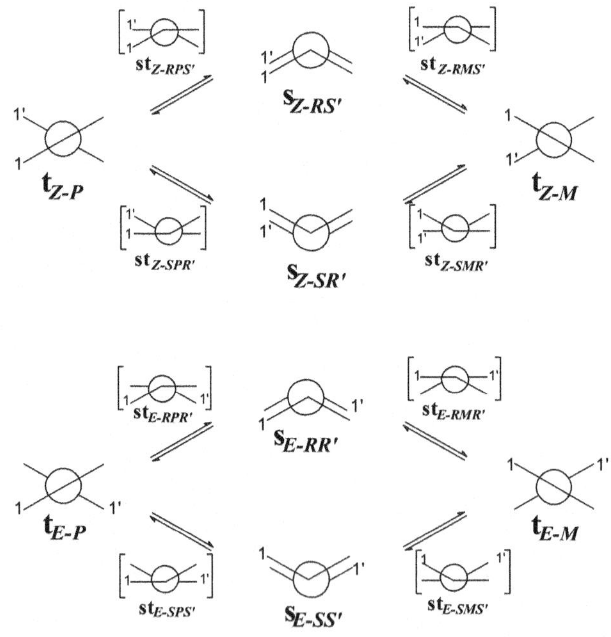

Fig. 37 Schematic mechanism for the interconversion of the **t**-D_2 twisted and **s**-C_{2v} (x) *syn*-folded conformations via a transition state **st**-C_2 (x)

conformations or as a two-step conformational inversion mechanism of the *syn*-folded conformations. In Figs. 37 and 38, enantiomeric versions of the conformations and transition states are found on opposite sides. *E*- and *Z*-configurations are not interconverted.

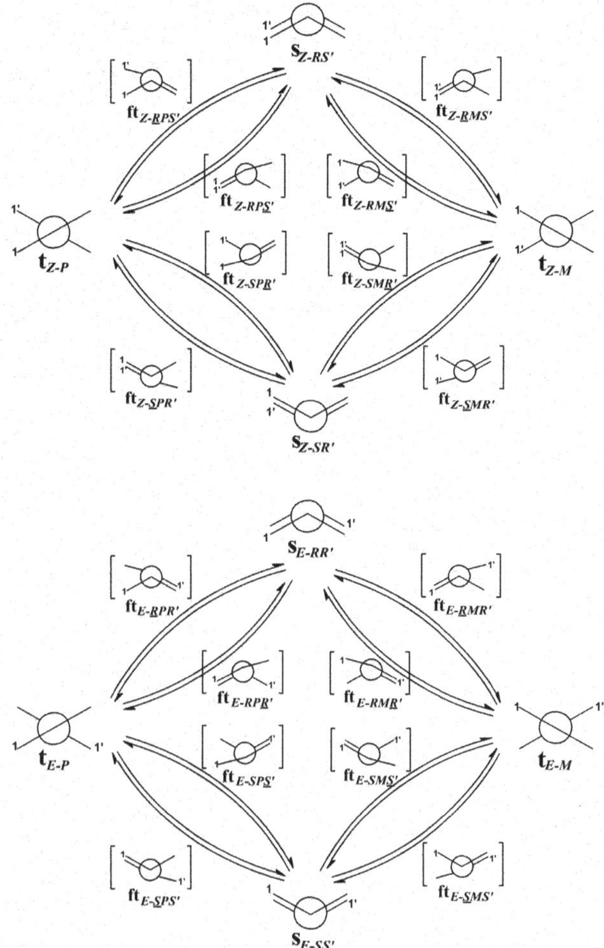

Fig. 38 Schematic mechanism for the interconversion of the **t**-D_2 twisted and **s**-C_{2v} (x) *syn*-folded conformations via a transition state **ft**-C_1

Table 23 Symmetry analysis of the *anti*-folded to *syn*-folded conformational isomerization

	Permutation-inversion operators
Symmetry operators of **a**-C_{2h} (y)	E, (11′)(88′)(99′), (18)(1′8′)*, (18′)(81′)(99′)*
Symmetry operators of **s**-C_{2v} (x)	E, (18′)(81′)(99′), (18)(1′8′)*, (11′)(88′)(99′)*
Largest common subgroup	E, (18)(1′8′)*

4.3.6 Interconversion of the *anti*-Folded and *syn*-Folded Conformations

The permutation-inversion operators of the *anti*-folded and the *syn*-folded conformations, **a**-C_{2h} (y) and **s**-C_{2v} (x), and their largest common subgroup are shown in Table 23.

Table 24 Possible point groups and conformations of the transition state for interconversion of the **a**-C_{2h} (y) *anti*-folded and the **s**-C_{2v} (x) *syn*-folded conformations

Group of permutation-inversion operators	TS	h_{TS}	n_{TS}	[a]	[b]
$\{E, (18)(1'8')^*\}$	**f**-C_s (xz)	2	8	C_s (xz)	A′
$\{E\}$	**ft**-C_1	1	16	C_1	A

[a]Point group symmetry along pathway from transition state to reactant or product, i.e., maximum common subgroup of transition state and reactant or product

[b]Symmetry species of the mode of the transition vector (using the conventional setting of the transition state point group [279])

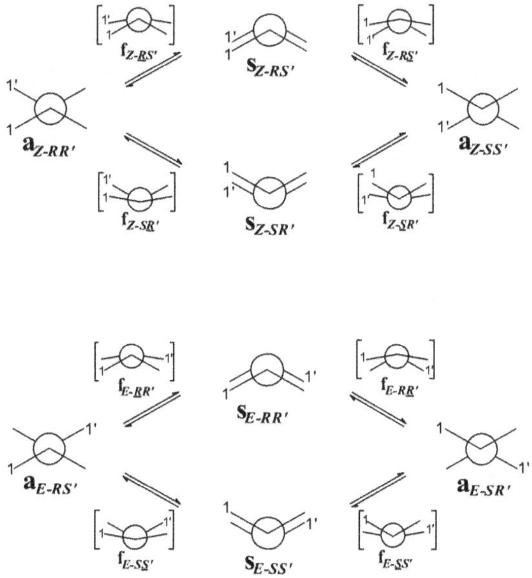

Fig. 39 Schematic mechanism for the interconversion of the **a**-C_{2h} (y) *anti*-folded and **s**-C_{2v} (x) *syn*-folded conformations via a transition state **f**-C_s (xz)

The largest common subgroup, $\{E, (18)(1'8')^*\}$, corresponds to the point group C_s (xz). According to Table 4, conformations with C_s (xz) symmetry may be classified as unequally folded: **f**-C_s (xz) (not specifying the relative direction of folding (*syn*/*anti*); cf. Sect. "Folded Conformations"). An **ft**-C_1 transition state is also possible, as shown in Table 24.

In conformations with point group C_s (xz), the two moieties may have different degrees of folding; however, twisting of the central ethylene group is not allowed. There are $n_{TS} = 8$ versions of the **f**-C_s (xz) transition state: four with Z-configuration and four with E-configuration. The four Z-versions convert the two Z-versions of the **a**-C_{2h} (y) conformation into the two Z-versions of the **s**-C_{2v} (x) conformation and vice versa as outlined Fig. 39. The corresponding E-versions interconvert via an analogous mechanism.

It should be noted that in Fig. 39 it is assumed that **s**-C_{2v} (x) has a higher energy than **a**-C_{2h} (y) and thus, according to the Hammond principle, the transition state **f**-C_s (xz) is *syn*-folded (**su**-C_s (xz)). Symmetry predicts unequal degrees of folding

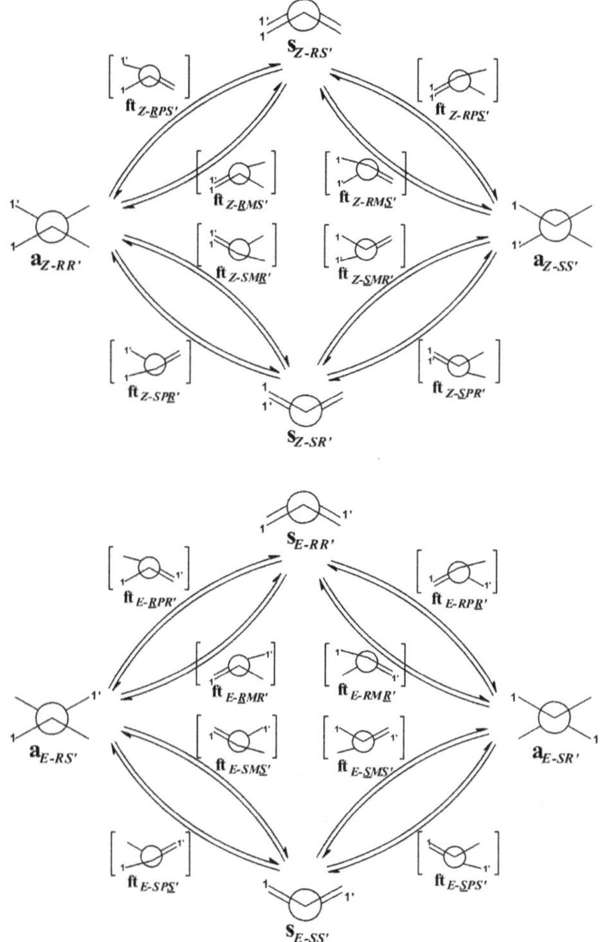

Fig. 40 Schematic mechanisms for the interconversion of the **a**-C_{2h} (y) *anti*-folded and **s**-C_{2v} (x) *syn*-folded conformations via a transition state **ft**-C_1

(C_s (xz) symmetry) but does not indicate whether the two moieties have the same (*syn*) or opposite (*anti*) direction of folding.

In the lower symmetry transition state, **ft**-C_1, the molecule may also be twisted in addition to the unequal degrees of folding. Twisting the central double bond in opposite directions gives rise to a doubling of all pathways ($\boldsymbol{p} = 2$) in this mechanism (Fig. 40). Each transition state (and pathway) of Fig. 39 is replaced by two transition states, one with positive (P) and one with negative (M) twist. The twist breaks the C_s symmetry.

The more folded moiety of the **ft**-C_1 transition state is indicated by underlining in the labels identifying the versions. In addition to interconverting the *anti*-folded and *syn*-folded conformations, the above two mechanisms may also serve as two-step mechanisms for conformational inversion of the *anti*-folded and *syn*-folded conformations.

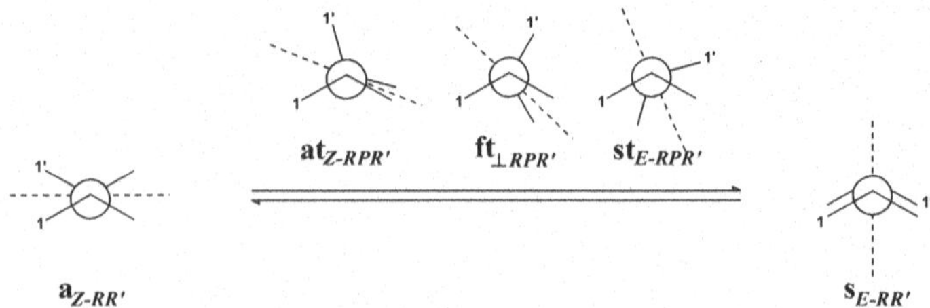

Fig. 41 Interconversion of an *anti*-folded and a *syn*-folded conformation with simultaneous *E*,*Z*-isomerization

4.3.7 Interconversion of the *anti*-Folded and *syn*-Folded Conformations with Simultaneous *E*,*Z*-Isomerization

The point groups C_{2h} (*y*) and C_{2v} (*x*) of the *anti*-folded and the *syn*-folded conformations, respectively, both include one C_2 axis as symmetry element. However, the axes C_2 (*y*) and C_2 (*x*) have a different orientation in space and correspond to different permutation-inversion operators: (11′)(88′)(99′) and (18′)(81′)(99′), respectively. Thus, they are distinct symmetry elements, in contrast to the mirror plane σ (*xz*), which belongs to the common subgroup (cf. Sect. 4.3.6). However, the two permutation operators may be interconverted by interchanging atom labels 1′ and 8′, indicating the possibility of a pathway converting an *anti*-folded conformation into a *syn*-folded conformation with a simultaneous *E*,*Z*-isomerization. This process may conserve C_2 symmetry. Figure 41 schematically shows the conversion of $\mathbf{a}_{Z\text{-}RR'}$ into $\mathbf{s}_{E\text{-}RR'}$. Intermediate C_2 symmetric *anti*-folded/twisted, folded/orthogonally twisted, and *syn*-folded/twisted transient structures are shown above the reaction arrow.

All intermediate structures along the pathway have C_2 symmetry, including the transition state. Thus, the transition state is not defined by a point of higher symmetry and does not necessarily correspond to the structure with exactly 90° twist of the central double bond. However, the orthogonally twisted and folded structure is exceptional with respect to the stereochemical nomenclature: it is neither *Z* nor *E*, and the borderline case between *anti*- and *syn*-folded conformations. It may be labeled $\mathbf{ft}_{\perp RPR'}$, where *P* refers to the sign of the torsion angle $\tau(C_1\text{-}C_9\text{-}C_{9'}\text{-}C_{1'})$ (Sect. 2.2). In the example of Fig. 41, the second moiety is rotated clockwise by 180° relative to the first moiety. Note that the direction of folding of both moieties does not change throughout the process. The degree of folding may change in a synchronous manner, in accord with the C_2 symmetry. In general, the *anti*-folded conformation will have a different degree of folding as compared to the *syn*-folded conformation. Intermediate structures along the pathway may have considerably higher or lower degrees of folding than the end points. A dashed line indicates the orientation of the C_2 axis, which changes with the rotation of the second moiety. Due to this change in orientation of the C_2 axis during the process, it cannot be labeled consistently. Also, the corresponding permutation-inversion operator changes at the point of *E*,*Z*-isomerization.

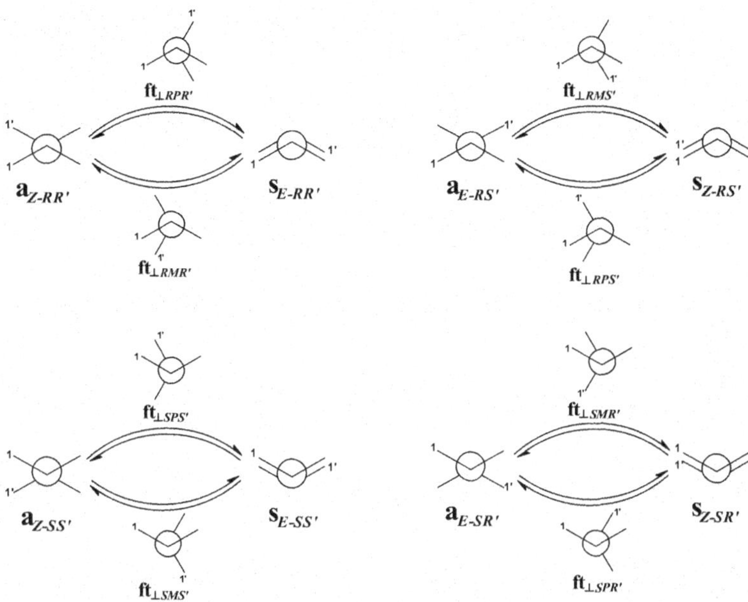

Fig. 42 Schematic mechanism for the interconversion of the **a**-C_{2h} (y) *anti*-folded and **s**-C_{2v} (x) *syn*-folded conformations with simultaneous *E*,*Z*-isomerization

In addition to the clockwise rotation of the second moiety relative to the first moiety (Fig. 41), it may also be rotated counterclockwise. Thus, there are two pathways interconverting $\mathbf{a}_{Z\text{-}RR'}$ and $\mathbf{s}_{E\text{-}RR'}$. The connectivity of the four versions of the *anti*-folded conformation with the four versions of the *syn*-folded conformation by this mechanism of conformational isomerization with simultaneous *E*,*Z*-isomerization is shown in Fig. 42.

Note that the orthogonally twisted/folded conformations shown above and below the reaction arrows in Fig. 42 correspond to a conformation half way along the reaction coordinate (90° twist). There are $\boldsymbol{n} = 8$ versions of this conformation with twofold symmetry ($h = 2$). However, this midpoint of the reaction does not necessarily correspond to the transition state of the reaction (see above).

This dynamic process performs an *E*,*Z*-isomerization, while converting an *anti*-folded conformation into the *syn*-folded conformation that has the same folding directions (*R*/*S*) for both moieties. Since folding directions are not changed, it cannot serve as a mechanism of conformational inversion. It connects each version of the *anti*-folded conformation with one and only one *syn*-folded conformation in a pairwise manner.

It should be noted that the peculiarities in deriving the mechanism for interconversion of the *anti*- and *syn*-folded conformations with simultaneous *E*,*Z*-isomerization are due to the choice of a common orientation and reference framework for *E*- and *Z*-isomers and of pure permutation-inversion operators to represent the molecular symmetry group. Using differently oriented frameworks for *E*- and

Z-isomers and pairs of permutation operators and rotation/reflection operators results in a molecular symmetry group of larger size, however, may allow one to derive the above mechanism in a more consistent manner.

4.3.8 Summary of the Dynamic Stereochemistry of Homomerous BAEs

Symmetry considerations, e.g., the molecular symmetry group approach, are very powerful tools to derive and classify all possible types of conformations for minima, transition states, and higher order saddle points. Permutation and permutation-inversion operators may also be used to derive the mechanisms of automerization and isomerization reactions of these conformations and the connectivity of the various labeled versions of the minima and transition states. Thus, the molecular symmetry group allows a systematic analysis of the conformational space and dynamic stereochemistry of molecules. However, symmetry considerations alone can only predict stationary points [280]. They cannot distinguish minima, transition states, and higher order saddle points. Such a classification requires calculations of the actual geometries and their vibrational frequencies or experimental results on structures and their relative energies.

In theoretical studies, prior knowledge about symmetry allows one to introduce symmetry constraints that reduce the degrees of freedom and therefore the number of coordinates that have to be optimized. In the optimization of a transition state, symmetry rules allow one to distinguish cases where a symmetry constrained energy minimization will converge to the transition state from cases where a full transition state search is necessary. A symmetry-constrained minimization is sufficient when the transition vector is not totally symmetric, i.e., the transition state has a higher symmetry than the structures along the pathways to reactants and products [270]. This greatly facilitates calculations.

In view of the large number of possible mechanisms for automerizations and isomerizations of BAEs predicted by the symmetry considerations, an illustrative example based on a scenario of plausible assumptions about the conformational type and symmetry of the minima and transition states may be instructive.

The following conformations have been observed experimentally in BAEs and may be assumed to be (local) minima:

- Twisted conformations **t**-D_2
- *anti*-Folded conformations **a**-C_{2h} (y)
- *syn*-Folded conformations **s**-C_{2v} (x)

Twisted conformations have been observed, e.g., in the X-ray crystal structures of bifluorenylidene (**2**) [47], and have been identified with the thermochromic and photochromic **B** forms. *anti*-Folded conformations have been observed, e.g., in the X-ray crystal structures of bianthrone (**3**) [56–60], and dixanthylene (**4**) [70, 71]. *syn*-Folded conformations have been observed in the X-ray crystal structure of tetrabenzo[7,7′]fulvalene (**8**) and related BAEs [61, 76], and have been identified as the photochromic **E** form. The relatively small deviations of the conformations

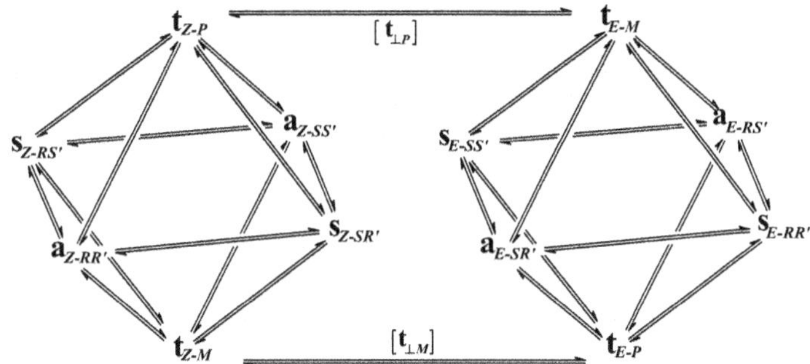

Fig. 43 Network of isomerization pathways for homomerous BAEs with twisted, *anti*-folded, and *syn*-folded minimum energy conformations

observed in the X-ray crystal structures from the symmetry indicated above may be due to crystal packing effects.

Planar conformations of BAEs are extremely overcrowded and thus may be expected to have very high energy [1, 13, 112, 285, 286]. For this reason, planar conformations are excluded from consideration as (local) minima or transition states.

In the isomerization and automerization mechanisms, the option with the highest symmetry transition state may be selected (principle of maximum symmetry, PMS) [252]. Transition states with lower symmetry would lead to similar results, but the mechanisms would be more complicated due to the parallel pathways.

Excluding the planar conformation as transition states or intermediates, the following conformational isomerization processes may be expected:

1. **t**-D_2 **a**-C_{2h} (y) via [**at**-C_2 (y)] (cf. Sect. 4.3.4, Fig. 35)
2. **t**-D_2 **s**-C_{2v} (x) via [**st**-C_2 (x)] (cf. Sect. 4.3.5, Fig. 37)
3. **a**-C_{2h} (y) **s**-C_{2v} (x) via [**f**-C_s (xz)] (cf. Sect. 4.3.6, Fig. 39)
4. *E*,*Z*-isomerization of **t**-D_2 via [**t**$_\perp$-D_{2d}] (cf. Sect. "*E*,*Z*-Isomerization with Simultaneous Inversion of Helicity of the Twisted Conformation", Fig. 33)

This scenario leads to the network of isomerization pathways shown in Fig. 43.

Conformational inversions of the *anti*- and *syn*-folded conformations and enantiomerizations of the twisted conformations proceed in two steps via an intermediate, using one of the three conformational isomerization mechanisms 1–3. *E*,*Z*-isomerization of the twisted conformation may proceed directly in one step via mechanism 4. Note that simultaneously with the *E*,*Z*-isomerization, the helicity of the central double bond (*P*/*M*) is inverted. For an *E*,*Z*-isomerization, the *anti*-folded and *syn*-folded conformations first have to be converted into a twisted conformation (alternatively, an *E*,*Z*-isomerization with simultaneous interconversion of *syn*- and *anti*-folded conformations may be considered; cf. Sect. 4.3.7, Fig. 42). According to the network of pathways shown in Fig. 43, twisted conformations have a special role as hub for *E*,*Z*-isomerizations. BAEs with twisted global

Table 25 Conformations and point groups of heteromerous BAEs

Conformations	Point groups
Planar	**p**-C_{2v} (z) **p**-C_s (yz)
Orthogonally twisted	$\mathbf{t}_\perp$-C_{2v} (d) $\mathbf{t}_\perp$-C_s (d) [a]
Twisted	$\mathbf{t_P}$-C_2 (z)
Unequally folded	**au**-C_s (xz) **su**-C_s (xz) **f**-C_s (xz) [b]
Unequally folded/twisted	**ft**-C_1 [b]

[a]The reflection planes of C_{2v}, and C_s in this orientation are the planes $x = y$ and/or $x = -y$
[b]Unequally folded conformations may have *anti*-folded, *syn*-folded, or mono-folded character

minima always have distinct barriers for enantiomerization and for *E,Z*-isomerization. On the other hand, in BAEs with *anti*-folded global minima, the interconversion of the *anti*-folded and twisted conformations necessary before an *E,Z*-isomerization may also serve as a mechanism for conformational inversion. In the case where the transition state for the *anti*-folded to twisted conformational isomerization [**at**-C_2 (y)] is higher than the transition state for *E,Z*-isomerization [$\mathbf{t}_\perp$-D_{2d}], the former is the highest transition state for both processes. Such a common barrier was observed in several BAEs with central six-membered rings [73, 170, 187, 188].

4.4 Symmetry of Heteromerous BAEs and Disubstituted Homomerous BAEs

In the heteromerous BAEs there is no symmetry relation between the two moieties due their different constitution. This drastically reduces the number of symmetry operators and available point groups. The possible conformations and their point groups are listed in Table 25 (cf. Table 4).

Note that there is no point group distinctly defining an *anti*-folded or *syn*-folded conformation. All pure folded conformations (not twisted) have the point group C_s (xz), irrespective of the relative direction of folding of the two moieties.

In *E*- and *Z*-disubstituted homomerous BAEs, the left and right side of the tricyclic moieties are different and thus cannot be symmetry related. The available conformations and point groups are listed in Table 26. Their number is smaller than in unsubstituted homomerous BAEs (cf. Table 4).

Note that there is no symmetry defined orthogonally twisted conformation. A conformation with planar tricyclic moieties twisted by 90° about the central double bond would have C_2 symmetry. There are no symmetry constraints on the planarity of the tricyclic moieties and on the twist of the double bond of such a C_2 symmetric conformation. On the contrary, there is a continuum of twisted/*anti*-folded conformations of the *Z*-isomer ($\mathbf{ta}_{Z\text{-}RPR'}$-$C_2$ (y) and $\mathbf{at}_{Z\text{-}RPR'}$-$C_2$ (y)) and a continuum of twisted/*syn*-folded conformations of the *E*-isomer ($\mathbf{ts}_{E\text{-}RR'}$-$C_2$ (x) and $\mathbf{st}_{E\text{-}RR'}$-$C_2$ (x)). However, $\mathbf{a}_{E\text{-}RS'}$-$C_i$, and $\mathbf{s}_{Z\text{-}RS'}$-$C_s$ (xy) still have unique point groups.

Table 26 Conformations and point groups of *Z*- and *E*-disubstituted homomerous BAEs

	Z-Disubstituted	E-Disubstituted
Planar	$\mathbf{p}_{Z}$-C_{2v} (y), $\mathbf{p}_{Z}$-C_s (yz)	$\mathbf{p}_{E}$-C_{2h} (x), $\mathbf{p}_{E}$-C_s (yz)
Twisted	$\mathbf{ta}_{Z\text{-}RPR'}$-$C_2$ (y)	$\mathbf{ts}_{E\text{-}RR'}$-$C_2$ (x)
anti-Folded	$\mathbf{at}_{Z\text{-}RPR'}$-$C_2$ (y)	$\mathbf{a}_{E\text{-}RS'}$-$C_i$
syn-Folded	$\mathbf{s}_{Z\text{-}RS'}$-$C_s$ (xy)	$\mathbf{st}_{E\text{-}RR'}$-$C_2$ (x)
Unequally folded/twisted	$\mathbf{ft}_{Z\text{-}RPR'}$-$C_1$ [a] $\mathbf{ft}_{Z\text{-}RPS'}$-$C_1$ [b]	$\mathbf{ft}_{E\text{-}RPS'}$-$C_1$ [a] $\mathbf{ft}_{E\text{-}RPR'}$-$C_1$ [b]

[a]The relative direction of folding of the two moieties is *anti*
[b]The relative direction of folding of the two moieties is *syn*

In disubstituted heteromerous BAEs all non-planar conformations necessarily have C_1 symmetry. In monosubstituted BAEs, all non-planar conformations have C_1 symmetry, with the exception of the orthogonally twisted conformation $\mathbf{t}_{\perp}$-C_s (d). Planar conformations of disubstituted heteromerous BAEs and monosubstituted BAEs have C_s symmetry (cf. Table 4).

Because of the reduced symmetry, an analysis based on the molecular symmetry groups of heteromerous and/or (di)substituted BAEs would yield substantially less information as compared to the above analysis of unsubstituted homomerous BAEs. Therefore, it is not developed here. In many cases, a qualitative approach may be more promising. A substitution of a homomerous BAE, e.g., in 2- and 2′-positions, should lead only to small changes in energy and geometry. Such a situation has been termed ‘slightly distorted skeleta’ [260]. Mislow et al. introduced the concept of pseudo-symmetry for substituted systems [287]. All conformations should remain similar (in energy as well as in geometry) to those of the unsubstituted parent system. In particular, the mechanisms of the conformational isomerizations and the corresponding transition states should remain similar. Thus considerably more information may be gained by studying the dynamic stereochemistry based on the molecular symmetry group of the unsubstituted parent system and considering the substitution in a second step as small “perturbation” causing minor changes in the energies and geometries. Likewise, in heteromerous BAEs the dynamic stereochemistry may be considered similar to that of homomerous BAEs, as long as the steric and conformational properties of the two moieties are not too different. Such an approximate application of symmetry [268] assuming an analogous dynamic stereochemistry will yield more powerful predictions.

5 Conclusions

Bistricyclic aromatic enes (BAEs) and the related polycyclic systems are a class of molecular materials that displays a rich variety of conformations, dynamic stereochemistry and switchable chirality, color, and spectroscopic properties. This is due to the a subtle interplay of the inherent preference for planarity of aromatic systems and the competing necessity of non-planarity due to intramolecular overcrowding in the fjord regions built into the general molecular structure of BAEs (**1**).

The conformational, dynamic, and spectroscopic properties may be designed and fine-tuned, e.g., by variation of the bridging groups X and Y, the overcrowding in the fjord regions, extensions of the aromatic system, or modifications of the general BAE structure **1**, based on the fundamental understanding of the structure–property relationships (SPR).

The symmetry analysis presented in Sect. 4 allows deeper insight into the conformations, chirality, and the mechanisms of the dynamic stereochemistry.

The objective of the present work was to evaluate the conformational spaces and the dynamic stereochemistry of overcrowded bistricyclic aromatic enes (**1**), applying fundamental symmetry considerations.

The conformations of BAEs may be described by the predominant mode(s) of out-of-plane deformation. The following conformational types were defined: planar (**p**), orthogonally twisted ($\mathbf{t}_{\perp}$), twisted (**t**), *anti*-folded (**a**), *syn*-folded (**s**), and folded conformations with one planar (or nearly planar) tricyclic moiety (**f**) (Sect. 4.1.2). In addition, combined modes were considered. The definitions of these conformational types are empirical and qualitative. An exact classification scheme may be derived from the point group symmetries (Sect. 4.1.3).

Combination of the qualitative description by conformational types with the classification according to point groups allows one to derive possible mechanisms for dynamic stereochemical processes in unsubstituted homomerous BAEs based on the molecular symmetry group formalism [246, 247] and the symmetry theorems for transition states [268, 270]. The principle of maximum symmetry [252] allows one to predict mechanistic scenarios for the enantiomerizations, conformational inversions, *E*,*Z*-isomerizations, and interconversions of conformations, based on plausible assumptions about the relative stability of the conformational types. The exact symmetry analysis for unsubstituted homomerous BAEs may be generalized to less symmetric systems following the notion of slightly distorted skeleta [260] and the concept of pseudo-symmetry for substituted systems [287].

Hopefully, the present symmetry analysis of the conformational spaces of BAEs will serve as a basis for future experimental and computational studies of the dynamic stereochemistry of BAEs, with special emphasis on the transition states for interconversions of conformers, *E*,*Z*-isomerizations, and enantiomerizations. Prior knowledge of the symmetry of the transition states, as derived in Sect. 4.3, is highly instrumental in finding their geometries, as in many cases the geometry optimization reduces to a simple symmetry constrained energy minimization rather than a much more demanding transition state search.

The rapidly increasing understanding of the fundamental principles underlying thermochromism, photochromism, piezochromism, and electrochromism in BAEs and of related molecular switches and motors will hopefully lead to useful applications in the future. Many fields in modern chemical science have certainly profited from the quest towards elucidating the fascinating enigmas of BAEs and related aromatic compounds.

References

1. Shoham G, Cohen S, Suissa MR, Agranat I (1988) Stereochemistry of strained, overcrowded bistricyclic ethylenes. In: Stezowski JJ, Huang J-L, Shao MC (eds) Molecular structure: chemical reactivity and biological activity. IUCr Crystallographic Symposia 2, Oxford University Press, Oxford, pp 290–312
2. Biedermann PU, Stezowski JJ, Agranat I (1998) Overcrowded polycyclic aromatic enes. In: Thummel RP (ed) Advances in theoretically interesting molecules, vol 4. JAI Press, Stamford, pp 245–322
3. Biedermann PU, Stezowski JJ, Agranat I (2001) Conformational space and dynamic stereochemistry of overcrowded homomerous bistricyclic aromatic enes – a theoretical study. Eur J Org Chem 15–34
4. de la Harpe C, Van Dorp WA (1875) Über die Einwirkung des erhitzten Bleioxids auf Fluoren. Ber Dtsch Chem Ges 8:1048–1050
5. Meyer H (1909) Über neue Reduktionsprodukte des Anthrachinons. Ber Dtsch Chem Ges 42:143–145
6. Meyer H (1909) Über neue Derivate des Anthrachinons. Monatsh Chem 30:165–177
7. Gurgenjanz G, Kostanecki S (1895) Über ein neues Reduktionsprodukt des Xanthons. Ber Dtsch Chem Ges 28:2310–2311
8. Schönberg A, Schütz O (1928) Über das Verhalten des Dixanthylens in der Wärme. Ber Dtsch Chem Ges 61:478–479
9. Bell F, Waring DH (1949) The symmetrical dianthryls. Part III. J Chem Soc 2689–2693
10. Harnik E, Herbstein FH, Schmidt GMJ (1951) X-Ray crystallographic measurements on some compounds showing intramolecular overcrowding. Nature 168:158–160
11. Theilacker W, Kortüm G, Elliehausen H, Wilski H (1956) Zur Frage der Thermochromie, IV. Mitteilung: Dibenzo-dehydrodianthrone und Dehydrodianthron-carbonsäure-(3). Chem Ber 89:1578–1592
12. Schönberg A, Sidky MM (1959) Syntheses of thermochromic ethylenes. A study of the relationship between consitiution and thermochromism. J Am Chem Soc 81:2259–2262
13. Mustafa A, Asker W, Sobhy MEED (1960) Thermochromism of dixanthylenes. Reactions with substituted xanthones. III. J Org Chem 25:1519–1525
14. Day JH (1963) Thermochromism. Chem Rev 63:65–80
15. Jørgensen M, Lerstrup K, Frederiksen P, Bjørnholm T, Sommer-Larsen P, Schaumburg K, Brunfeldt K, Bechgaard K (1993) Synthesis and electrochemical properties of 1,4-dithiafulvenyl-substituted bianthrone molecules with potential application as molecular switches. J Org Chem 58:2785–2790
16. Sommer-Larsen P, Bjørnholm T, Jørgensen M, Lerstrup K, Frederiksen P, Schaumburg K, Brunfeldt K, Bechgaard K, Roth S, Poplawski J, Byrne H, Anders J, Eriksson L, Wilbrandt R, Frederiksen J (1993) A molecular switch involving large conformational changes. A theoretical study. Mol Cryst Liq Cryst 234:89–96
17. Feringa BL, Schoevaars AM, Jager WF, de Lange B, Huck NPM (1996) Switching of chirality by light. Enantiomer 1:325–335
18. Feringa BL (2001) In control of motion: from molecular switches to molecular motors. Acc Chem Res 34:504–513
19. Koumura N, Geersema EM, van Gelder MB, Meetsma A, Feringa BL (2002) Second generation light-driven molecular motors. Unidirectional rotation controlled by a single stereogenic center with near-perfect photoequilibria and acceleration of the speed of rotation by structural modification. J Am Chem Soc 124:5037–5051
20. Hagen S, Nuechter U, Nuechter M, Zimmermann G (1995) High temperature synthesis of bowl-shaped subunits of fullerenes. II From 9,9′-bi-9*H*-fluorenylidene to diindeno[1,2,3,4-*defg*; 1′,2′,3′,4′-*mnop*]chrysene. Polycycl Aromat Compd 4:209–217

21. Biederman PU, Luh TY, Weng DTC, Kuo CH, Stezowski JJ, Agranat I (1996) From overcrowded polycyclic aromatic enes to semifullerenes. Polycycl Aromat Compd 8:167–175
22. Rabideau PW, Sygula A (1996) Buckybowls: polynuclear aromatic hydrocarbons related to the buckminsterfullerene surface. Acc Chem Res 29:235–242
23. Scott LT (1996) Fragments of fullerenes: novel syntheses, structures and reactions. Pure Appl Chem 68:291–300
24. Mehta G, Rao HSP (1998) Synthetic studies directed towards bucky-balls and bucky-bowls. Tetrahedron 54:13325–13370
25. Pogodin S, Biedermann PU, Agranat I (1998) From twisted and folded overcrowded polycyclic aromatic enes to planar fullerene fragments towards bowl-shaped PAHS. In: Kadish KM, Ruoff RS (eds) Recent advances in the chemistry of fullerenes and related materials, vol 6. The Electrochemical Society, Pennington, pp 1110–1116
26. Avoyan RL, Struchkov YT, Dashevskii VG (1966) Steric hindrance and conformation in aromatic molecules. J Struct Chem 7:283–320
27. Sandström J (1983) Static and dynamic stereochemistry of push-pull and strained ethylenes. Top Stereochem 14:83–181
28. Biedermann PU, Levy A, Stezowski JJ, Agranat I (1995) Aspects of chirality in overcrowded bistricyclic enes. Chirality 7:199–205
29. Luef W, Keese R (1991) Strained olefins: structure and reactivity of nonplanar carbon–carbon double-bonds. Top Stereochem 20:231–318
30. Sandström J (1997) Strained olefins. In: Patai S (ed) The chemistry of double-bonded functional groups, Supplement A3. Wiley, New York, pp 1253–1280
31. Gutman I, Cyvin SJ (1989) Introduction to the theory of benzenoid hydrocarbons. Springer-Verlag, Berlin Heidelberg, pp 20–25
32. Rieger R, Mullen K (2010) Forever young: polycyclic aromatic hydrocarbons as models cases for structural and optical studies. J Phys Org Chem 23:315–325
33. Allen FH, Kennard O, Watson DG, Brammer L, Orpen AG, Taylor R (1987) Tables of bond lenths determined by X-ray and neutron diffraction. Part 1. Bond lengths in organic compounds. J Chem Soc Perkin Trans 2 Suppl S1–S19
34. Ermer O (1974) Conformation of trans-cyclooctene. Angew Chem Int Ed Engl 13:604–606
35. Bürgi HB, Shefter E (1975) Out-of-plane deformations of cyclic polyenes. A force-field interpretation of structural data. Tetrahedron 31:2976–2981
36. Ermer O (1976) Calculation of molecular properties using force fields. Applications in organic chemistry. Struct Bond 27:161–211
37. Haddon RC (1990) Measure of nonplanarity in conjugated organic molecules: which structurally characterized molecule displays the highest degree of pyramidalization? J Am Chem Soc 112:3385–3389
38. Radhakrishnan TP, Agranat I (1991) Measures of pyramidalization. Struct Chem 2:107–115
39. Haddon RC (1986) GVB and POAV analysis of rehybridization and π-orbital misalignment in non-planar conjugated systems. Chem Phys Lett 125:231–234
40. Haddon RC (1987) Rehybridization and π-orbital overlap in nonplanar conjugated organic molecules: π-orbital axis vector (POAV) analysis and three-dimensional Hückel molecular orbital (3D-HMO) theory. J Am Chem Soc 109:1676–1685
41. Helmchen G (1995) Nomenclature and vocabulary of organic stereochemistry. In: Büchel KH, Falbe J, Hagemann H, Hanack M, Klamann D, Kreher R, Kropf H, Regitz M, Schaumann E (eds) Stereoselective synthesis, Houben–Weyl, methods of organic chemistry, additional and supplementary volumes to the 4th edn, vol E21a. Thieme Verlag, Stuttgart/New York, pp 10 and 22
42. Klyne W, Prelog V (1960) Description of steric relationships across single bonds. Experientia 16:521–523
43. Cahn RS, Ingold CK (1951) Specification of configuration about quadricovalent asymmetric atoms. J Chem Soc (Lond) 612–622

44. Herbstein FH (1991) The crystal and molecular structures of overcrowded halogenated compounds IX. Stereochemistry of molecules containing the perchlorofluorenyl and perchlorofluorenylidene moieties and of their hydrogen analogues. Acta Cryst B 47:288–298
45. Kanawati B, Genest A, Schmitt-Kopplin P, Lenoir D (2012) Bis-dibenzo[a.i]fluorenylidene, does it exist as stable 1,2-diradical? J Mol Model 18:5089–5095
46. Allen FH, Kennard O (1993) 3D search and research using the Cambridge Structural Database. Chem Design Automat News 8:31–37
47. Lee JS, Nyburg SC (1985) Refinement of the α-modification of 9,9′-bifluorenylidene, $C_{26}H_{16}$, and structure analyses of the β-modification, the 2:1 pyrene complex, $2(C_{26}H_{16})\bullet C_{16}H_{10}$, and the 2:1 perylene complex, $2(C_{26}H_{16}).C_{20}H_{12}$. Acta Cryst C 41:560–567
48. Lee JS, Chen IY (1989) Structural and physical characterization of 1:1 9,9′-bifluorenylidene complex with 7,7,8,8-tetracyanoquinodimethane, BFL-TCNQ. J Chin Chem Soc (Taipei) 36:123–130
49. Pogodin S, Agranat I (2003) Biphenalenylidene: the forgotten bistricyclic aromatic ene. A theoretical study. J Am Chem Soc 125:12829–12835
50. Pogodin S, Agranat I (2007) Theoretical notions of aromaticity and antiaromaticity: phenalenyl ions versus fluorenyl ions. J Org Chem 72:10096–10107
51. Bergmann ED (1955) The Fulvenes. Prog Org Chem 3:81–171
52. Rabinovitz M, Agranat I, Bergmann ED (1965) Fulvenes and thermochromic ethylenes. Part 35. The NMR spectrum and the spatial structure of dibiphenyleneethene. Tetrahedron Lett 6:1265–1269
53. Bailey NA, Hull SE (1978) The structures of diisopropyl-9,9′-bifluorenylidene-1,1′-dicarboxylate and 9,9′-bifluorenylidene. Acta Cryst B 34:3289–3295
54. Biedermann PU, Levy A, Suissa MR, Stezowski JJ, Agranat I (1996) The controversial dynamic stereochemistry of bifluorenylidenes. Enantiomer 1:75–80
55. Bergmann E (1935) Dipole moment and molecular structure. Part XIV. 2:2′-Difluorobis diphenylene-ethylene. J Chem Soc (Lond) 987–989
56. Harnik E, Schmidt GMJ (1954) The structure of overcrowded aromatic compounds. Part II. The crystal structure of dianthronylidene. J Chem Soc (Lond) 3295–3302
57. Apgar PA, Wasserman E (1987) Unpublished results. Quoted in [1].
58. Wolstenholme DJ, Cameron TS (2006) Comparative study of weak interactions in molecular crystals: H–H bonds vs hydrogen bonds. J Phys Chem A 110:8970–8978
59. Naumov P, Ishizawa N, Wang J, Pejov L, Pumera M, Lee SC (2011) On the origin of the solid-state thermochromism and thermal fatigue of polycyclic overcrowded enes. J Phys Chem A 115:8563–8570
60. Johnstone RDL, Allan D, Lennie A, Pidcock E, Valiente R, Rodríguez F, Gonzalez J, Warren J, Parsons S (2011) The effect of pressure on the crystal structure of bianthrone. Acta Cryst B 67:226–237
61. Dichmann KS, Nyburg SC, Pickard FH, Potworowski JA (1974) The crystal and molecular structures of two stereoisomers of Δ5,5′-bi-5H-dibenzo[a, d]cycloheptene (2,3;6,7;2′,3′;6′,7′-tetrabenzoheptafulvalene). Acta Cryst B 30:27–36
62. Levy A, Biedermann PU, Agranat I (2000) Interplay of twisting and folding in overcrowded heteromerous bistricyclic aromatic enes. Org Lett 2:1811–1814
63. Biedermann PU, Stezowski JJ, Agranat I (2006) Polymorphism versus thermochromism: interrelation of color and conformation in overcrowded bistricyclic aromatic enes. Chem Eur J 12:3345–3354
64. Riklin M, von Zelewsky A, Bashall A, McPartlin M, Baysal A, Connor JA, Wallis JD (1999) Synthesis, structure and chemistry of a twisted olefinic bis-dentate proligand: 5,5′-bi-5*H*-cyclopenta[2,1-*b*:3,4-*b*′]dipyridinylidene. Helv Chim Acta 82:1666–1680
65. Pogodin S, Cohen S, Biedermann PU, Agranat I (2002) Stereochemistry of (E)- and (Z)-1,1′-dichlorobifluoroenylidenes, substituted overcrowded fullerene fragments. Enantiomer 7:261–269

66. Molins E, Miravitlles C, Espinosa E, Ballester M (2002) 1,1′,3,3′,6,6′,8,8′-Octachloro-9,9′-bifluorenylidene and perchloro-9,9′-bifluorenylidene, two exceedingly twisted ethylenes. J Org Chem 67:7175–7178
67. Assadi N, Pogodin S, Cohen S, Agranat I (2013) Naphthologs of overcrowded bistricyclic aromatic enes: (*E*)-bisbenzo[*a*]fluorenylidene. Struc Chem 24:1229–1240
68. Stezowski JJ, Hildenbrand T, Suissa MR, Agranat I (1990) Overcrowded twisted bi-4*H*-cyclopenta[*def*]phenanthren-4-ylidene versus bi-9*H*-fluorenylidene, syn pyramidalization. Struct Chem 1:123–126
69. Lee-Ruff E, Grant A, Stynes DV, Vernik I (2000) Crystal structure and molecular modeling of 4,4′-bi-4H-cyclopenta[def]phenanthrenylidene and related derivatives. Struct Chem 11:245–248
70. Mills JFD, Nyburg SC (1963) Thermochromism and related effects in bixanthenylidenes and bianthronylidenes. Part I. Crystal structure analyses. J Chem Soc (Lond) 308–321
71. Shi J, Chang N, Li C, Mei J, Deng C, Luo X, Liu Z, Bo Z, Dong YQ, Tang BZ (2012) Locking the phenyl rings of tetraphenylethene step by step: understanding the mechanism of aggregation-induced emission. Chem Commun 48:10675–10677
72. Pogodin S, Suissa MR, Levy A, Cohen S, Agranat I (2008) The tetracyanoquinodimethane motif in overcrowded bistricyclic aromatic enes: avoiding thermochromism. Eur J Org Chem 2887–2894
73. Tapuhi Y, Suissa MR, Cohen S, Biedermann PU, Levy A, Agranat I (2000) The stereochemistry of overcrowded homomerous bistricyclic aromatic enes with alkylidene bridges. J Chem Soc Perkin Trans 2:93–100
74. Geertsema EM, Meetsma A, Feringa BL (1999) Asymmetric synthesis of overcrowded alkenes by transfer of axial single bond chirality to axial double bond chirality. Angew Chem Int Ed Engl 38:2738–2741
75. Levy A, Biedermann PU, Cohen S, Agranat I (2000) Selenium- and tellurium-bridged overcrowded homomerous bistricyclic aromatic enes. J Chem Soc Perkin Trans 2:725–735
76. Luo J, Song K, Gu FL, Miao Q (2011) Switching of non-helical overcrowded tetrabenzoheptafulvalene derivatives. Chem Sci 2:2029–2034
77. Takezawa H, Murase T, Fujita M (2012) Temporary and permanent trapping of the metastable twisted conformer of an overcrowded chromic alkene via encapsulation. J Am Chem Soc 134:17420–17423
78. Levy A, Pogodin S, Cohen S, Agranat I (2007) Thermochromism at room temperature in overcrowded bistricyclic aromatic enes: closely populated twisted and folded conformations. Eur J Org Chem 5198–5211
79. Levy A, Biedermann PU, Cohen S, Agranat I (2001) Stereochemistry of selenium- and tellurium-bridged heteromerous bistricyclic aromatic enes. The fluorenylidene-chalcoxanthene series. J Chem Soc Perkin Trans 2 2329–2341
80. Levy A, Cohen S, Agranat I (2003) Overcrowded 1,8-diazafluorenylidene-chalcoxanthenes. Introducing nitrogens at the fjord regions of bistricyclic aromatic enes. Org Biomol Chem 1:2755–2763
81. Assadi N, Pogodin S, Cohen S, Levy A, Agranat I (2009) Overcrowded naphthologs of mono-bridged tetraarylethylenes: analogs of bistricyclic aromatic enes. Struc Chem 20:541–556
82. Zhou J, Chang Z, Jiang Y, He B, Du M, Lu P, Hong Y, Kwok HS, Qin A, Qiu H, Zhao Z, Tang BZ (2013) From tetraphenylethene to tetranaphthylethene: structural evolution in AIE luminogen continues. Chem Commun 49:2491–2493
83. Pogodin S, Assadi N, Agranat I (2013) Conformational space of tetrakis(2-naphthalenyl) ethene: a naphthologous AIE luminogen. Struct Chem 24:1747–1757
84. Zhang X, Jiang X, Luo J, Chi C, Chen H, Wu J (2010) A cruciform 6,6′-dipentacenyl: synthesis, solid-state packing and applications in thin-film transistors. Chem Eur J 16:464–468
85. deGunst GP (1969) The photochemistry of bifluorenylidene. Rec Trav Chim Pays Bas 88:801–804

86. Hagen S, Nuechter U, Nuechter M, Zimmermann G (1994) High-temperature synthesis towards bowl-shaped subunits of fullerenes. III. From 4,4′-bi-4H-cyclopenta[def]phenanthrenylidene towards [5.5]circulene. Tetrahedron Lett 35:7013–7014
87. Pogodin S, Biedermann PU, Agranat I (1997) Facile aromatization reactions of overcrowded polycyclic aromatic enes leading to fullerene fragments. J Org Chem 62:2285–2287
88. Mehta G, Rao HSP (1997) Synthetic studies related to fullerenes and fullerence fragments. In: Halton B (ed) Advances in strain in organic chemistry, vol 6. JAI Press, London, pp 139–187
89. Rabideau PW, Sygula A (1995) Polynuclear aromatic hydrocarbons with curved surfaces – hydrocarbons possesing carbon frameworks related to Buckminsterfullerene. In: Thummel RP (ed) Advances in theoretically interesting molecules, vol 3. JAI Press, Greenwich, p 1230
90. Bronstein HE, Choi N, Scott LT (2002) Practical synthesis of an open geodesic polyarene with a fullerene-type 6:6-double bond at the center: diindeno[1,2,3,4-*defg*;1′,2′,3′,4′-*mnop*] chrysene. J Am Chem Soc 124:8870–8875
91. Brockmann H, Mühlmann R (1949) Über die photochemische Cyclisierung des Helianthrons und Dianthrons zum meso-Naphthodianthron. Chem Ber 82:348–357
92. Filipescu N, Avram E, Welk KD (1977) Photooxidative transformations of anthrone, bianthronyl, and bianthrone in acid solution. J Org Chem 42:507–512
93. Kazama S, Kamiya M, Akahori Y (1991) Mechanism of the photochemical reaction of 10-(9H-xanthenylidene)-9(10H)-anthracenone. Chem Pharm Bull 39:3103–3105
94. Müller U, Enkelmann V, Adam M, Müllen K (1993) Synthesis of alkyl-substituted helianthrones and mesonaphthobianthrone by highly regioselective photocyclization of bianthronylidenes. Chem Ber 126:1217–1225
95. Yamada S, Kubo M, Fuke H, Tsubaki N, Maeda K (1993) Photochemical ring closure of 10,10′-disubstituted 9,9′(10*H*,10′*H*)-biacridinylidenes followed by dehydrogenation. Bull Chem Soc Jpn 66:1834–1836
96. Nicodem DE, da Cunha MFV (1997) Bianthrone photophysics and photochemistry: solvent heavy-atom-induced triplet state formation. J Photochem Photobiol A 107:165–167
97. Linde K, Berner MM, Kriston L (2008) St. John's wort for major depression. Cochrane Database Systematic Reviews CD000448. doi:10.1002/14651858.CD000448.pub3
98. Lavie G, Mazur Y, Lavie D, Meruelo D (1995) The chemical and biological properties of hypericin – a compound with a broad spectrum of biological activities. Med Res Rev 15:111–119
99. Ciogli A, Bicker W, Lindner W (2010) Determination of enantiomerization barriers of hypericin and pseudohypericin by dynamic HPLC on immobilized polysaccaride-type chiral stationary phases and off-column racemization experiments. Chirality 22:463–471
100. Zhang X, Li J, Qu H, Chi C, Wu J (2010) Fused bispentacenequinone and its unexpected Michael addition. Org Lett 12:3946–3949
101. Luo J, Xu X, Mao R, Miao Q (2012) Curved polycyclic aromatic molecules that are pi-isoelectronic to hexabenzocoronene. J Am Chem Soc 134:13796–13803
102. Pogodin S, Agranat I (2002) Overcrowding motifs in large PAHs. An ab initio study. J Org Chem 67:265–270
103. Zhang H, Wu D, Li SH, Yin J (2012) Star-shaped polycyclic aromatic hydrocarbons: design and synthesis of molecules. Curr Org Chem 16:2124–2158
104. Ruiz M, Gomez-Lor B, Santos A, Echavarren AM (2004) Overcrowded 5,10,15-trisubstituted derivatives: synthesis of 5,10,15-tri(fluorenylidene)truxene. Eur J Org Chem 858–866
105. Iyoda M, Otani H, Oda M (1988) Tris(fluorenylidene)cyclopropane, a novel [3]radialene. Angew Chem Int Ed 27:1080–1081
106. Iyoda M, Mizusuna A, Kurata H, Oda M (1989) Novel cyclo-oligomerization of nickel carbenoids: a simple and efficient synthesis of radialenes. J Chem Soc Chem Commun 1690–1692
107. Sugimoto T, Misaki Y, Kajita T, Nagatomi T, Yoshida Z, Yamauchi J (1988) Tris (thioxanthe-9-ylidene)cyclopropane, and its radical cation and dication. Angew Chem Int Ed 27:1078–1080

108. Kortüm G (1974) Die Thermochromie, Piezochromie und Photochromie der Dehydrodianthrone. Ber Bunsenges Phys Chem 78:391–403
109. Grubb WT, Kistiakowsky GB (1950) On the nature of thermochromism. J Am Chem Soc 72:419–424
110. Theilacker W, Kortüm G, Friedheim G (1950) Zur Frage der Thermochromie; ein Beispiel für eine echte Valenztautomerie. Chem Ber 83:508–519
111. Theilacker W (1953) Die Dehydrodianthrone und ihr thermisches und optisches Verhalten. Angew Chem 65:216–216
112. Kortüm G (1958) Thermochromie, Piezochromie, Photochromie, und Photomagnetismus. Angew Chem 70:14–20
113. Kortüm G, Theilacker W, Schreyer G (1957) Zur Frage der Thermochromie IV. Mitteilung: Die Thermochromie adsorbierter Molekeln; Identität von Thermochromie und Piezochromie. Z Physik Chem NF 11:182–195
114. Grabowski ZR, Balasiewicz MS (1968) Polarography and kinetics of the thermochromic system of dehydrodianthrone. Trans Faraday Soc 64:3346–3353
115. Tapuhi Y, Kalisky O, Agranat I (1979) Thermochromism and thermal E, Z-isomerizations in bianthrones. J Org Chem 44:1949–1952
116. Kortüm G, Zoller W (1967) Piezochromie und Thermochromie der Dehydrodianthrone. Chem Ber 100:280–292
117. Kortüm G, Krieg P (1969) Spektroskopische Untersuchungen zur Thermo-, Piezo- und Photochromie von Bixanthyliden-(9,9′) und Biflavyliden-(4,4′). Chem Ber 102:3033–3045
118. Hirshberg Y, Fischer E (1953) Low-temperature photochromism and its relation to thermochromism. J Chem Soc (Lond) 629–636
119. Matsue T, Evans DH (1984) Electron-transfer reactions and associated conformational changes. The effect of disproportionation on the reduction of bianthrone at elevated temperatures. J Electroanal Chem 168:287–298
120. Fanselow DL, Drickamer HG (1974) High pressure studies of the electronic behavior of bianthrones. J Chem Phys 61:4567–4574
121. Wasserman E, Davis RE (1959) On the identity of the green forms of bianthrone. J Chem Phys 30:1367–1367
122. Theilacker W, Kortüm G, Elliehausen H (1956) Zur Frage der Thermochromie, V. Mitteilung: Optische Aktivierung der Dehydrodianthron-carbonsäure-(3). Die Ursache der Thermochromie. Chem Ber 89:2306–2318
123. Hirshberg Y (1950) Photochromie dans la série de la bianthrone. Compt Rend hebd Séances Acad Sci 231:903–904
124. Hirshberg Y (1951) The mechanism of low temperature photochromy in the bianthrone series. Bull Res Council Israel 1:123–125
125. Fischer E (1984) Photochromism and other reversible photoreactions in the dianthrylidenes and implications regarding environmental control of photoreactions, colour-structure correlations, and other points. Rev Chem Intermed 5:393–422
126. Muszkat KA (1988) Photochromism and thermochromism in bianthrones and bianthrylidenes. In: Patai S, Rappoport Z (eds) The chemistry of quinonoid compounds, vol 2. Wiley, Chichester, pp 203–224
127. Lorenz R, Wild U, Huber JR (1969) The photochromism of the dehydrodianthrones. Pariser–Parr–Pople calculations. Photochem Photobio 10:233–242
128. Bercovici T, Korenstein R, Muszkat KA, Fischer E (1970) Dianthrone photochromism 1950–1970. Pure Appl Chem 24:531–565
129. Kortüm G, Zoller W (1970) Photochromie benzo-substituierter Xanthylidenanthrone. Chem Ber 103:2062–2076
130. Fischer E (1975) Photochromie und photochrome Verbindungen. Chem Unserer Zeit 9:85–95
131. Laarhoven WH (1990) 4n+2 systems: molecules derived from Z-hexa-1,3,5-triene / cyclohexa-1,3-diene. In: Dürr H, Bouas-Laurent H (eds) Photochromism, molecules and systems. Elsevier, Amsterdam, pp 270–313

132. Kortüm G, Bayer G (1963) Zur Thermochromie und Photochromie der Dehydrodianthrone. Ber Bunsenges Phys Chem 67:24–28
133. Kortüm G, Bach H (1965) Fluorescenz- und Absorptionsmessungen am 1,3,6′,8′-Tetramethyl-Dehydrodianthron. Z Physik Chem NF 46:305–319
134. Kortüm G, Theilacker W, Braun V (1954) Die Photochromie der Dehydrodianthrone. Z Physik Chem NF 2:179–196
135. Bercovici T, Fischer E (1969) Photochromism in dianthrone and related compounds. Part III. Solutions in aliphatic hydrocarbons. Israel J Chem 7:127–133
136. Korenstein R, Muszkat KA, Fischer E (1975) Large enhancement of photoisomerization quantum yields by external perturbing agents. Chem Phys Lett 36:509–513
137. Korenstein R, Muszkat KA, Fischer E (1976) Reversible photoisomerization and fluorescence of bithioxanthene: temperature and spin-orbit perturbation effects. J Photochem 5:345–353
138. Korenstein R, Muszkat KA, Slifkin MA, Fischer E (1976) Reversible photochemistry, photooxidation and fluorescence of dixanthylidene: temperature and external spin-orbit perturbation effects. J Chem Soc Perkin Trans 2:438–443
139. Stegemeyer H (1969) Elektronenspektren und Molekülstruktur von Di-biphenylen-äthylen und vinyloger Verbindungen. Ber Bunsenges Phys Chem 73:612–619
140. Fabian J, Zahradnik R (1989) The search for highly colored organic-compounds. Angew Chem Int Ed 28:677–694
141. Biedermann PU, Stezowski JJ, Agranat I (2001) Thermochromism of overcrowded bistricyclic aromatic enes (BAEs). A theoretical study. Chem Commun 954–955
142. Wizinger R (1927) Über Halochromie und tieffarbige Ketone. Angew Chem 40:939–945
143. Hirschberg Y (1956) Reversible formation and eradication of colors by irradiation at low temperatures. A photochemical memory model. J Am Chem Soc 78:2304–2312
144. Schönberg A, Ismail AFA, Asker W (1946) Reactions of thermochromic ethylenes. J Chem Soc (Lond) 442–446
145. Schönberg A, Asker W (1945) Dipyrylenes, dichromylenes, dixanthylenes, and their sulfur analogues. Chem Rev 37:1–14
146. Kortüm G, Theilacker W, Zeininger H, Elliehausen H (1953) Zur Frage der Thermochromie. II. Mitteilung: Methylsubstituierte Dehydrodianthrone. Chem Ber 86:294–307
147. Schönberg A, Nickel S (1931) Über die Dichromylene und die Valenz-tautomerie ungesättigter Systeme. Ber Dtsch Chem Ges 64:2323–2327
148. Schönberg A, Kaltschmitt H, Schulten H (1933) Über die Einwirkung von Lithium-Phenyl auf Trimethylen-1.3-disulfide und über die Synthese von Di-chromylenen (3. Mitteil. über metallorganische Verbindungen). Ber Dtsch Chem Ges 66:245–250
149. Bergmann E, Engel L (1930) Über die Bedeutung von Dipolmessungen für die Stereochemie des Kohlenstoffs. Z Phys Chem B 8:111–137
150. Bergmann E, Corte H (1933) Über thermochrome Äthylene. (Beiträge zur Kenntnis der doppelten Bindung, V. Mitteilung). Ber Dtsch Chem Ges 66:39–43
151. Schönberg A, Nickel S (1934) Über Chromon-chloride und Dichromylene. Ber Dtsch Chem Ges 67:1795–1798
152. Bergmann ED (1948) Isomerism and isomerization of organic compounds. Interscience Publishers, New York
153. Nielsen WG, Fraenkel GK (1953) Paramagnetism in a thermochromic compound. J Chem Phys 21:1619–1619
154. Wasserman E (1959) The electron spin resonance of the thermochromic form of bianthrone. J Am Chem Soc 81:5006–5007
155. Mills JFD, Nyburg SC (1963) Thermochromism and related effects in bixanthenylidenes and bianthronylidenes. Part II. General observations. J Chem Soc (Lond) 927–935
156. Matlow SL (1955) LCAO MO study of the phenomenon of thermochromism. J Chem Phys 23:152–154

157. Schönberg A, Junghans K (1965) Photochemische Reaktionen, XXI. Photosensibilisierte Dehydrierung des Bixanthylens und die Konstitution seiner thermochromen Form. Chem Ber 98:2539–2544
158. Korenstein R, Muszkat KA, Seger G (1976) Structure of the short wavelength photoisomers of bianthrone analogues. J Chem Soc Perkin Trans 2:1536–1540
159. Woodward RB, Wasserman E (1959) The structure of the thermochromic form of bianthrone. J Am Chem Soc 81:5007–5007
160. Schönberg A, Elkaschef M, Nosseir M, Sidky MM (1958) Experiments with 4-thiopyrones and with 2,2′-6,6′-tetraphenyl-4,4′-dipyrylene. The piezochromism of diflavylene. J Am Chem Soc 80:6312–6315
161. Schönberg A, Mustafa A, Sobhy MEED (1953) Thermochromism of dixanthylenes. J Am Chem Soc 75:3377–3378
162. Mustafa A, Sobhy MEED (1955) Thermochromism of dixanthylenes. II. J Am Chem Soc 77:5124–5126
163. Bergmann ED, Berthier G, Hirschberg Y, Loewenthal E, Pullman B, Pullman A (1951) Fulvènes et éthylènes thermochromes, 9e partie. Influence du mèthyle sur les spectres des benzo- et dibenzofulvènes. Bull Chem Soc France 669–681
164. Hirshberg Y, Loewenthal E, Bergmann ED (1951) The cause of the thermochromy of bianthrone. Bull Res Council Israel 1:139–141
165. Schönberg A, Mustafa A, Asker W (1954) New types of thermochromic substances. The stereochemical aspect of thermochromism; thermochromism and vinylogy. J Am Chem Soc 76:4134–4136
166. Harnik E (1956) On chromic isomerism in dianthrone and its analogs. J Chem Phys 24:297–299
167. Korenstein R, Muszkat KA, Sharafy-Ozeri S (1973) Photochromism and thermochromism through partial torsion about an essential double bond. Structure of the B colored isomers of bianthrones. J Am Chem Soc 95:6177–6181
168. Korenstein R, Muszkat KA (1976) Environmental effects on processes of excited molecules. External spinorbit coupling effects on photochemical reactions of bianthrone-like molecules. In: Pullman B (ed) Environmental effects on molecular structure and properties. Riedel, Doertrecht, pp 561–571
169. Kortüm G, Koch K-W (1967) Magnetische Untersuchungen zur Photochromie und Thermochromie der Dehydrodianthrone. Chem Ber 100:1515–1520
170. Agranat I, Tapuhi Y (1979) Dynamic stereochemistry and steric effects in overcrowded ethylenes. Conformational behavior of dixanthylenes. J Am Chem Soc 101:665–671
171. Agranat I, Tapuhi Y (1979) Dynamic stereochemistry in overcrowded ethylenes. Conformational behavior of bianthrones. J Org Chem 44:1941–1948
172. Agranat I, Tapuhi Y (1978) Steric effects in fast thermal E, Z isomerization of overcrowded ethylenes. Conformational behavior of N, N′-dimethylbiacridans. J Am Chem Soc 100:5604–5609
173. Kortüm G, Theilacker W, Littmann G (1957) Zur Photochromie der Dehydrodianthrone. Naturwiss 44:114–115
174. Hammerich O, Parker VD (1981) Electron transfer reactions accompanied by large structural changes. II. Rates and activation parameters for the interconversion of bianthrone isomers. Acta Chem Scand B 35:395–402
175. Evans DH, Busch RW (1982) Electron-transfer reactions and associated conformational changes. Extended redox series for some bianthrones, lucigenin, and dixanthylene. J Am Chem Soc 104:5057–5062
176. Evans DH, Hu K (1996) Inverted potentials in two-electron processes in organic electrochemistry. J Chem Soc Faraday Trans 92:3983–3990
177. Macías-Ruvalcaba NA, Evans DH (2006) Electron-transfer reactions with significant changes in structure. Unsymmetrical crowded ethylenes. J Phys Chem B 110:24786–24795

178. Macías-Ruvalcaba NA, Evans DH (2007) Electron transfer and structural change: distinguishing concerted and two-step processes. Chem Eur J 13:4386–4395
179. Ahlberg E, Hammerich O, Parker VD (1981) Electron-transfer reactions accompanied by large structural changes. 1. Lucigenin – 10,10′-dimethyl-9,9′-biacridylicene redox system. J Am Chem Soc 103:844–849
180. Mills NS, Benish M (2006) The aromaticity/antiaromaticity continuum. 1. Comparison of the aromaticity of the dianion and the antiaromaticity of the dication of tetrabenzo[5,5]fulvalene via magnetic measures. J Org Chem 71:2207–2213
181. Eliel EL, Wilen SH, Mander LN (1994) Stereochemistry of organic compounds. Wiley, New York, pp 546–548
182. Oki M (1985) Olefins in which the transition state for rotation is selectively stabilized. In: Oki M (ed) Methods in stereochemical analysis, vol 4, Applications of dynamic NMR spectroscopy to organic chemistry. VCH, Deerfield Beach, pp 108–111
183. Sutherland IO (1971) The Investigation of the kinetics of conformational changes by nuclear magnetic resonance spectroscopy. Annu Rep NMR Spectrosc 4:71–235
184. Gault IR, Ollis WD, Sutherland IO (1970) Conformational behavior of bisfluorenylidenes. Chem Commun 269–271
185. Agranat I, Tapuhi Y (1976) Fast thermal *E*,*Z* isomerization in symmetrical overcrowded ethylenes. Low barriers due to ground-state destabilization. J Am Chem Soc 98:615–616
186. Kuhn R, Zahn H, Scholler KL (1953) Über stereoisomere 2,2′-Diamino-bisdiphenylen-äthylene. Justus Liebigs Ann Chem 582:196–217
187. Jager WF, de Lange B, Feringa BL (1992) Chiroptical molecular switches 2; Resolution, properties and applications of inherent dissymmetric alkenes. Mol Cryst Liq Cryst 217:133–138
188. Feringa BL, Jager WF, de Lange B (1992) Resolution of sterically overcrowded ethylenes; a remarkable correlation between bond lengths and racemization barriers. Tetrahedron Lett 33:2887–2890
189. Agranat I, Suissa MR (1993) Syn, anti isomerization of 5,5′-bis-5H-dibenzo[a, d] cyclohepten-5-ylidene. Struct Chem 4:59–66
190. Kalinowski HO, Kessler H (1973) Fast isomerizations about double bonds. Top Stereochem 7:295–383
191. Doering WvE, Roth WR, Bauer F, Breuckmann R, Ebbrecht T, Herbold M, Schmidt R, Lennartz HW, Lenoir D, Boese R (1989) Rotationsbarrieren gespannter Olefine. Chem Ber 122:1263–1275
192. Leigh WJ, Arnold DR (1981) Merostabilization in radical ions, triplets, and biradicals. 5. The thermal cis-trans isomerization of para-substituted tetraphenylethylenes. Can J Chem 59:609–620
193. Agranat I, Tapuhi Y (1977) Double edged steric effects in fast thermal *Z*, *E*-isomerization of overcrowded ethylenes. Evidence against an orthogonal biradical transition state. Noveau J Chim 1:361–362
194. Schönberg A, Sodtke U, Praefcke K (1969) Darstellung, Reaktionen und Stereochemie der 2.3;6.7;2′.3′;6′.7′-tetrabenzo-heptafulvalene. Chem Ber 102:1453–1467
195. Schönberg A, Sodtke U, Praefcke K (1968) Über Stereoisomere und Photochemie des 2.3;6.7;2′.3′;6′.7′-Tetrabenzo-heptafulvalens. Tetrahedron Lett 9:3253–3256
196. Bergmann ED, Rabinovitz M, Agranat I (1968) The spatial structure of tetrabenzohepta-fulvalene. Chem Commun 334–335
197. Pullman B, Pullman A, Berthier G, Points J (1952) Fulvènes et éthylènes thermochromes (21e parite). Structure et spectres d'absorption de la bianthrone, de la benzhydrylidène-anthrone et de la fluorénylidène-anthrone. Bull Soc Chim France 271–275
198. Kazama S, Kamiya M, Akahori Y (1981) Study of conformation of anion radicals of some thermochromic ethylenes by electron spin resonance spectroscopy and semi-empirical molecular orbital calculations. Chem Pharm Bull 29:3375–3378

199. Meyer AY, Yinnon H (1972) Planar and non-planar unsaturation – II. Twisting-distortions in fulvalenic structures. Tetrahedron 28:3915–3928
200. Lenoir D, Lemmen P (1980) Reaktion von 1-Indanonen mit niedrigwertigen Titanverbindungen. Sterischer Einfluß von Methylsubstituenten bei 1-(1-Indanyliden)indan und 9-(9-Fluorenyliden)fluoren. Chem Ber 113:3112–3119
201. Lemmen P, Lenoir D (1980) Kraftfeldrechnungen an Fluorenylidenfluorenen. Vergleich der berechneten Geometrie eines stark tordierten Olefins mit den Röntgenstrukturdaten. Fresenius' Z Anal Chem 304:284–285
202. Kikuchi O, Kawakami Y (1986) An MO study of conformational behavior of bianthrone and its ions. J Mol Struct (Theochem) 137:365–372
203. Kikuchi O, Matsushita KI, Morihashi K, Nakayama M (1986) A molecular orbital study of conformational behavior of 9,9′-bifluorenylidene. Bull Chem Soc Jpn 59:3043–3046
204. Filatov M (2011) Theoretical study of the photochemistry of a reversible three-state bis-thiaxanthylidene molecular switch. ChemPhysChem 12:3348–3353
205. Pogodin S, Rae ID, Agranat I (2006) The effects of fluorine and chlorine substituents across the fjords of bifluorenylidenes: overcrowding and stereochemistry. Eur J Org Chem 5059–5068
206. Krcsmar L, Grunenberg J, Dix I, Jones PG, Ibrom K, Ernst L (2005) Spin–spin interactions across the "cove" in the (*Z*) and (*E*) isomers of 1,1′-difluoro-9,9′-bifluorenylidene. Eur J Org Chem 5306–5312
207. Agranat I, Rabinovitz M, Gosnay I, Weitzen-Dagan A (1972) "Through-space" coupling between bucking fluorine and hydrogen nuclei in trans-1,1′-difluorotetrabenzopentafulvalene. J Am Chem Soc 94:2889–2891
208. Assadi N, Pogodin S, Agranat I (2011) Peterson olefination: unexpected rearrangement in the overcrowded polycyclic aromatic ene series. Eur J Org Chem 6773–6780
209. Irie M (2000) Photochromism: memories and switches – introduction. Chem Rev 100:1683
210. Feringa BL, van Delden RA, ter Wiel MKJ (2001) Chiroptical molecular switches. In: Feringa BL (ed) Molecular switches. Wiley-VCH, Weinheim, pp 123–161
211. Feringa BL, Jager WF, de Lange B (1993) Organic materials for reversible optical data storage. Tetrahedron 49:8267–8310
212. Anders J, Byrne HJ, Poplawski J, Roth S, Sommer-Larsen P, Björnholm T, Jörgensen M, Schaumburg K (1993) Time resolved excited state spectroscopy of anthracene based photochromic systems. Mol Cryst Liq Cryst Sci Technol Sect A 235:231–236
213. Anders J, Byrne HJ, Poplawski J, Roth S, Sommer-Larsen P, Björnholm T, Jörgenson M, Schaumburg K (1993) Influence of substitution on the electronic properties of bianthrones. Synth Met 61:177–180
214. Lara-Avila S, Danilov A, Geskin V, Bouzakraoui S, Kubatkin S, Cornil J, Bjørnholm T (2010) Bianthrone in a single-molecule junction: conductance switching with a bistable molecule facilitated by image charge effects. J Phys Chem C 114:20686–20695
215. de Lange B, Jager WF, Feringa BL (1992) Chiroptical molecular switches 1: principles and syntheses. Mol Cryst Liq Cryst 217:129–132
216. Jager WF, de Lange B, Schoevaars AM, van Bolhuis F, Feringa BL (1993) Sterically overcrowded alkenes; synthesis, resolution and circular dichroism studies of substituted bithioxanthylidenes. Tetrahedron Asym 4:1481–1497
217. Feringa BL, Jager WF, de Lange B (1993) Sterically overcrowded alkenes; a stereospecific photochemical thermal isomerization of a benzoannulated bithioxanthylidene. J Chem Soc Chem Commun 288–290
218. Huck NPM, Jager WF, de Lange B, Feringa BL (1996) Dynamic control and amplification of molecular chirality by circular polarized light. Science 273:1686–1688
219. van Delden RA, van Gelder MB, Huck NPM, Feringa BL (2003) Controlling the color of cholesteric liquid-crystalline films by photoirradiation of a chiroptical molecular switch used as dopant. Adv Funct Mater 13:319–324

220. ter Wiel MKJ, van Delden RA, Meetsma A, Feringa BL (2003) Increased speed of rotation for the smallest light-driven molecular motor. J Am Chem Soc 125:15076–15086
221. Klok M, Walko M, Geertsema EM, Ruangsupapichat N, Kistemaker JCM, Meetsma A, Feringa BL (2008) New mechanistic insight in the thermal helix inversion of second-generation molecular motors. Chem Eur J 14:11183–11193
222. Pollard MM, Klok M, Pijper D, Feringa BL (2007) Rate acceleration of light-driven rotary molecular motors. Adv Funct Mater 17:718–729
223. Landaluce TF, London G, Pollard MM, Rudolf P, Feringa BL (2010) Rotary molecular motors: a large increase in speed through a small change in design. J Org Chem 75:5323–5325
224. Ruangsupapichat N, Pollard MM, Harutyunyan SR, Feringa BL (2011) Reversing the direction in a light-driven rotary molecular motor. Nat Chem 3:53–60
225. Huck NPM, Feringa BL (1995) Dual-mode photoswitching of luminescence. J Chem Soc Chem Commun 1095–1096
226. Browne WR, Pollard MM, de Lange B, Meetsma A, Feringa BL (2006) Reversible three-state switching of luminescence: a new twist to electro- and photochromic behavior. J Am Chem Soc 128:12412–12413
227. Oosterling MLCM, Schoevaars AM, Haitjema HJ, Feringa BL (1996) Polymer-bound chiroptical molecular switches; photochemical modification of the chirality of thin films. Isr J Chem 36:341–348
228. Ivashenko O, Logtenberg H, Areephong J, Coleman AC, Wesenhagen PV, Geertsema EM, Heureux N, Feringa BL, Rudolf P, Browne WR (2011) Remarkable stability of high energy conformers in self-assembled monolayers of a bistable electro- and photoswitchable overcrowded alkene. J Phys Chem C 115:22965–22975
229. Aleksiuk O, Biali SE (1996) Dixanthylene double calix[6]arene. J Org Chem 61:5670–5673
230. Yip YC, Wang XJ, Ng DKP, Mak TCW, Chiang P, Luh TY (1990) (*Z*)-2,2′-Disubstituted bifluorenylidenes by intramolecular desulfurdimerization reactions. J Org Chem 55:1881–1889
231. Schoevaars AM, Hulst R, Feringa BL (1994) Synthesis and ^{1}H NMR complexation studies of alkalimetal bithioxanthylidene crown ether complexes. Tetrahedron Lett 35:9745–9748
232. Geertsema EM, Schoevaars AM, Meetsma A, Feringa BL (2006) Bisthioxanthylidene biscrown ethers as potential stereodivergent chiral ligands. Org Biomol Chem 4:4101–4112
233. Querol M, Stoekli-Evans H, Belser P (2002) 4,5-Diazafluorene-based overcrowded alkene: a new ligand for transition metal complexes. Org Lett 4:1067–1070
234. Sans MQ, Belser P (2002) Towards new chiroptical switches. Coord Chem Rev 229:59–66
235. Smid WI, Schoevaars AM, Kruizinga W, Veldman N, Smeets WJJ, Spek AL, Feringa BL (1996) Synthesis and dynamic behaviour of new metallo-based sterically overcrowded alkenes. Chem Commun 2265–2266
236. ter Wiel MKJ, Meetsma A, Feringa BL (2002) Synthesis and molecular structure of chiral metallo-based sterically overcrowded alkenes. Tetrahedron 58:2183–2188
237. Decker H, Petsch W (1935) Biacridyl und die sich von ihm ableitenden Radikale und Leuchtsalze, die Luzigenine. J Prakt Chem 143:211–235
238. Albert A (1966) The acridines: their preparation, physical, chemical, and biological properties and uses, 2nd edn. Eduard Albert, London
239. Papadopoulos K, Spartalis S, Nikokavouras J, Mitsoulis K, Dimotikali D (1994) Chemiluminescence of N, N′-dialkyl-9,9′-biacridylidenes in homogeneous and micellar media. J Photochem Photobiol A Chem 83:15–19
240. Hong Y, Lam JWL, Tang BZ (2011) Aggregation-induced emission. Chem Soc Rev 40:5361–5388
241. Hong Y, Lam JWY, Tang BZ (2009) Aggregation-induced emission: phenomenon, mechanism and applications. Chem Commun 4332–4353

242. Hoekstra A, Vos A (1975) The crystal and molecular structures of tetraphenylhydrazine and related compounds at −160°C. III. Discussion of the conformation of the molecules. Acta Cryst B 31:1722–1729
243. Hoekstra A, Vos A (1975) The crystal and molecular structures of tetraphenylhydrazine and related compounds at −160°C. II. The crystal structures of tetraphenylethylene (TPE) and diphenylaminotriphenylmethane (DTM). Acta Cryst B 31:1716–1721
244. Brunetti FG, Gong X, Tong M, Heeger AJ, Wudl F (2010) Strain and Hückel aromaticity: driving forces for a promising new generation of electron acceptors in organic electronics. Angew Chem Int Ed 49:532–536
245. Sun GY, Li HB, Geng Y, Su ZM (2012) Potential of bifluorenylidene derivatives as nonfullerene small-molecule acceptor for heterojunction organic photovoltaics: a density functional theory study. Theor Chem Acc 131:1099
246. Longuet-Higgins HC (1963) The symmetry groups of non-rigid molecules. Mol Phys 6:445–460
247. Hougen JT (1986) Symmetry beyond point groups in molecular spectroscopy. J Phys Chem 90:562–568
248. Pechukas P (1976) On simple saddle points of a potential surface, the conservation of nuclear symmetry along paths of steepest descent, and the symmetry of transition states. J Chem Phys 64:1516–1521
249. Rodger A, Schipper PE (1986) Symmetry selection rules for reaction mechanisms. Chem Phys 107:329–342
250. Rodger A, Schipper PE (1987) Symmetry selection rules for reaction mechanisms: a practical formulation for the generation of symmetry-allowed mechanisms and applications. J Phys Chem 91:189–195
251. Rodger A, Schipper PE (1988) Symmetry selection rules for reaction mechanisms: application to metal-ligand isomerizations. Inorg Chem 27:458–466
252. Metropoulos A (1987) Permutation-inversion symmetry considerations in the allowed geometry of transition-state complex. J Phys Chem 91:2233–2234
253. Bone RGA, Rowlands TW, Handy NC, Stone AJ (1991) Transition states from molecular symmetry groups: analysis of non-rigid acetylene trimer. Mol Phys 72:33–73
254. Bone RGA (1992) Deducing the symmetry operations generated at a transition state. Chem Phys Lett 193:557–564
255. Ruch E, Hässelbarth W, Richter B (1970) Doppelnebenklassen als Klassenbegriff und Nomenklaturprinzip für Isomere und ihre Abzählung. Theoret Chim Acta (Berlin) 19:288–300
256. Hässelbarth W, Ruch E (1973) Classifications of rearrangement mechanisms by means of double cosets and counting formulas for the numbers of classes. Theoret Chim Acta (Berlin) 29:259–268
257. Klemperer WG (1972) Enumeration of permutational isomerization reactions. J Chem Phys 56:5478–5489
258. Klemperer WG (1973) Steric courses of chemical reactions. I. J Am Chem Soc 95:380–396
259. Klemperer WG (1973) Steric courses of chemical reactions. II. J Am Chem Soc 95:2105–2120
260. Brocas J, Gielen M, Willem R (1983) The permutational approach to dynamic stereochemistry. McGraw-Hill, New York
261. Schaad LJ, Hu J (1998) Symmetry rules for transition structures in degenerate reactions. J Am Chem Soc 120:1571–1580
262. Bytautas L, Klein DJ (1998) Symmetry aspects of nonrigid molecules and transition structures in chemical reactions. Int J Quantum Chem 70:205–217
263. Barron LD (1986) True and false chirality and parity violation. Chem Phys Lett 123:423–427
264. Avalos M, Babiano R, Cintas P, Jiménez JL, Palacios JC (1998) Absolute asymmetric synthesis under physical fields: facts and fictions. Chem Rev 98:2391–2404
265. Eyring H (1935) The activated complex in chemical reactions. J Chem Phys 3:107–115
266. Wynne-Jones WFK, Eyring H (1935) The absolute rate of reactions in condensed phases. J Chem Phys 3:492–502

267. Barrow GM (1988) Physical chemistry, 5th edn. McGraw-Hill, New York, pp 751–758
268. McIver JW, Stanton RE (1972) Symmetry selection-rules for transition-states. J Am Chem Soc 94:8618–8620
269. Murrell JN, Laidler KJ (1968) Symmetries of activated complexes. Trans Faraday Soc 64:371–377
270. Stanton RE, McIver JW Jr (1975) Group theoretical selection-rules for transition-states of chemical-reactions. J Am Chem Soc 97:3632–3646
271. Nourse JG (1980) Self-inverse and non-self-inverse degenerate isomerizations. J Am Chem Soc 102:4883–4889
272. Brocas J, Willem R (1983) Chiral or achiral paths of steepest descent and transition-states – a permutational approach. J Am Chem Soc 105:2217–2220
273. Salem L (1971) Narcissistic reactions – synchronism vs nonsynchronism in automerizations and enantiomerizations. Acc Chem Res 4:322–328
274. Fukui K (1970) A formulation of reaction coordinate. J Phys Chem 74:4161–4163
275. Fukui K (1981) The path of chemical-reactions – the IRC approach. Acc Chem Res 14:363–368
276. Murrell JN, Pratt GL (1970) Statistical factors and symmetry of transition states. Trans Faraday Soc 66:1680–1684
277. Minyaev RM (1994) Reaction path as a gradient line on a potential energy surface. Int J Quantum Chem 49:105–127
278. Valtazanos P, Ruedenberg K (1986) Bifurcations and transition-states. Theor Chim Acta 69:281–307
279. Altmann SL, Herzig P (1994) Point - group theory tables. Clarendon Press, Oxford
280. Mezey PG (1990) A global approach to molecular symmetry: theorems on symmetry relations between ground- and excited-state configurations. J Am Chem Soc 112:3791–3802
281. Bunker PR, Schutte CJH, Hougen JT, Mills IM, Watson JKG, Winnewisser BP (1997) Notations and conventions in molecular spectroscopy: part 3. Permutation and permutation-inversion symmetry notation (IUPAC Recommendations 1997). Pure Appl Chem 69:1651–1657
282. Schutte CJH, Bertie JE, Bunker PR, Hougen JT, Mills IM, Watson JKG, Winnewisser BP (1997) Notations and conventions in molecular spectroscopy: part 2. Symmetry notation (IUPAC Recommendations 1997). Pure Appl Chem 69:1641–1649
283. Hougen JT (1962) Classification of rotational energy levels for symmetric-top molecules. J Chem Phys 37:1433–1441
284. Hougen JT (1963) Classification of rotational energy levels. II. J Chem Phys 39:358–365
285. Nyburg SC (1954) Steric hindrance in the dibiphenylene-ethylene molecule. Acta Cryst 7:779–780
286. Harnik E, Herbstein FH, Schmidt GMJ, Hirschfeld FL (1954) The structure of overcrowded aromatic compounds. Part I. A preliminary survey. J Chem Soc (Lond) 3288–3294
287. Hutchings MG, Nourse JG, Mislow K (1974) A first approach to the stereochemical analysis of tetraarylmethanes. Tetrahedron 30:1535–1549

Index

Lightning Source UK Ltd.
Milton Keynes UK
UKOW06n1819071016

284715UK00003B/102/P

9 783319 073019